AF292277

Probability and Its Applications

Series Editors
Thomas Liggett
Charles Newman
Loren Pitt

Robert M. Blumenthal

Excursions of
Markov Processes

Birkhäuser
Boston · Basel · Berlin

Robert M. Blumenthal
Department of Mathematics
University of Washington
Seattle, WA 98193

Library of Congress Cataloging-in-Publication Data

Blumenthal, R. M. (Robert McCallum), 1931-
    Excursions of Markov processes / R. M. Blumenthal.
         p.   cm. -- (Probability and its applications)
    Includes bibliographical references and indexes.
    ISBN 978-1-4684-9414-3
    1. Markov processes.   I. Title. II. Series.
  QA274.7.B58   1992   92-38702
  519.2'33--dc20   CIP

Printed on acid-free paper.

ISBN 978-1-4684-9414-3   ISBN 978-1-4684-9412-9 (eBook)
DOI 10.1007/978-1-4684-9412-9

Camera-ready copy prepared in TeX by the author.

9 8 7 6 5 4 3 2 1

Table of Contents

Preface

Let $\{X_t; t \geq 0\}$ be a Markov process in R^1, and break up the path X_t into (random) component pieces consisting of the zero set $(\{t | X_t = 0\})$ and the "excursions away from 0," that is pieces of path $X_s : r \leq s \leq t$, with $X_{r-} = X_t = 0$, but $X_s \neq 0$ for $r < s < t$. When one measures the time in the zero set appropriately (in terms of the local time) the excursions acquire a measure theoretic structure practically identical to that of processes with stationary independent increments, except the values of the process are paths rather than real numbers. And there is a measure on path space that helps describe the measure theoretic properties of the excursions in the same way that the Lévy measure describes the jumps of a process with independent increments. The entire circle of ideas is called excursion theory.

There are many attractive things about the subject: it is an area where one can use to advantage general probabilistic potential theory to make quite specific calculations, it provides a natural setting for applying esoteric things like David Williams' path decomposition, it provides a method for constructing processes whose description in terms of an infinitesimal generator or some such analytic object would be complicated. And the ideas seem to be closely related to a good deal of current research in probability.

The purpose of this book is to set down in an organized way a few items from excursion theory that might attract a probabilist to the subject or make it easier for someone interested in learning the subject to get started.

The prerequisites are a working knowledge of measure theory and enough background in probability to include concepts like independence and conditioning and their use in calculating specific probabilities and expectations. In order to have an official list of prerequisites to rely on I will proceed as if the reader knows all the notation, definitions and basic facts from Chapter 0 and 1 of [BG, 1]. However that is a bit of a burden, so I have included a brief review at the beginning. The reader should be able to proceed from there, filling in from [BG, 1] as the need arises.

On two levels the presentation has less generality than is possible. Firstly I take the state space to be locally compact with a countable base and the basic Markov process to be standard, (rather than U-space and "right" process that is the basic data for today's research). This allows one to refer to the textbook literature for background. And since the major emphasis is on computations and special cases this specialization doesn't represent any significant loss. More serious is the fact that at times I have dodged an issue that basically is measure theoretic by using a simplifying analytic hypothesis. Itô's synthesis theorem will be a case in point.

The general form of excursion theory goes under the labels "regenerative systems" or "homogeneous random sets." The cleanest, most general and most usable presentation of this theory always is to be found in the papers of Maisonneuve. The development in Chapter III comes directly from [Ma, 1]. As with all theories one finds that many people have recognized the excursion structure and introduced important techniques long before the emergence of a general theory. But I have made no effort to give the history of the subject or to survey the literature.

The application of general excursion theory to the analysis and synthesis of Markov processes through their excursions away from the boundary of the state space is what provides the principal focus of this book. Without question the most important papers on this aspect of things are Itô's [I,1] and Motoo's [Mo,1]. Chapter V is devoted to Itô's work and Chapter VII to Motoo's. As to the rest of the book, Chapter I gives an introduction to those aspects of Markov processes relevant to probabilistic potential theory. Chapter II treats examples, general and specific, that will reappear throughout the rest of the book. Also it includes a detailed solution of the analysis and synthesis problem in an important special case. Chapter III presents a special case (generalized in Chapter VII) of excursion theory and applies it to giving Itô's theory of the Poisson point process of excursions away from a point. Chapter IV exploits the wealth of specific formulas in the Brownian motion case to illustrate Itô's synthesis theorem and David Williams' path decomposition, and to develop excursion aspects of Brownian motion that reappear in Chapter VI. And Chapter VI gives successful applications of excursion theory to a pair of difficult problems.

Exercises are rather few and far between, especially for a theory which can be applied so successfully to making computations. Most of the Exercises contain material that is vital to subsequent developments; we urge readers to at least understand their statements. On the other hand the text omits full treatment of many measure theoretic and analytic details; and the reader can get plenty of exercise just filling the gaps.

Much current work using the concept of excursions is not touched on in this book. For example the monograph by Burdzy [Bu,1] deals entirely with a specific application of Brownian motion excursion theory to a problem in complex analysis. Other current research studies delicate local properties of the excursions of Brownian motion in R^n away from a surface. The surface might even be random, for example the convex hull of a piece of the path. The problems are specific and difficult, and there is not much general theory to go by.

I will take this opportunity to thank a few of the many people who helped me with this book. As he always does, my colleague Bruce Erickson provided guidance on a number of points of mathematics and exposition; and he participated with an interesting piece of research [E,1] dealing exactly with the subject of this book. Pat Fitzsimmons and Tom Salisbury read large parts of the book. Their comments helped me with technical points and matters of content and general purpose (of which I was not always sure myself.) Frances Chen and Mary Sheetz of the University of Washington Department of Mathematics prepared the manuscript for reproduction, always displaying skill at transcription and expertise about format. And their good-natured and enthusiastic manner was an added bonus. My wife, Sarah, encouraged me to take up this project; and then she applied the appropriate mix of cheerfulness and tolerance to my grapplings with each new mathematical crisis. I dedicate this book to her as an expression, however insufficient, of my appreciation.

Seattle, August 1991

I Markov Processes

0. Introduction.

We will assume the reader is familiar with the concepts of a probability triple $(\Omega, \mathcal{F}, P)$ and conditional expectation relative to a sub σ-algebra $\mathcal{G}$ of $\mathcal{F}$. If $(E, \mathcal{E})$ is a measurable space and $\{X_t; t\epsilon T\}$ is an indexed family of functions from Ω to E such that $X_t^{-1}(A)\epsilon\mathcal{F}$ for each $A\epsilon\mathcal{E}$ we say that $\{X_t; t\epsilon T\}$ is a *stochastic process* (defined over Ω) *with state space E*. We call Ω the *sample space*. If $\{t_1, \cdots, t_n\}$ is an ordered finite subset of T the measure $\mu_{t_1,\ldots,t_n}$ on $\mathcal{E}^n$ defined by $\mu_{t_1,\ldots,t_n}(D) = P\{(X_{t_1}, \cdots, X_{t_n})\epsilon D\}$ is called the distribution in E^n of $(X_{t_1}, \cdots, X_{t_n})$; and the collection of all these is called the set of *finite dimensional distributions* of the stochastic process. Two stochastic processes, even defined over different probability spaces are called *equal in law* if their finite dimensional distributions agree. In some parts of stochastic process theory it is reasonable to regard two processes as equivalent if they are equal in law, and to admit only concepts which are dependent only on the finite-dimensional distributions. If T is a subset of the real line then a process $\{X_t; t\epsilon T\}$ is said to be a *Markov process* if for every n and every choice of $t_1 < \cdots < t_{n+1}$ from T we have

$$(0.1) \qquad P\{X_{t_{n+1}}\epsilon A \mid X_{t_1}, \cdots, X_{t_n}\} = P\{X_{t_{n+1}}\epsilon A \mid X_{t_n}\}$$

for all A in $\mathcal{E}$. It is clear that with this definition if two processes are equal in law then if one is Markovian so is the other.

However, in the present day treatment a Markov process is a rather elaborate object, $X = (\Omega, \mathcal{F}, \mathcal{F}_t, X_t, \theta_t, P^x)$ where the underlying Ω and the functions X_t are of themselves important, where the totality of possible initial distributions is labelled explicitly by the measures P^x, where shift operators $\{\theta_t\}$ compatible with the $\{X_t\}$ are featured as are σ-algebras $\mathcal{F}_t$ potentially containing more information than what is learned from just the behavior of X_r for $r \leq t$. And these objects are in fact essential even in very specific situations where one is dealing with familiar objects like Brownian motion. There is nothing basically unfamiliar here, but the set-up takes some getting used to. So it seems wise to devote several pages to elaborating on the various components.

1. Basic terminology.

A *measurable space* is a pair $(A, \mathcal{A})$ consisting of a set A and a σ-algebra $\mathcal{A}$ of subsets of A. If in addition μ is a measure on $\mathcal{A}$ then the triple $(A, \mathcal{A}, \mu)$ is a *measure space*. If $(A, \mathcal{A})$ and $(B, \mathcal{B})$ are measurable spaces and X is a function from A to B we say that X is *measurable relative to* $\mathcal{A}$ and $\mathcal{B}$ if $X^{-1}(D)\epsilon\mathcal{A}$ for all $D\epsilon\mathcal{B}$. Often we write $X\epsilon\mathcal{A}/\mathcal{B}$ in this situation. When B is the real line and $\mathcal{B}$ is the σ-algebra of topological Borel sets often we omit mention of $\mathcal{B}$ and simply say "X is $\mathcal{A}$ measurable" or write $X\epsilon\mathcal{A}$. If μ is a measure on $\mathcal{A}$ and $X\epsilon\mathcal{A}/\mathcal{B}$ then the formula $\nu(D) = \mu\big(X^{-1}(D)\big)$ defines a measure ν on $\mathcal{B}$ called the *distribution* (relative to μ) of X in B." The phrase in parentheses is dropped if μ is the only measure on $\mathcal{A}$ to be considered over the course of the discussion. Let Ω be a set, $(B, \mathcal{B})$ a measurable space, and $\{X_t; t\epsilon T\}$ a family of functions from Ω to B. We denote by $\sigma\{X_t; t\epsilon T\}$ the σ-algebra of subsets of Ω generated by the sets $X_t^{-1}(B)$ with t and B ranging over T and $\mathcal{B}$. Clearly $\mathcal{B}$ should be displayed in the notation, but the viewpoint is that $\mathcal{B}$ is known and fixed throughout the discussion. If $\{\mathcal{F}_t; t\epsilon T\}$ is a family of subclasses of subsets of Ω then $\sigma\{\mathcal{F}_t; t\epsilon T\}$ denotes the smallest σ-algebra of subsets of Ω containing all the $\mathcal{F}_t$. If $(\Omega, \mathcal{F}, P)$ is a probability measure space then sub σ-algebras, $\mathcal{F}_1, \cdots, \mathcal{F}_n$ of $\mathcal{F}$ are said to be *independent* or *mutually independent* if $P(B_1 \cap \cdots \cap B_n) = P(B_1) \cdots P(B_n)$ for all choices of the B's with $B_i\epsilon\mathcal{F}_i$. Obviously there are grammatical problems here but the usage is traditional. An indexed family $\{\mathcal{F}_t; t\epsilon T\}$ of sub σ-algebras is called mutually independent if every finite sub family is so.

For a probability space $(\Omega, \mathcal{F}, P)$ and an $\mathcal{F}$ measurable integrable real valued function X one writes $E(X; A)$ for the integral $\int_A X\, dP$ of X over a set $A\epsilon\mathcal{F}$, or just $E(X)$ when $A = \Omega$. The notation $P(X)$ or $P(X; A)$ is common also. We will use both.

A *monotone class theorem* is, more or less, any general theorem that says that a relationship (typically the equality of two measures) which holds over all sets in a certain class $\mathcal{A}$ automatically holds as well for all sets in the σ-algebra generated by $\mathcal{A}$. The one that seems most easy to apply is Dynkin's: he calls a collection $\mathcal{S}$ of subsets of Ω a π-system if $\mathcal{S}$ is closed under finite intersections; and he calls a collection $\mathcal{D}$ a d-system if $\Omega\epsilon\mathcal{D}$ and $\mathcal{D}$ is closed under proper differences and countable monotone unions.

His theorem is that if $\mathcal{D}$ is a d-system, $\mathcal{S}$ is a π-system and $\mathcal{D} \supset \mathcal{S}$ then $\mathcal{D} \supset \sigma\{\mathcal{S}\}$. We will assume that the reader is familiar with monotone class arguments and will not give details when such an argument must be invoked to complete a proof.

2. Stationary transition functions.

Let $(E, \mathcal{E})$ be a measurable space. A real valued function $P(t, x, A)$ defined for $t \geq 0$, $x \epsilon E$ and $A \epsilon \mathcal{E}$ is called a *stationary transition function* (over $(E, \mathcal{E})$) if for fixed t and A it is $\mathcal{E}$ measurable in x, for fixed t and x it is a probability measure in A and if it satisfies the Chapman-Kolmogorov equation;

$$(2.1) \qquad P(t + s, x, A) = \int_E P(t, x, dy) P(s, y, A)$$

for all t, s, x and A. There is a more general notion of (not necessarily stationary) transition function, $P(s, x; t, A)$ defined for $0 \leq s < t$, $x \epsilon E$ and $A \epsilon \mathcal{E}$ which is $\mathcal{E}$ measurable in x and a probability measure in A and satisfies

$$P(s, x; t, A) = \int_E P(s, x; r, dy) P(r, y; t, A)$$

for all s, r, t, x and A with $s < r < t$.

If $\{X_t; t \geq 0\}$ is a Markov process defined over a probability space $(\Omega, \mathcal{F}, P)$ we say that $P(s, x; t, A)$ is a transition function for the process if

$$(2.2) \qquad P(s, X_s; t, A) = P\{X_t \epsilon A \mid \sigma\{X_r; r \leq s\}\}$$

for all $A \epsilon \mathcal{E}$ and s and t with $0 \leq s < t$. In most cases of interest to us there will be a stationary transition function $P(r, x, A)$ such that the left side of (2.2) can be replaced by $P(t - s, X_s, A)$. Then we call the Markov process *time-homogeneous*. We use the term "transition function" in either situation and often drop the qualification "stationary".

Let $P(t, x, A)$ be a transition function. The formula

$$P_t f(x) = \int_E f(y) P(t, x, dy)$$

defines a transformation of those real valued $\mathcal{E}$ measurable functions for which the integral makes sense. Certainly this will include those that are bounded or positive. The fact that $P_t P_s f = P_s P_t f = P_{t+s} f$ is an immediate consequence of the Chapman-Kolmogorov equation, provided f is such that all the operations make sense. One viewpoint is that only the transformations are of basic interest and that a time homogeneous Markov process with the given stationary transition function is useful only as an auxiliary device for studying the semi-group $\{P_t; t > 0\}$. But this viewpoint is now out-of-date, and one considers measure-theoretic properties of the process to be of interest as well.

We will assume always that $\{x\}\epsilon\mathcal{E}$ for each $x\epsilon E$. Let $P(t, x, A)$ be a transition function over $(E, \mathcal{E})$. Given a point $x\epsilon E$ and a finite set of times $t_1, \cdots, t_n$ with $0 < t_1 < \cdots < t_n$ the expression

$$(2.3) \quad \varepsilon_x(dx_0)P(t_1, x_0, dx_1)P(t_2 - t_1, x_1, dx_2)\cdots\cdot P(t_n - t_{n-1}, x_{n-1}, dx_n)$$

defines a probability measure, let us call it $\mu_{0,t_1,\ldots,t_n}(x, \cdot)$, on $\mathcal{E}^{n+1}$. (The meaning of (2.3) is that if $A\epsilon\mathcal{E}^{n+1}$ then one first calculates

$$\int P(t_n - t_{n-1}, x_{n-1}, dx_n)I_A(x_0, \cdots, x_{n-1}, x_n)$$

for fixed $(x_0, \cdots, x_{n-1})$ the integration producing a function of $(x_0, \cdots, x_{n-1})$ which is $\mathcal{E}^n$ measurable and which then is integrated in x_{n-1} relative to $P(t_{n-1} - t_{n-2}, x_{n-2}, dx_{n-1})$, and so on). It is clear that for fixed $t_1, \cdots, t_n$ and A, $\mu_{0,t_1,\ldots,t_n}(x, A)$ defines an $\mathcal{E}$ measurable function of x.

Now suppose $\{X_t; t \geq 0\}$ is a time homogeneous Markov process with $P(t, x, A)$ as transition function and initial distribution ν (that is $\nu(B) = P\{X_0\epsilon B\}$ for $B\epsilon\mathcal{E}$). Then the finite dimensional distributions of the process are given by

$$(2.4) \qquad P\{(X_0, \cdots, X_{t_n})\epsilon A\} = \int_E \mu_{0,t_1,\ldots,t_n}(x, A)\nu(dx)$$

that is, they are determined by the initial distribution and the transition function. Given our time points $t_1, \cdots, t_n$ pick one of them, t_m. Suppose $A \in \mathcal{E}^{n+1}$ is the intersection of two sets of the form $\{(x_0, \cdots, x_n) \mid (x_0, \cdots, x_m)\epsilon C\}$ and $\{(x_0, \cdots, x_n) \mid (x_{m+1}, \cdots, x_n)\epsilon D\}$ with $C\epsilon\mathcal{E}^{m+1}$ and $D\epsilon\mathcal{E}^{n-m}$. Since the expression (2.3) can be split apart at any place it is clear that

$$(2.5) \quad \mu_{0,t_1,\ldots,t_n}(x, A) = \int_C \mu_{0,t_1,,\ldots,t_m}(x, dy)\mu_{0,t_{m+1}-t_m,\ldots,t_n-t_m}(y_m, D).$$

For the Markov process this says that

$$\mu_{0,t_{m+1}-t_m,\ldots,t_n-t_m}(X_{t_m}, D) = P\{(X_{t_{m+1}}, \cdots, X_{t_n})\epsilon D \mid X_0, X_{t_1}, \cdots, X_{t_m}\}.$$

By the usual measure theory arguments one can replace $\sigma\{X_0, X_{t_1}, \cdots, X_{t_m}\}$ with $\sigma\{X_r; r \leq t_m\}$ on the right of this equality.

3. Time homogeneous Markov processes.

Let $P(t, x, A)$ be a stationary transition function over a measurable space $(E, \mathcal{E})$. We would like to know that for each $y \epsilon E$ there is a time-homogeneous Markov process which has $P(t, x, A)$ as transition function and unit mass at y as initial distribution.

An initial step toward this goal always can be taken: let Ω denote the set of all functions $t \to \omega(t)$ from $[0, \infty)$ to E, and consider the coordinate process $\{x_t; t \geq 0\}$ where $x_t(\omega) = \omega(t)$. The finite dimensional σ-algebras in Ω are by definition those of the form $\sigma\{x_r; r \epsilon \Lambda\}$ where Λ denotes a finite subset $\{0, t_1, \cdots, t_n\}$ of $[0, \infty)$. (To avoid extra words we will assume always that $0 < t_1 < \cdots < t_n$). The union, over all Λ, of these is an algebra, $\mathcal{F}'$, of subsets of Ω and the σ-algebra generated by $\mathcal{F}'$ is $\sigma\{x_t; t \geq 0\}$. We will denote this by $\mathcal{F}^0$. Any set, Γ, in $\mathcal{F}'$ is of the form $\Gamma = \{\omega \mid (x_0(\omega), \cdots, x_{t_m}(\omega)) \epsilon D\}$ for some m, choice of $t_1, \cdots, t_m$ and $D \epsilon \mathcal{E}^{m+1}$, but a given set can be represented in more than one way. Let us define, for each $y \epsilon E$, a set function P^y on $\mathcal{F}'$ by $P^y(\Gamma) = \mu_{0, t_1, \cdots, t_m}(y, D)$ for Γ as above. Since Γ has more than one representation one must check that this is indeed a definition, but this follows easily from the Chapman-Kolmogorov equation. The set function P^y is finitely additive on $\mathcal{F}'$ and is a probability measure when restricted to any σ-algebra $\sigma\{x_r; r \epsilon \Lambda\}$ with Λ finite. If we can argue that P^y can be extended so as to be a measure on $\mathcal{F}^0$ then over $(\Omega, \mathcal{F}^0, P^y)$ the coordinate process will be a Markov process with unit mass at y as initial distribution and the given $P(t, x, A)$ as transition function. The extension will be possible if and only if $\lim_{n \to \infty} P^y(\Gamma_n) = 0$, whenever $\{\Gamma_n\}$ is a decreasing sequence of $\mathcal{F}'$ sets with empty intersection. A condition for this general enough to cover all cases of interest to us is that E be locally compact Hausdorff with a countable base and $\mathcal{E}$ be the Borel sets of the topology, for then one can use compactness considerations and reason by contradiction (Kolmogorov's Consistency Theorem.) Of course even in the absence of some such topological assistance it might be that some special feature of the situation allows us to reach the same conclusion.

So for this paragraph and the next one we will assume that we have succeeded somehow in arguing that for each y in E there is a probability measure (necessarily unique) on $\mathcal{F}^0$ agreeing with P^y as defined on $\mathcal{F}'$. Clearly, from the way the P^y are constructed, if Γ is a set in $\mathcal{F}'$ then $P^y(\Gamma)$

defines an $\mathcal{E}$ measurable function of y. By a monotone class argument this holds also for T in $\mathcal{F}^0$. If μ is a probability measure on $\mathcal{E}$ and P^μ is the measure on $\mathcal{F}^0$ given by

$$(3.1) \qquad P^\mu(\Gamma) = \int_E P^y(\Gamma)\mu(dy)$$

then over $(\Omega, \mathcal{F}^0, P^\mu)$ the coordinate process is Markov with $P(t, x, A)$ as transition function, but now the initial distribution is μ.

For t positive define the transformation θ_t from Ω to Ω by

$$(\theta_t\omega)(r) = \omega(r+t).$$

Then $x_r \circ \theta_t = x_{r+t}$. Suppose B is a set from $\mathcal{E}^{m+1}$ and J is the indicator of $\{\omega \mid (x_0(\omega), x_{t_1}(\omega), \cdots, x_{t_m}(\omega))\epsilon B\}$ where as usual $0 < t_1 < \cdots < t_m$. Then $J \circ \theta_t$ is the indicator of $\{\omega \mid (x_t(\omega), x_{t+t_1}(\omega), \cdots, x_{t+t_m}(\omega))\epsilon B\}$. It follows that θ_t as a mapping from $(\Omega, \mathcal{F}^0)$ to $(\Omega, \mathcal{F}^0)$ is measurable, and in the domain side $\mathcal{F}^0$ may be replaced by $\sigma\{x_r; r \geq t\}$. In particular (2.6) may be rewritten as

$$(3.2) \qquad P^{x_t}(J) = P^\mu\{J \circ \theta_t \mid \sigma(x_r; r \leq t)\},$$

P^μ in the conditional probability expression standing for the fact that the underlying measure on $(\Omega, \mathcal{F}^0)$ is P^μ. Obviously we may replace J by any positive $\mathcal{F}^0$ measurable function. Usually one takes equation (3.2) as the best expression of the Markov property of the process and of the use of the measures P^y in expressing the conditional expectation of a function $J \circ \theta_t$, dependent on the future given the past, $\sigma\{x_r; r \leq t\}$, in terms of the present, x_t.

But for adequate flexibility and generality one must give up reliance on the specific sample space and process variables we have just considered and instead axiomatize these objects. Coming to specifics, we assume as before that we are given a stationary transition function $P(t, x, A)$ on a state space $(E, \mathcal{E})$. Suppose we are given a measurable space $(\Omega, \mathcal{F})$ and a family $\{\mathcal{F}_t; t \geq 0\}$ of sub σ-algebras of $\mathcal{F}$ with $\mathcal{F}_s \subset \mathcal{F}_t$ whenever $s \leq t$. (Such a family is called a *filtration* of $(\Omega, \mathcal{F})$.) Suppose we are given also a family $\{X_t; t \geq 0\}$ of functions from Ω to E, such that $\sigma\{X_t\} \subset \mathcal{F}_t$ for each t, and for each $y \epsilon E$ a probability measure P^y on $\mathcal{F}$ such that

$$(3.3) \qquad \int_\Lambda P(t - s, X_s, A) dP^y = P^y \{X_t \epsilon A; \Lambda\}$$

for all $A \epsilon \mathcal{E}$, $0 \leq s < t$, $\Lambda \epsilon \mathcal{F}_s$. One phrases (3.3) by saying that over $(\Omega, \mathcal{F}, P^y)$, $\{X_t, \mathcal{F}_t; t \geq 0\}$ is a time homogeneous Markov process with $P(t, x, A)$ as transition function. The validity of (3.3) just for Λ from $\sigma\{X_r; r \leq t\}$ is what we assumed previously in defining Markov processes with a given transition function. We will suppose in addition that we have for each positive t a transformation $\theta_t : \Omega \to \Omega$ such that $X_r \circ \theta_t = X_{r+t}$ for all positive r. We say that $(\Omega, \mathcal{F}, \mathcal{F}_t, X_t, \theta_t, P^x)$ is a *time homogeneous Markov process*. We will make the additional assumption that for each y in E, $P^y(X_0 = y) = 1$. This is a restriction, but for our purposes it is a very slight one. We will derive some simple consequences of these assumptions and give some examples to justify the generality.

From (3.3) and our last assumption we have that $P(t, x, A) = P^x(X_t \epsilon A)$ for all x, t and A, and defining $P(0, x, \cdot)$ to be unit mass at x is consistent with our other assumptions. Let $\mathcal{F}^0$ be the σ-algebra $\sigma\{X_r; r \geq 0\}$. For a basic set in $\mathcal{F}^0$,

$$\Gamma = \{\omega | (X_0(\omega), X_{t_1}(\omega), \cdots, X_{t_m}(\omega)) \epsilon B\},$$

with $B \epsilon \mathcal{E}^{m+1}$, we have

$$P^y(\Gamma) = \mu_{0, t_1, \cdots, t_m}(y, B)$$

in the notation of section 2, so as a function of y, $P^y(\Gamma)$ is $\mathcal{E}$ measurable. By a monotone class argument this holds for any $\Gamma \epsilon \mathcal{F}^0$. The measure P^y restricted to $\mathcal{F}^0$ is determined by the transition function and the requirement that $P^y(X_0 = y) = 1$. For the set Γ above, $\theta_t^{-1}\Gamma = \{\omega | (X_t(\omega), \cdots, X_{t+t_m}(\omega)) \epsilon B\}$, so $\theta_t^{-1}\Gamma$ is an $\mathcal{F}^0$ set. The discussion which led to (3.2) allows us to conclude that for all y

$$(3.4) \qquad P^{X_t}(J) = P^y(J \circ \theta_t \mid \mathcal{F}_t)$$

for all t and positive $\mathcal{F}^0$ measurable functions J. In (3.4) we have to write P^y instead of P^μ because (3.1) makes sense only for $\Gamma\epsilon\mathcal{F}^0$ and hence the measure P^μ is defined there only. Before going on we should remind ourselves that from the viewpoint of basic stochastic process theory all that is required to specify a Markov process is enough information to specify the finite dimensional distributions. Usually this means giving an initial distribution and a transition function, and defining the finite dimensional distributions by way of (2.3) and (2.4). One can, with reasonable accuracy, reconcile this with our more elaborate set-up by thinking of the P^y as an unequivocal determination of "conditional probability given $X_0 = y$", the θ_t as a useful device for describing "events dependent only on the process beyond time t", and the $\mathcal{F}_t$ as being enlargements of $\sigma\{X_r; r \leq t\}$ necessary to exploit the sort of "zero-one law" considerations that are critical to probabilistic potential theory. We will take a flexible viewpoint; so, for example, in a situation where the effort in defining the θ_t exceeds the profit from doing so we will drop these transformations from our consideration.

To illustrate the axioms let $(E, \mathcal{E})$ be the real line and Borel sets, and let $P(t, x, A)$ be the *Brownian motion transition function*,

$$(3.5) \qquad\qquad P(t, x, A) = \int_A p(t, x, y)dy$$

where dy denotes lebesgue measure and $p(t, x, y)$ is the Gauss *kernel*

$$p(t, x, y) = \frac{1}{\sqrt{2\pi t}}\, e^{-\frac{1}{2t}(x-y)^2}.$$

One checks easily the transition function requirements and so given any probability measure μ on $\mathcal{E}$ there is a time-homogeneous Markov process with μ as initial distribution and $P(t, x, A)$ as transition function. We call any such process *Brownian motion with initial distribution μ*. We will consider several descriptions that specify the underlying space, basic variables, translation operators and so forth. Take W to consist of all functions $t \to w(t)$ from $[0, \infty)$ to $R, x_t(w) = w_t$ for $t \geq 0$, $\mathcal{F}_t^0 = \sigma\{x_r; r \leq t\}$, $\mathcal{F}^0 = \sigma\{x_r; r \geq 0\}, \theta_t^0 w(s) = w(t+s)$ and P^y the measure on $\mathcal{F}^0$ relative to which $\{x_t; t \geq 0\}$ is Brownian motion starting at y. Then $(W, \mathcal{F}^0, \mathcal{F}_t^0, x_t, \theta_t^0, P^y)$ is a Markov process in our restricted sense — we are reserving the letters Ω, X_t, and θ_t for use later on. Now given any stochastic process $\{X_t; t \geq 0\}$

over a sample space Ω the functions $t \to X_t(w)$ defined for each $w\epsilon\Omega$ are called the *sample functions* of the process. One wishes to apply real variables techniques to these functions and for this reason, as well as some others to be considered later, it is desirable to have them be as regular as restrictions inherent in the finite dimensional distributions will allow. In this light the coordinate representation of Brownian motion is quite useless as is any description which specifies only the finite dimensional distributions. To correct this, let Ω denote the subset of W consisting of all functions, w, which when restricted to the rationals, Q, agree with the restriction to Q of a continuous function. The continuous function is of course uniquely determined by w if w is in Ω; we will denote it by $\varphi(w)$. The description of Ω involves conditions on $x_t(w)$ only for t in a fixed countable set and so clearly $\Omega\epsilon\mathcal{F}^0$. In Chapter II we will show that $P^y(\Omega) = 1$ for each y. Let us take this as known. Let $\mathcal{F}$ denote the σ-algebra in Ω consisting of all $\mathcal{F}^0$ sets which are contained in Ω and, there being little chance of confusion, use P^y to denote the original P^y restricted to the measurable space $(\Omega, \mathcal{F})$. Define a process $\{X_t; t \geq 0\}$ over $(\Omega, \mathcal{F})$ by

$$X_t(w) = x_t\big(\varphi(w)\big).$$

For $w\epsilon\Omega$ and any t

$$X_t(w) = \lim_{\varepsilon \to 0} \sup_{|q-t|<\varepsilon, q\epsilon Q} x_q(w)$$

and so each X_t is $\mathcal{F}$ measurable, that is $\{X_t; t \geq 0\}$ is a stochastic process over $(\Omega, \mathcal{F})$. Clearly $x_t = X_t$ if t is rational. We can't define θ_t as θ_t^0 restricted to Ω, but if for $w\epsilon\Omega$ we set $\theta_t w(r) = \varphi(w)(r + t)$ then θ_t maps Ω into Ω, and $X_{r+t} = X_r \circ \theta_t$, as desired. Coming to the necessary filtration $\{\mathcal{F}_t; t \geq 0\}$ it is tempting to say that $\mathcal{F}_t$ consists of all sets Γ which can be written as $\Delta \cap \Omega$ for some $\Delta\epsilon\mathcal{F}_t^0$. Instead call that σ-algebra $\mathcal{G}_t$ and set $\mathcal{F}_t = \bigcap_{\varepsilon>0} \mathcal{G}_{t+\varepsilon}$, (usually denoted $\mathcal{G}_{t+}$). Clearly $\{\mathcal{G}_t; t \geq 0\}$ is an increasing family of sub σ-algebras of $\mathcal{F}$ and hence so is $\{\mathcal{F}_t; t \geq 0\}$. For any positive number t, real number b and $w\epsilon\Omega$ we have that $X_t(w) < b$ if and only if for some strictly positive ε and δ, $x_q(w) < b - \varepsilon$ for all positive rational q with $t - \delta \leq q \leq t$. Hence each X_t is measurable relative to $\mathcal{G}_t$ and thus relative to $\mathcal{F}_t$ as well. Finally we have to check the Markov property, that is (3.3), for $\{X_t, \mathcal{F}_t; t \geq 0\}$. Let t be positive, s be strictly positive and let t_n and s_n

be rationals decreasing to t and s from strictly above. Let φ be a bounded continuous function on R and $\Lambda \epsilon \mathcal{F}_t$. Thus for each n, $\Lambda = \Omega \cap \Lambda_n$ for some $\Lambda_n \epsilon \mathcal{F}^0_{t_n}$. Then using the fact that the coordinate process is Brownian motion and $x_t = X_t$ for rational t,

$$P^y\big(\varphi(X_{s_n+t_n}); \Lambda\big) = P^y\big(\varphi(x_{s_n+t_n}); \Lambda_n\big)$$

(3.6)
$$= \int_{\Lambda_n} P(s_n, x_{t_n}, \varphi) dP^y = \int_{\Lambda} P(s_n, X_{t_n}, \varphi) dP^y$$

(In (3.6) we used the notation $P(s, x, \varphi)$ for $\int \varphi(y) P(s, x, dy)$). Now let n approach infinity. On the left side $\varphi(X_{s_n+t_n})$ converges to $\varphi(X_{s+t})$. Also it is easy to see from the Gauss kernel that for a bounded Borel function φ, the function $(s, x) \to P(s, x, \varphi)$ is jointly continuous on $(0, \infty) \times R$. So the last integrand on the right converges to $P(s, X_t, \varphi)$. Clearly this implies (3.3), which is what we needed to check. We have now set up a Brownian motion $(\Omega, \mathcal{F}, \mathcal{F}_t, X_t, \theta_t, P^y)$ so that the sample functions are continuous. Except for some measure theoretic completions, the σ-algebras $\mathcal{F}_t$ are the ones required for a potential-theoretic analysis of Brownian motion.

Some processes closely related to Brownian motion illustrate other aspects of the general set-up: let $(\Omega, \cdots)$ be the sextuple we have just constructed for defining Brownian motion with continuous sample functions and right continuous filtration and now use ω to denote points of Ω. Define the density $p^+(t, x, y)$ for $x \geq 0, y \geq 0$ only by

$$p^+(t, x, y) = p(t, x, y) + p(t, -x, y),$$

with $p(t, x, y)$ the Gauss kernel, and set

$$P^+(t, x, A) = \int_A p^+(t, x, y) dy$$

for A a Borel subset of $[0, \infty)$. It is easy to check that $P^+(t, x, A)$ is a transition function over the state space $\Big([0, \infty), \mathcal{B}([0, \infty))\Big)$. The probability measures P^y on $(\Omega, \mathcal{F})$ will be those from our Brownian motion construction, but only for y positive. For process variables and translation operators we will take $\{|X_t|; t \geq 0\}$ and $\{|\theta_t|; t \geq 0\}$ ($|\theta_t|$ transforms the function $\omega \epsilon \Omega$ into $|\theta_t \omega|$, which is again in Ω). It is clear that $|X_t|$ is $\mathcal{F}_t$ measurable and that $|X_t| \circ |\theta_r| = |X_{t+r}|$. As to the Markov property for $\{|X_t|, \mathcal{F}_t; t \geq 0\}$, suppose we are given positive numbers t and s, a positive Borel function φ on $[0, \infty)$ and a set $\Lambda \epsilon \mathcal{F}_t$. Then

$$P^y\left(\varphi(|X_{t+s}|); \Lambda\right) = P^y\left(\overline{\varphi}(X_{t+s}); \Lambda\right)$$

where $\overline{\varphi}$ is the Borel function on R defined by $\overline{\varphi}(x) = \varphi(|x|)$. The right side in the display is given by

$$P^y\left(P(s, X_t, \overline{\varphi}); \Lambda\right)$$

since the X_t process is Brownian motion. It is easy to see that $P(s, x, \overline{\varphi}) = P^+(s, |x|, \varphi)$ for all s and x, and so

$$P^y\left(\varphi(|X_{t+s}|); \Lambda\right) = P^y\left(P^+(s, |X_t|, \varphi); \Lambda\right).$$

This shows that $(\Omega, \mathcal{F}, \mathcal{F}_t, |X_t|, |\theta_t|, P^y)$ with $y \geq 0$ is a time-homogeneous Markov process with transition function $P^+(t, x, A)$. Any Markov process with this transition function is called Brownian motion on $[0, \infty)$ reflected at 0 or simply *reflecting Brownian motion*. This version has all of its sample functions continuous. The σ-algebras $\mathcal{F}_t$ are in fact much larger than $\sigma\{|X_r|; r \leq t\}$, but the only part of this that is of any importance is the fact that $\mathcal{F}_t$ contains $\bigcap\limits_{\varepsilon > 0} \sigma\{|X_r|; r \leq t + \varepsilon\}$.

There is a different description of reflecting Brownian motion that is useful for calculations: let $(\Omega, \mathcal{F}, \mathcal{F}_t, X_t, \theta_t, P^x)$ be the description of Brownian motion given above. Once again consider the measures P^x only for x positive. For $\omega \epsilon \Omega$ define

$$m_t(\omega) = \min(0, \min_{s \leq t} X_s(\omega))$$

and

$$X_t^+(\omega) = X_t(\omega) - m_t(\omega).$$

In words, the path of X_t^+ is just the path of X_t until that path reaches $(-\infty, 0]$ and after that, the current minimum is subtracted off, so that in particular X_t^+ is always positive. It turns out that $\{X_t^+; t \geq 0\}$ is reflecting Brownian motion: that is, it is a time homogeneous Markov process and its transition function is $P^+(t, x, A)$. We will argue now for the Markovian character but will postpone the identification of the transition function to

II-2. To start with $m_t(\omega)$ is the minimum of 0 and the infimum of $X_s(\omega)$ as s ranges over the rationals less than t; so clearly X_t^+ is measurable relative to $\sigma\{X_r; r \leq t\}$, which is contained in $\mathcal{F}_t$. Coming to the Markov property calculation, let $G(\omega, \omega')$ be a positive function on $\Omega \times \Omega$ of the form $f(\omega)g(\omega')$ where f and g are positive functions on Ω with f being $\mathcal{F}_t$ measurable and g being $\sigma\{X_r; r \geq 0\}$ measurable. Then one way of expressing the Markov property equation, (3.4), is

$$(3.7) \qquad \int_\Omega G(\omega, \theta_t\omega) P^y(d\omega) = \int_\Omega \left\{ \int_\Omega G(\omega, \omega') P^{X_t(\omega)}(d\omega') \right\} P^y(d\omega).$$

By a monotone class argument (3.7), along with the measurability assertions needed to guarantee that it is meaningful, holds for any positive function $G(\omega, \omega')$ which is measurable in (ω, ω') relative to the σ-algebra $\mathcal{F}_t \times \sigma\{X_r; r \geq 0\}$. In particular let h be a positive Borel function on $[0, \infty)$, t and r positive numbers and Λ a set in $\mathcal{F}_t$. The function

$$G(\omega, \omega') = h\big(X_r^+(\omega' - m_t(\omega))\big) I_\Lambda(\omega)$$

is suitably measurable, and $G(\omega, \theta_t\omega)$ is equal to $I_\Lambda h(X_{r+t}^+)$. Applying (3.7) we have

$$(3.8) \qquad \int_\Lambda h(X_{r+t}^+) dP^y$$

$$= \int_\Lambda \left\{ \int_\Omega h\left(X_r^+(\omega' - m_t(\omega)) \right) P^{X_t(\omega)}(d\omega') \right\} P^y(d\omega).$$

Now it is quite clear that given any real number x, the mapping from Ω to Ω that carries the function ω to the function $\omega - x$ carries the measure P^y to the measure P^{y-x}, for any y. Applying this to the inner integral on the right of (3.8) we see that the expression in braces can be written

$$\int h(X_r^+) dP^{X_t^+(\omega)}.$$

Now take t and x positive and set

$$\overline{P}(t, x, A) = P^x(X_t^+ \epsilon A)$$

for Borel sets $A \subset [0, \infty)$. Then we have established

$$\int_\Lambda h(X_{r+t}^+) dP^y = \int_\Lambda \overline{P}(r, X_t^+, h) dP^y.$$

This completes the proof, subject to our postponed verification that $\overline{P}(t, x, A)$ and $P^+(t, x, A)$ are the same. It is worth noting that in this example the appropriate translation operators will have to be defined with a little more care: given a function $\omega \epsilon \Omega$ take $\theta_t^+ \omega$ to be the function

$$(\theta_t^+ \omega)(r) = \omega(t + r) - m_t(\omega).$$

Then $X_r^+ \circ \theta_t^+ = X_{r+t}^+$ as required. From these equalities we conclude that $(\Omega, \mathcal{F}, \mathcal{F}_t, X_t^+, \theta_t^+, P^x)$ is reflecting Brownian motion.

4. The strong Markov property.

Let $X = (\Omega, \mathcal{F}, \mathcal{F}_t, X_t, \theta_t, P^x)$ be a time homogeneous Markov process with transition function $P(t, x, A)$. Equation (3.4) expresses the fact that for each fixed t the process $\{X_{t+s}; s \geq 0\}$ is a Markov process with $P(s, x, A)$ as transition function and initial position X_t, and that once X_t is fixed this process is independent of the enlarged past history as embodied in $\mathcal{F}_t$.

For establishing analytic properties of the transition function by means of probabilistic properties of the process it is useful to know that (3.4) holds also when the fixed times t are replaced by certain random times T, that is by positive functions on Ω not necessarily constant. Naturally it will be necessary to interpret symbols like $\mathcal{F}_T$ and to impose proper restrictions on the T. Specifically, it is clear that the values of T must not be based on a look into the future; that is any event of the form $\{T \leq b\}$ should not involve the behavior of the variables X_s for s larger than b. Positive random variables with this property are called stopping times, and the validity of (3.4) for all stopping times is called the strong Markov property. We will devote the next few paragraphs to a more careful definition of stopping times and a discussion of when the strong Markov property holds.

Let $(\Omega, \mathcal{F})$ be a measurable space and $\{\mathcal{F}_t; t \geq 0\}$ be a filtration of $(\Omega, \mathcal{F})$. A function T from Ω to $[0, \infty]$ is called an $\{\mathcal{F}_t\}$ -*stopping time* if for each positive number b, the event $\{T \leq b\}$ is in the σ-algebra $\mathcal{F}_b$. The specific filtration is part of the definition, but usually it is the filtration in a Markov process $X = (\cdots, \mathcal{F}_t, \cdots)$ under discussion. Hence it is regarded as part of the basic set up and without mentioning the filtration one can say "let $S, T, \cdots$ be stopping times..." without much fear of causing confusion. The basic properties of stopping times follow easily from the definition: any stopping time is $\mathcal{F}$ measurable. The maximum, minimum and sum of two stopping times are stopping times. The supremum of a countable family of stopping times is a stopping time; but if S denotes the infimum of a countable family of $\{\mathcal{F}_t\}$ stopping times all we can say is that for each t, $\{S < t\} \in \mathcal{F}_t$ or, equivalently, for each t, $\{S \leq t\} \in \mathcal{F}_{t+}$, where $\mathcal{F}_{t+}$ denotes the intersection of $\mathcal{F}_{t+\varepsilon}$ over all ε strictly positive. Usually we can set up things so that the *filtration is right continuous* in the sense that $\mathcal{F}_t = \mathcal{F}_{t+}$ for all t. Then S will indeed be a stopping time.

Given a stopping time T the σ-algebra, $\mathcal{F}_T$ *of events depending on* $\{\mathcal{F}_t\}$

only up to time T is defined by saying that an event Λ is in $\mathcal{F}_T$ provided

$$\Lambda \cap \{T \le b\} \in \mathcal{F}_b$$

for all positive b. It is easy to check that $\mathcal{F}_T$ is a sub σ-algebra of $\mathcal{F}$, that T is $\mathcal{F}_T$-measurable and that the notation is consistent in the sense that $\mathcal{F}_T = \mathcal{F}_t$ if T is identically equal to t. Also $\mathcal{F}_T \subset \mathcal{F}_S$ if T and S are stopping times with $T \le S$.

We will be concerned only with the situation set forth at the beginning of this section where X is a Markov process and $\{\mathcal{F}_t\}$ is its filtration. We say that X is *progressively measurable* if for each positive t the mapping from $[0, t] \times \Omega$ to the state space E that sends (s, ω) to $X_s(\omega)$ is measurable relative to $\mathcal{E}$ in E and $\mathcal{B}[0, t] \times \mathcal{F}_t$ in $[0, t] \times \Omega$. (A sufficient condition for this is that E is a metric space with $\mathcal{E}$ the Borel sets and $X_s(\omega)$ is for each ω a right continuous function of s. Usually we are able to arrange things so that this holds.) If T is a finite stopping time and X is progressively measurable then X_T is measurable relative to $\mathcal{F}_T$ and $\mathcal{E}$. If T can take on the value ∞ we regard X_T as being defined only on $\{T < \infty\}$ but it is still $\mathcal{F}_T$ measurable. So if X is progressively measurable the expression $P^{X_T}(J)$ defines, for any set J from $\mathcal{F}^0$, a function on $\{T < \infty\}$ which is $\mathcal{F}_T$ measurable.

We say that the Markov process X has the *strong Markov property* if for each stopping time T, X_T is $\mathcal{F}_T$ measurable and,

$$(4.1) \qquad E^x(f(X_{t+T}); \Gamma \cap \{T < \infty\}) = E^x(P^{X_T}(f(X_t)); \Gamma \cap \{T < \infty\})$$

for each $\Gamma \in \mathcal{F}_T$, positive $\mathcal{E}$ measurable function f, positive t and point $x \in E$. The analogy with (3.4) is better displayed if we express (4.1) as

$$(4.2) \qquad\qquad E^x\big(f(X_{t+T})|\mathcal{F}_T\big) = P^{X_T} f(X_t),$$

which is the same thing once we introduce conventions to deal with the event $\{T = \infty\}$.

For any positive function T on Ω we define the transformation θ_T at least on $\{T < \infty\}$ by

$$\theta_T \omega = \theta_{T(\omega)} \omega$$

so that $f(X_{t+T})$ is written also as $f(X_t) \circ \theta_T$. By iterating (4.1) or (4.2) in the usual way one shows that if X has the strong Markov property then

$$(4.3) \qquad\qquad E^x(J \circ \theta_T | \mathcal{F}_T) = P^{X_T}(J)$$

for all positive $\mathcal{F}^0$ measurable functions J. At this point we have not introduced conventions for dealing with the event that T is infinite, so an accurate interpretation of (4.3) requires an equality between two expectations as in (4.1).

The following theorem will allow us to conclude that practically any transition function that we want to consider is the transition function of a strong Markov process and in fact of one where the filtration is rich enough for the development of a statisfactory potential theory. Suppose we are given a transition function $P(t, x, A)$ on a metric space E. Recall the associated transformation $P_t f(x) = \int f(y) P(t, x, dy)$ of bounded or positive real valued $\mathcal{E}$ measurable functions. (Here $\mathcal{E} = \mathcal{B}(E)$).

(4.4) Theorem. *Suppose that $\{X_t; t \geq 0\}$ is a time homogeneous Markov process defined over a probability space $(\Omega, \mathcal{F}, P)$ with transition function $P(t, x, A)$, that for each ω the sample function $t \to X_t(\omega)$ is right continuous on $[0, \infty)$ and that for each bounded continuous real valued function f on E, $P_t f(x)$ is continuous in x for each positive t. Define the filtration $\{\mathcal{F}_t; t \geq 0\}$ by $\mathcal{F}_t = \sigma\{X_r; r \leq t\}_+$. Then for each $\{\mathcal{F}_t\}$ stopping time T, $X_T I_{T < \infty} \in \mathcal{F}_T$ and*

$$(4.5) \qquad E\big(g(X_{t+T}); \Gamma \cap \{T < \infty\}\big) = E\big(P_t g(X_T); \Gamma \cap \{T < \infty\}\big)$$

for every $\Gamma \in \mathcal{F}_T$, positive t and positive $\mathcal{E}$ measurable function g.

Remark: Clearly this says that the process $\{X_t, \mathcal{F}_t; t \geq 0\}$ is strong Markov, but it is phrased in terms of one measure on Ω rather than a family of them. Also the transformations associated with Brownian motion or reflecting Brownian motion obviously transform bounded continuous functions into continuous functions so these processes are strong Markov, even relative to the somewhat enlarged filtration provided we take versions with continuous sample functions.

Proof. The right continuity property makes the process progressively measurable and so on $\{T < \infty\}$ $X_T \in \mathcal{F}_T$. Define random variables T_n by

$$T_n = \frac{k+1}{2^n} \qquad \text{if} \qquad \frac{k}{2^n} \leq T < \frac{k+1}{2^n},$$

$k = 0, 1, 2, \cdots$, and $T_n = \infty$ if $T = \infty$. One sees easily that each T_n is a stopping time relative to the filtration $\sigma\{X_r; r \leq t\}_+$ and that if Γ is in $\mathcal{F}_T$ then Γ is in $\mathcal{F}_{T_n}$. Since T_n is countably valued, integrals can be broken down to the sets where T_n is constant and then the simple Markov property implies that

$$E\big(g(X_{t+T_n}); \Gamma \cap \{T_n < \infty\}\big) = E\big(P_t g(X_{T_n}); \Gamma \cap \{T_n < \infty\}\big)$$

whenever g is bounded and $\mathcal{E}$ measurable. Now suppose g is continuous and bounded. Clearly T_n decreases to T and $T_n + t$ decreases to $T + t$, and so $g(X_{t+T_n})$ converges to $g(X_{t+T})$ and $P_t g(X_{T_n})$ converges to $P_t g(X_T)$ all convergences taking place boundedly. It follows that (4.5) holds at least for continuous g, but then it holds also for all bounded or positive $\mathcal{E}$ measurable g by the usual measure theory arguments. This completes the proof.

In the applications of interest to us practically always we start with a transition function such that the semigroup $\{P_t; t \geq 0\}$ acts on bounded continuous functions, and an associated process having right continuous paths. So we have the strong Markov property. We might consider other processes obtained by making various analytic or probabilistic transformations of these; but the transformations generally are such that the strong Markov property is easily seen to be preserved even if the sufficient conditions of (4.4) are destroyed.

(4.6) Exercise: For a strong Markov process generalize (3.7) as follows: if T is a stopping time and $G : \Omega \times \Omega \to R$ is bounded and measurable relative to $\mathcal{F}_T \times \mathcal{F}$ then

$$\int_\Omega G(\omega, \theta_T \omega) P^y(d\omega) = \int_\Omega \left\{ \int_\Omega G(\omega, \omega') P^{X_T(\omega)}(d\omega') \right\} P^y(d\omega).$$

(4.7) Exercise: If X is Brownian motion with continuous sample functions, if b and t are strictly positive take $T = \inf\{r | X_r(\omega) = b\}$, $G(\omega, \omega') = I_{\{T(\omega) \leq t, X_{t-T(\omega)}(\omega') \leq b\}}$ in (4.6) to conclude that

$$P^0(\max_{s \leq t} X_s \geq b) = 2P^0(X_t \geq b).$$

(4.8) Remark. Note that in (4.4), and (5.1) of the next section, when E is locally compact separable metric it is enough that the conditions: $P_t f(x)$ is continuous in x, and $P_t f(x) \to f(x)$ as $t \to 0$ hold only for f continuous and vanishing at infinity. This is needed in (6.1).

5. Hitting Times.

The most important random time is the time T_A of hitting a subset A of E. The proper definition is

$$T_A(\omega) = \inf\{t > 0 | X_t(\omega) \in A\},$$

with $T_A(\omega) = \infty$ if for no strictly positive t is $X_t(\omega)$ in A. To ensure that, at least when A is a Borel set, T_A is a stopping time the filtration must have the correct properties. And to benefit from probability arguments, which establish relationships only with probability one, we must augment the σ-algebras with sets of measure 0. Also we need to know that in some sense T_A varies continuously with A, and this leads to a property called quasi-left continuity. We will discuss these matters in this section. Proper choice of the σ-algebras and quasi-left-continuity have deep implications for the subject as a whole; but merely the need to deal properly with the hitting times is motivation enough.

Let $(\Omega, \mathcal{H}, P)$ be a probability space, and suppose $\mathcal{H}$ is complete, that is any subset of an $\mathcal{H}$ set of P measure 0 is already in $\mathcal{H}$. For subsets A and B of E set $A \sim B$ if $A \vartriangle B$ is in $\mathcal{H}$ and $P(A \vartriangle B) = 0$. If $\mathcal{G}$ is a sub σ-algebra of $\mathcal{H}$ define $\overline{\mathcal{G}}$ by saying that $\Gamma \in \overline{\mathcal{G}}$ if there is a set $\Delta \in \mathcal{G}$ with $\Gamma \sim \Delta$. Obviously $\overline{\mathcal{G}}$ is a sub σ-algebra of $\mathcal{H}$ containing $\mathcal{G}$ and if $\mathcal{G}$ is contained in another sub σ-algebra $\mathcal{F}$ then $\overline{\mathcal{G}}$ is contained in $\overline{\mathcal{F}}$. Thus if $\{\mathcal{F}_t; t \geq 0\}$ is a filtration of $(\Omega, \mathcal{H})$ then $\{\overline{\mathcal{F}_t}; t \geq 0\}$ is a filtration also, and, as one can verify easily, if the original filtration is right continuous then so is the $\{\overline{\mathcal{F}_t}\}$ filtration. So suppose we are given a right continuous filtration $\{\mathcal{F}_t; t \geq 0\}$. We will show in a moment that if T is a stopping time relative to $\{\overline{\mathcal{F}_t}\}$ then there is a stopping time T' relative to $\{\mathcal{F}_t\}$ such that $P(T' \neq T) = 0$ and that if $\Lambda \in \overline{\mathcal{F}}_T$ then there is a set $\Lambda' \in \mathcal{F}_{T'}$ such that $\Lambda \sim \Lambda'$. The implication of this for processes is that, if the filtration is right continuous, then once a relationship like (4.1) is established, it continues to make sense and to be valid for the larger class of stopping times one obtains when we let $\mathcal{H}$ denote the P^x completion of $\sigma\{\mathcal{F}_t; t \geq 0\}$ and replace $\mathcal{F}_t$ by $\overline{\mathcal{F}_t}$. The processes covered by Theorem (4.4) are instances of this. To show the existence of T' and Λ' let $\Gamma_{kn} = \left\{\frac{k-1}{2^n} \leq T < \frac{k}{2^n}\right\}$ and define an $\{\overline{\mathcal{F}_t}\}$ stopping time T_n by setting $T_n(\omega) = k/2^n$ if $\omega \in \Gamma_{k,n}$. (Let us assume that T is finite, to save a few extra sentences.) The set Γ_{kn} is in $\overline{\mathcal{F}_{k/2^n}}$

and so there is a set Δ_{kn} in $\mathcal{F}_{k/2^n}$ such that $\Delta_{kn} \sim \Gamma_{kn}$. We may assume that the sets $\Delta_{1n}, \Delta_{2n}, \cdots$ are pairwise disjoint. Then if we define R_n by $R_n(\omega) = k/2^n$ if $\omega \in \Delta_{kn}$, and $R_n(\omega) = \infty$ if $\omega \in (\bigcup_k \Delta_{kn})^c$ we obtain a sequence $\{R_n\}$ of $\{\mathcal{F}_t\}$ stopping times and for each n, $P(R_n \neq T_n) = 0$. As n increases to infinity T_n decreases to T. If $T' = \inf_n R_n$ then T' is an $\{\mathcal{F}_t\}$ stopping time since the $\mathcal{F}_t$ filtration is right continuous. Clearly $P(T' \neq T) = 0$. Let $\Delta'_{kn} = \{(k-1)/2^n \leq T' < k/2^n\}$ and define $T'_n(\omega)$ to be $k/2^n$ if $\omega \in \Delta'_{kn}$. If Λ is a set in $\overline{\mathcal{F}}_T$ then $\Lambda \cap \Gamma_{kn} \in \overline{\mathcal{F}_{k/2^n}}$ and so there is a set $\Lambda_{kn} \in \mathcal{F}_{k/2^n}$ with $\Lambda_{kn} \sim \Lambda \cap \Gamma_{kn}$, and we may assume $\Lambda_{kn} \subset \Delta'_{kn}$ since $\Delta'_{kn} \sim \Gamma_{kn}$. Let $\Lambda_n = \bigcup_{k=1}^{\infty} \Lambda_{kn}$. Since $\Lambda_n \cap \{T'_n = k/2^n\}$ is in $\mathcal{F}_{k/2^n}$ for all k, Λ_n is a set in $\mathcal{F}_{T'_n}$. Regardless of n, $\Lambda_n \sim \Lambda$ so the various Λ_n are essentially the same. Set $\Lambda' = \lim \inf \Lambda_n$. Then $\Lambda' \sim \Lambda$. The set Λ' is in $\mathcal{F}_{T'_n}$ for each n and hence is in $\mathcal{F}_{T'}$ since T'_n decreases to T' and the right continuity is preserved when one passes from fixed times to stopping times.

Let $X = (\Omega, \mathcal{F}, \mathcal{F}_t, X_t, \theta_t, P^x)$ be a time homogemeous Markov process. We will suppose also that X is strong Markov. Then X is said to be *quasi-left-continuous* provided that whenever $\{T_n\}$ is an increasing sequence of stopping times (relative to $\{\mathcal{F}_t\}$, of course) with limit T it follows that X_{T_n} converges to X_T on $\{T < \infty\}$ almost surely P^x for each x. The assumption that X is strong Markov has no obvious purpose in this definition except to ensure that X_{T_n} and X_T are measurable, but quasi-left-continuity is useful only in conjunction with the strong Markov property. For a process like Brownian motion where we realize the transition function with a process having continuous sample functions the convergence of X_{T_n} to X_T is automatic. However many important transition functions can be realized by processes whose sample functions are regular only to the extent that they are right continuous and have left hand limits on $[0, \infty)$; and for these quasi-left-continuity turns out to be an adequate substitute for the missing left continuity. One obvious consequence of quasi-left-continuity is that if $\{T_n\}$ is a sequence of stopping times increasing to T then T_n must in fact equal T for all large n if T is finite and the path is discontinuous there. This is the starting point for Paul-André Meyer's remarkable theory of accessibility of stopping times, which is basic to the subject as a whole. But for us it will be enough to give the standard theorem showing that most processes of interest are quasi-left-continuous.

(5.1) Theorem. *Let $\{X_t, \mathcal{F}_t; t \geq 0\}$ be a time homogeneous Markov process over $(\Omega, \mathcal{F}, P)$ with transition function $P(t, x, A)$. Suppose that for each ω the sample function $t \to X_t(\omega)$ is right continuous and has left hand limits everywhere in $[0, \infty)$. Suppose also that for each bounded continuous function f on E, $P_t f(x)$ defines for each t a continuous function of x and that for such f, $P_t f(x) \to f(x)$ for each x as $t \to 0$. Then whenever $\{T_n\}$ is an increasing sequence of stopping times with limit T, $\lim\limits_{n \to \infty} X_{T_n} = X_T$ on $\{T < \infty\}$ P-almost surely.*

Remark: As with Theorem (4.4) we have phrased the theorem in terms of one measure on a probability space rather than a family of them.

Proof: Since we can first take the minimum of the stopping times with a constant and then let the constant increase to infinity it will suffice to prove the theorem assuming T is finite or even bounded. Note that the hypotheses of Theorem (4.4) are satisfied so our process has the strong Markov property. Since left hand limits exist, X_{T_n} approaches an E-valued random variable, Z, and for each strictly positive t, $X_{T_n + t}$ approaches a value Z_t. By the right continuity Z_t approaches X_T as t decreases to zero. Let f and g be bounded continuous functions on E. Then

$$
\begin{aligned}
E\big(f(Z)g(X_T)\big) &= \lim_{t \to 0} \lim_{n \to \infty} E\big(f(X_{T_n})g(X_{T_n + t})\big) \\
&= \lim_{t \to 0} \lim_{n \to \infty} E\big(f(X_{T_n})P_t g(X_{T_n})\big) \\
&= \lim_{t \to 0} E\big(f(Z)P_t g(Z)\big) \\
&= E\big(f(Z)g(Z)\big)
\end{aligned}
$$

and by a monotone class assignment this implies that P-almost surely $X_T = Z$ as was required.

Next we turn to the hitting times, T_A, and to the manner in which they vary with A. Hunt established the properties of T_A by an imaginative use of Choquet's capacitability theorem. The modern début and section theorems of Dellacherie and Meyer put capacitability into a more measure theoretic setting, and allow one to weaken the hypotheses on the state space. We will content ourselves with sketching Hunt's approach. It is adequate for our needs and spares us investing time in understanding the background for the section theorem.

Let $\{X_t, \mathcal{F}_t; t \geq 0\}$ be a Markov process over a probability space $(\Omega, \mathcal{F}, P)$. We will assume that the state space E is a locally compact metric space with a countable base for the topology. Also we will assume that for each $\omega \in \Omega$ the sample function $t \to X_t(\omega)$ is right continuous on $[0, \infty)$ and has left-hand limits (in E) on $(0, \infty)$. We will assume that the filtration $\{\mathcal{F}_t\}$ is right continuous and finally that whenever T_n is an increasing sequence of $\{\mathcal{F}_t\}$ stopping times with limit T then X_{T_n} converges to X_T, P almost surely on $\{T < \infty\}$ that is quasi-left-continuity of the fixed process at hand. The properties of hitting times we wish to establish require the adjoining of sets of probability zero. But by the discussion at the beginning of this section none of the above hypotheses are lost if we complete $\mathcal{F}$ relative to P and then complete each $\mathcal{F}_t$ within $\mathcal{F}$. We will assume that this has already been done, that is that $\Gamma \in \mathcal{F}$ and $P(\Gamma) = 0$ implies that $\Delta \in \mathcal{F}$ whenever $\Delta \subset \Gamma$, and that $\Gamma \in \mathcal{F}_t$, $\Delta \in \mathcal{F}$ and $P(\Gamma_\Delta \Delta) = 0$ implies that $\Delta \in \mathcal{F}_t$.

Let $[a, b]$ be a fixed closed interval in $[0, \infty)$. If A is a subset of E let

$$\widetilde{A} = \{\omega | X_t(\omega) \in A \quad \text{for some } t \in [a, b]\},$$

let $\mathcal{A}$ denote the class of subsets A of E such that $\widetilde{A} \in \mathcal{F}_b$, and define a set function φ on $\mathcal{A}$ by $\varphi(A) = P(\widetilde{A})$. If A is an open set then by the right continuity of the paths $\omega \in \widetilde{A}$ if and only if $X_b(\omega) \in A$ or $X_q(\omega) \in A$ for some rational $q \in [a, b]$, so any open set is in $\mathcal{A}$. Let A be a compact set and let O_n be a decreasing sequence of open sets with compact closures such that A is the intersection of the closures of the O_n. Define random times T_n by

$$\begin{aligned}
T_n(\omega) &= \inf \ \{t \in [a, b] | X_t(\omega) \in O_n\} \qquad \omega \in \widetilde{O}_n \\
&= b + 1 \qquad\qquad\qquad\qquad\qquad\ \omega \in (\widetilde{O}_n)^c.
\end{aligned}$$

Then T_n is a stopping time, even relative to the filtration $\sigma\{X_r; r \leq t\}_+$, and $\{T_n\}$ is an increasing sequence. If T denotes the limit then the fact that X_{T_n} is in the closure of O_n if $T_n \leq b$ implies that if $X_{T_n} \to X_T$ then $\widetilde{A} = \bigcap_n \widetilde{O}_n$. By quasi-left-continuity this holds almost surely and so $A \in \mathcal{A}$ and $\varphi(A) = \lim_n \varphi(O_n)$. Now consider the set function φ restricted to the compact subsets of E. It satisfies the three hypotheses: (i) if $A \subset B$ then $\varphi(A) \leq \varphi(B)$, (ii) given any A and $\varepsilon > 0$ there is an open set G such that

$A \subset G$ and $\varphi(B) < \varphi(A) + \varepsilon$ whenever $B \subset G$, (iii) $\varphi(A \cup B) + \varphi(A \cap B) \leq \varphi(A) + \varphi(B)$. The first statement is obvious, the second follows from the argument that showed $\mathcal{A}$ to contain the compact sets, and the third follows from the fact that $\varphi(A \cup B) - \varphi(B)$ is the probability that the set $\{X_t; a \leq t \leq b\}$ meets $A \cup B$ but not A, while $\varphi(A) - \varphi(A \cap B)$ is the probability of the larger event that the random set meets A but not $A \cap B$. A set function on the compact subsets of a locally compact separable metric space which has these three properties is called a *Choquet capacity*. Choquet's capacitability theorem says that if we define an *inner capacity* φ_* and an *outer capacity* φ^* on all subsets of E by

$$\varphi_*(A) = \sup \varphi(K)$$
$$\varphi^*(A) = \inf \varphi_*(O)$$

where the supremum is over all compact K contained in A and the infimum is over all open O containing A then any analytic set, A, is *capacitable* in the sense that $\varphi^*(A)$ is equal to $\varphi_*(A)$.. Let us apply this to our situation. If O is open then it is the limit of an increasing sequence $\{K_n\}$ of compacts and it follows immediately that $\varphi^*(O)$ is the same as $\varphi(O)$. If A is analytic (in particular if A is a Borel set) the fact that A is capacitable implies that there is an increasing sequence $\{K_n\}$ of compact sets contained in A and a decreasing sequence $\{O_n\}$ of open sets containing A such that $P(\widetilde{O}_n) - P(\widetilde{K}_n)$ is less than $1/n$. Now $\widetilde{K}_n \subset \widetilde{A} \subset \widetilde{O}_n$ and it follows that $\widetilde{A} \in \mathcal{A}$ and that $\varphi(A)$ is well approximated by $\varphi(K)$ or $\varphi(O)$ as K ranges over compact subsets of A, and O ranges over open neighborhoods of A. This immediately implies that for any Borel set A, T_A is an $\{\mathcal{F}_t\}$ stopping time: specifically if t is strictly positive then $T_A(\omega) < t$ if and only if for some integer n, ω is in the set $\{X_s \in A, \text{ some } s \in [1/n, t - 1/n]\}$. Each of these sets is known already to be in $\mathcal{F}_t$. So $\{T_A < t\}$ is in $\mathcal{F}_t$ for $t > 0$, and since the filtration is right continuous this is enough. Notice that we weren't allowed to use the variable interval $[0, t - 1/n]$ since T_A is defined in terms of X_t for t strictly positive. The capacitability theorem also yields the following two important facts about approximation of hitting times: (i) given a Borel set A there is an increasing sequence $\{K_n\}$ of compact subsets of A such that with probability one T_{K_n} decreases to T_A, (ii) if $P(X_0 \in A) = 0$ then there is a decreasing sequence $\{O_n\}$ of open sets containing A such that with probability one T_{G_n} increases to T_A. The reader should be able to supply the proofs and to see the need for the extra condition in statement (ii).

6. Standard Processes.

Given a stationary transition function, in addition to knowing that it can be realized by a time homogeneous Markov process we would like to know that we can take the process to have regular sample functions and to have the strong Markov and quasi-left-continuity properties. We will state the standard theorem giving sufficient conditions for this and will sketch the proof.

Let E be a locally compact separable metric space with $\mathcal{E}$ the Borel sets of the topology. Let $P(t, x, A)$ be a stationary transition function on $(E, \mathcal{E})$. As usual we write $P_t f$ for the associated transformation, $P_t f(x) = \int f(y) P(t, x, dy)$, of bounded $\mathcal{E}$ measurable functions. Let C_0 denote the set of real valued continuous functions on E which vanish at ∞.

(6.1) Theorem. *Suppose that for each $f \in C_0$ and $t \geq 0$, $P_t f \in C_0$ and that $P_t f \to f$ uniformly as $t \to 0$. Then given any probability measure ν there is a time homogeneous Markov process $\{X_t, \mathcal{F}_t; t \geq 0\}$ having $P(t, x, A)$ as transition function and initial distribution ν, whose sample functions are right continuous and have left hand limits, whose filtration is right continuous and which has the strong Markov property and is quasi-left continuous.*

Proof: Really all that needs to be established is the regularity properties of the sample functions, because Theorem (4.4) says that then we will have the strong Markov property when the filtration is $\sigma\{X_r; r \leq t\}_+$, which is indeed right continuous, and the quasi-left-continuity will follow from (5.1). Let $\{x_t; t \geq 0\}$ denote a Markov process with initial distribution ν and transition function $P(t, x, A)$ defined over a probability space $(\Omega, \mathcal{F}, P)$. Certainly one exists because (2.4) specifies finite dimensional distributions in a consistent way, and the topological conditions on E allow one to apply Kolmogorov's consistency theory. We will argue that we can modify $\{x_t\}$ so as to obtain a process $\{X_t; t \geq 0\}$ whose sample functions are right continuous with left limits and such that for each t $P\{X_t = x_t\} = 1$. Let Q denote a countable dense subset of $[0, \infty)$. First we will argue that for any finite t the set $\{x_q; q \in Q \cap [0, t]\}$ is with probability one contained in a compact subset of E. Given any bounded $\mathcal{E}$ measurable function f and $\alpha > 0$ the "α-potential" $U^\alpha f(x) = \int_0^\infty e^{-\alpha t} P_t f(x) dt$ is defined and $\mathcal{E}$ measurable because $P_t f(x)$ is jointly measurable in (t, x). If f is also positive then obviously $e^{-\alpha t}P_t U^\alpha f \le U^\alpha f$ and it follows from this and the Markov property that $\{e^{-\alpha t}U^\alpha f(x_t); t \ge 0\}$ is a non-negative supermartingale relative to the filtration $\sigma\{x_r; r \le t\}$. From a standard martingale fact it follows that for any t the event

$$\{e^{-\alpha t}U^\alpha f(x_t) > 0, \inf_{s \in Q \cap [0,t]} e^{-\alpha s}U^\alpha f(x_s) = 0\}$$

has probability 0. If f is in C_0 and strictly positive then so is $U^\alpha f$. The function $U^\alpha f$ vanishes at ∞ and so our boundedness assertion follows. Next we will argue that with probability one the sample functions of x_t with t restricted to lie in Q have right and left hand limits everywhere. Another standard martingale fact asserts that with probability one the sample functions of a positive supermartingale are this regular. In particular if f is bounded and positive and $\mathcal{E}$ measurable then with probability one the sample functions of $\{e^{-\alpha t}U^\alpha f(x_t); t \in Q\}$ have right and left limits everywhere in $[0, \infty)$. For $f \in C_0$, $\alpha U^\alpha f \to f$ uniformly as $\alpha \to \infty$. Thus the countable family $\alpha U^\alpha f$, as α ranges over the integers and f ranges over a countable dense subset of C_0 is uniformly dense in C_0 and it follows that with probability one $\{f(x_t); t \in Q\}$ has right and left limits everywhere for all $f \in C_0$. Now if x_t denotes any function from Q to E which is bounded on finite intervals then oscillatory behavior of $t \to x_t$ will be detected by the compositions $f(x_t)$ as f ranges over C_0. Our initial assertion about the sample functions of $\{x_t; t \in Q\}$ follow immediately. Remove from Ω the P-null set where these limits don't exist and then set

$$X_t = \lim_{q \in Q, q \downarrow t} x_q.$$

The process $\{X_t\}$ is right continuous and has left hand limits. To show that $P\{x_t = X_t\} = 1$ for each fixed t take functions f and g in C_0. Then

$$\begin{aligned} E\big(f(x_t)g(X_t)\big) &= \lim_{q \downarrow t} E\big(f(x_t)g(x_q)\big) \\ &= \lim_{q \downarrow t} E\big(P_{q-t}g(x_t)f(x_t)\big) \\ &= E\big(f(x_t)g(x_t)\big), \end{aligned}$$

and by standard reasoning the assertion follows.

Let $X = (\Omega, \mathcal{F}, \mathcal{F}_t, X_t, \theta_t, P^x)$ be a time-homogeneous Markov process. Then X is called a *standard process* if in addition to the basic properties from section 3, the state space E is a locally compact separable metric space with $\mathcal{E}$ the Borel sets of the topology, the sample functions $t \to X_t$ are right continuous and have left hand limits on $[0, \infty)$, the strong Markov property and quasi-left continuity hold, and the filtration $\{\mathcal{F}_t\}$ is right continuous and the $\mathcal{F}_t$ have been completed in an appropriate way. The meaning of this last phrase is that whenever Γ is a subset of Ω having the property that for each x there is a set $\Gamma_1 \in \mathcal{F}_t$ and a set $\Gamma_2 \in \mathcal{F}$ with $\Gamma \triangle \Gamma_1 \subset \Gamma_2$ and $P^x(\Gamma_2) = 0$ then Γ itself is in $\mathcal{F}_t$. What this amounts to saying is that each $\mathcal{F}_t$ is for each x closed with respect to the operation of P^x completion within $\mathcal{F}$ as described in section 5. Any transition function satisfying the hypothesis of Theorem 6.1 can be realized by a standard process. To see this let Ω be the set of all functions from $[0, \infty)$ to E which are right continuous and have left limits on $[0, \infty)$ and let X_t denote the coordinate function, $X_t(\omega) = \omega(t)$. Let $\mathcal{G}_t = \sigma\{X_r; r \leq t\}$, $\mathcal{G} = \sigma\{X_r; r < \infty\}$ and θ_t be the usual shift, $(\theta_t\omega)(r) = \omega(t + r)$. Given a point $x \in E$ let $\{Y_t; t \geq 0\}$ be a time homogeneous Markov process over a probability space $(\Omega', \mathcal{F}', P')$ having ε_x as initial distribution, paths which are right continuous with left limits and the given $P(t, x, A)$ as transition function. Theorem 6.1 guarantees the existence of such a process. The mapping Y from Ω' to Ω defined by $(Y(\omega'))(t) = Y_t(\omega')$ is measurable relative to $\mathcal{F}'$ and $\mathcal{G}$, and if P^x denotes its distribution then $(\Omega, \mathcal{G}, \mathcal{G}_t, X_t, \theta_t, P^x)$ is a process having right continuous left limit paths and the desired transition function. According to Theorem 4.4 this process is strong Markov relative to the right continuous filtration $\{\mathcal{G}_{t+}\}$ and by Theorem 5.1 it is quasi left continuous. Let $\mathcal{F}^x$ denote the σ-algebra obtained by completing $\mathcal{G}$ relative to P^x and $\mathcal{F}_t^x$ denote the completion relative to P^x of $\mathcal{G}_{t+}$ within $\mathcal{F}^x$. Then let $\mathcal{F}_t = \cap\mathcal{F}_t^x$, $\mathcal{F} = \cap\mathcal{F}^x$ where the intersections are over all $x \in E$. The σ-algebra $\mathcal{F}$ and filtration $\{\mathcal{F}_t\}$ will have the desired properties.

Standard processes are the ones with which one can operate freely; but a number of measurability and technical issues will occur to the careful reader. As an example, let A be a Borel set in E. We know that T_A is a stopping time. In fact if we take a finite measure μ on $\mathcal{E}$, form the measure $P^\mu = \int P^x \mu(dx)$ on $\mathcal{G}$ and let $\mathcal{H}_t$ denote the P^μ completion of $\mathcal{G}_{t+}$ then T_A is an $\{\mathcal{H}_t\}$ stopping time. Now suppose we take a second Borel set B and form $T_B + T_A \circ \theta_{T_B}$, that is the first time of hitting A after hitting

B. To prove that this is an $\{\mathcal{F}_t\}$ stopping time we note that the event $\{T_B + T_A \circ \theta_{T_B} < t\}$ is the union of the events $\{T_B < q, T_A \circ \theta_{T_B} < r\}$ as q and r range over the rational pairs whose sum is less than t. Then we find ourselves having to show that $\theta_{T_B}^{-1}\{T_A < r\} \in \mathcal{F}_{T_B + r}$. This would be immediate if $\{T_A < r\}$ were in $\mathcal{G}_r$ rather than some completion. However given $x \in E$ let μ be the measure $\mu(C) = P^x(X_{T_B} \in C)$ and then let Γ_1 and Γ_2 be sets in $\mathcal{G}_r$ such that $\Gamma_1 \subset \{T_A < r\} \subset \Gamma_2$ and $P^\mu(\Gamma_2 - \Gamma_1) = 0$. Then $\theta_{T_B}^{-1}(\Gamma_1) \subset \theta_{T_B}^{-1}(\{T_A < r\}) \subset \theta_{T_B}^{-1}(\Gamma_2)$. The extremes are indeed in $\mathcal{F}_{T_B + r}$ and by the strong Markov property their difference has P^x measure 0, which is all we must show. The reader should be able to supply such arguments when intuition is regarded as suspect. All such issues are treated carefully in Chapter I of [BG,1].

7. Killed and Stopped Processes.

Let $X = (\Omega, \mathcal{F}, \mathcal{F}_t, X_t, \theta_t, P^x)$ be a standard process and let T be an $\{\mathcal{F}_t\}$ stopping time having the additional property that for each positive t and $x \in E$, $t + T \circ \theta_t = T$ on $\{t < T\}$ almost surely P^x. Such a stopping time is called a *terminal time*. The most obvious example is the hitting time of a set in $\mathcal{E}$, and in this case the defining almost everywhere equality holds as an identity. Suppose in addition that for each positive t and set $A \in \mathcal{E}$ the function $x \to P^x(X_t \in A, t < T)$ is an $\mathcal{E}$ measurable function. It is easy to see that this will be the case if T is the time of hitting an open subset of E, and an argument using the quasi-left-continuity as in section 5 shows that it is also the case if T is the time of hitting a closed set. For $x \in E$, $t > 0$, and $A \in \mathcal{E}$ set

$$(7.1) \qquad Q(t, x, A) = P^x(X_t \in A; t < T).$$

This is a measure in A and a Borel function in x, and from the simple Markov property and the terminal time property it follows that $Q(t, x, A)$ satisfies the Chapman-Kolmogorov equation; that is it has all the properties of a stationary transition function except that $Q(t, x, E)$ might be less than 1. We will call such a thing a *sub-Markov transition function*, regardless of whether or not it arises as above. Any sub-Markov transition function can be turned into a genuine transition function by the following standard device. Adjoin to E an additional point Δ, obtaining a new set $E_\Delta = E \cup \{\Delta\}$. The σ-algebra $\mathcal{E}_\Delta$ in E_Δ consists of all sets $A \in \mathcal{E}$ and all sets $A \cup \{\Delta\}$ with $A \in \mathcal{E}$. Define a transition function $\overline{P}(t, x, A)$ over $(E_\Delta, \mathcal{E}_\Delta)$ by

$$(7.2) \qquad \begin{aligned} \overline{P}(t, x, A) &= Q(t, x, A) & x \in E, A \in \mathcal{E} \\ \overline{P}(t, x, \{\Delta\}) &= 1 - Q(t, x, E) & x \in E \\ \overline{P}(t, \Delta, \{\Delta\}) &= 1. \end{aligned}$$

One checks immediately that $\overline{P}(t, x, A)$ is a transition function on $(E_\Delta, \mathcal{E}_\Delta)$. Suppose that T is a terminal time and in fact satisfies the stronger condition that $T = t + T \circ \theta_t$ on $\{t < T\}$ without exception. Suppose Q is defined by (7.1) and $\overline{P}$ is defined by (7.2). Define on Ω a family of functions $\{\overline{X}_t; t \geq 0\}$ by

$$\begin{aligned} \overline{X}_t &= X_t & t < T \\ &= \Delta & t \geq T. \end{aligned}$$

Then $\overline{X}_t$ takes values in E_Δ and it is immediate that for each t, $\overline{X}_t$ is measurable relative to $\mathcal{F}_t$ and $\mathcal{E}_\Delta$; that is $\{\overline{X}_t, \mathcal{F}_t; t \geq 0\}$ is a stochastic process with state space E_Δ. Let R be an $\{\mathcal{F}_t\}$ stopping time, t a positive number, Λ a set in $\mathcal{F}_R$, f a positive $\mathcal{E}_\Delta$ measurable function with $f(\Delta) = 0$ and x a point of E. Then

$$
\begin{aligned}
E^x\left(f(\overline{X}_{R+t}); \Lambda\right) &= E^x\left(f(X_{R+t})I_{R+t<T}; \Lambda\right) \\
&= E^x\left(f(X_{R+t})I_{R+t<T}; \Lambda \cap \{R < T\}\right) \\
&= E^x\left(\left(f(X_t)I_{t<T}\right) \circ \theta_R; \Lambda \cap \{R < T\}\right) \\
&= E^x\left(\overline{P}(t, \overline{X}_R, f); \Lambda\right)
\end{aligned}
$$

by the strong Markov property of X and the fact that $\overline{X}_R = \Delta$ if $R \geq T$. The hypothesis that $f(\Delta) = 0$ is easily dispensed with. The conclusion is that relative to P^x the process $\{\overline{X}_t, \mathcal{F}_t, t \geq 0\}$ is not only a Markov process with $\overline{P}(t, x, A)$ as transition function but is a strong Markov process. The process $\{\overline{X}_t; t \geq 0\}$ is called "X killed at time T", the idea being that at time T the path moves to a special point Δ where it remains forever. One would like to set up $\overline{X}$ so as to be a standard process. Up to a point this presents no problem: adjoin a point ω_Δ to Ω and extend the σ-algebra $\mathcal{F}_t$ and $\mathcal{F}$ so that $\{\omega_\Delta\}$ is also an element, calling the resulting objects $\overline{\Omega}, \overline{\mathcal{F}}_t$ and $\overline{\mathcal{F}}$. Define the measure P^Δ on $\overline{\mathcal{F}}$ by $P^\Delta(\{\omega_\Delta\}) = 1$ and extend the definition of $\overline{X}_t$ to $\overline{\Omega}$ by setting $\overline{X}_t(\omega_\Delta) \equiv \Delta$. Extend the operators θ_t by setting

$$
\begin{aligned}
\overline{\theta}_t(\omega) &= \theta_t(\omega) & \omega \in \Omega, t < T(\omega) \\
&= \omega_\Delta & \omega \in \Omega, t \geq T(\omega) \\
\overline{\theta}_t(\omega_\Delta) &= \omega_\Delta & t \geq 0.
\end{aligned}
$$

Then $\overline{X}_t \circ \overline{\theta}_h = \overline{X}_{t+h}$. Now it is easy to check that $(\overline{\Omega}, \overline{\mathcal{F}}, \overline{\mathcal{F}}_t, \overline{X}_t, \overline{\theta}_t, P^x)_{x \in E_\Delta}$ is a time homogeneous Markov process with state space E_Δ and $\overline{P}(t, x, A)$ as transition function. For considering the sample function properties assume that E_Δ is topologized so that the original topology on E is the subspace topology. Then the sample functions of $\overline{X}_t$ are right continuous and have left hand limits because those of X_t do. As to quasi-left-continuity, if $\{R_n\}$ is an increasing sequence of $\{\mathcal{F}_t\}$ stopping times with limit R then the quasi-left continuity of X will imply that if $R < \infty$ then $\overline{X}_{R_n} \to \overline{X}_R$ except perhaps if $R = T$ and $R_n < R$ for all n. This possibility can't be ruled out so we have to be more careful in using the property. Also some difficulty may arise concerning the state space. For example suppose T is the time of hitting a set A. It seems reasonable that we should remove from E all points x that are *regular* for A in the sense that $P^x(T_A = 0) = 1$, because for such points $P^x(\overline{X}_0 = \Delta) = 1$. The remaining subset of E might not be locally compact in the subspace topology, so another one of the conditions for a standard process could be lost.

For general probabilistic potential theory the existence of such situations can't be ignored and the attendant difficulties must be dealt with. But we will deal only with the case in which the killing time is the time of hitting a closed subset K of E. Then we will take $E - K$ as our state space and include Δ as an isolated point if $E - K$ is compact and as the point at ∞ for $E - K$ otherwise. If stopping times R_n increase to R then on $\{R = T_K < \infty\}$ either $R_n = T_K$ for all large n or $X_{R_n} \to \Delta$ in the topology of $(E - K) \cup \Delta$ and $\overline{X}_R$ will equal Δ.

In fact even the existence of this solution is not important. We definitely will be concerned with the sub-Markov transition function $Q(t, x, A)$ obtained by killing a standard process at a terminal time, but nothing we do will require the existence of a standard process which realizes the transition function $\overline{P}(t, x, A)$ which we obtain from Q by putting the deficient mass at the point Δ.

There is another variant of the killed process that will be important. Let K be a closed subset of E and define a stopping time T by

$$T = \inf\{t \geq 0 | X_t \in K\}.$$

The only difference between T and T_K is the inclusion of $t = 0$ in the set of admissible time points. Using arguments like those in section 5 one shows rather easily that for each positive $\mathcal{E}$ measurable function h the expressions $E^x(h(X_t); t < T)$, $E^x(h(X_T); T < \infty)$ and $E^x(h(X_T); t \geq T)$ define $\mathcal{E}$ measurable functions of x. Define $\mathcal{P}(t, x, A)$ by setting

$$\mathcal{P}(t, x, A) = P^x(X_{t \wedge T} \in A).$$

What we have just said implies that $\mathcal{P}(t, x, A)$ is $\mathcal{E}$ measurable in x. The fact that T is a terminal time and the Markov property imply that $\mathcal{P}(t, x, A)$ is a stationary transition function. The process $\{X_{t \wedge T}; t \geq 0\}$ is easily seen to be a time homogeneous Markov process and $\mathcal{P}(t,x,A)$ is its transition function. For obvious reasons it is called "X stopped at the time of hitting K", with the understanding that if $X_0 \in K$ then X is stopped immediately, even if T_K is strictly positive.

(7.3) **Exercise:** Let X be Brownian motion, with continuous sample functions, let σ be the time it hits $\{0\}$ and let h be a positive Borel function on R. Take $G(\omega,\omega') = h\big(X_{t-\sigma(\omega)}(\omega')\big) I_{t \geq \sigma(\omega)}$ and apply (4.6) to show that $P^x\big(h(X_t); \sigma \leq t\big)$ defines an even function of x. Note that if h vanishes on $(-\infty, 0]$ and x is negative then $P^x\big(h(X_t)\big) = P^x\big(h(X_t); \sigma \leq t\big)$ and conclude that $p^-(t,x,y) = p(t,x,y) - p(t,-x,y)$ with $x > 0, y > 0$ is the transition density for Brownian motion on $(0,\infty)$ killed when it reaches 0. (Here $p(t,x,y)$ is the Gauss kernel from section 3.)

8. Canonical realizations.

If $X = (\Omega, \mathcal{F}, \mathcal{F}_t, X_t, \theta_t, P^x)$ is a standard process then certain statements about X might be intuitively more clear or at least notationally simpler if one assumes that Ω is the space of functions from $[0, \infty)$ to E which are right continuous with left limits, X_t are the coordinate functions and so forth. We will point out in a moment that something on this order always can be done. But we do not want to insist that the framework of "path space" be used always, for doing so will deprive us of a lot of useful flexibility, as shown by the various presentations of reflecting Brownian motion, each with its own advantages.

So given a standard process X as above with state space E let U denote the set of all function u from $[0, \infty)$ to E which are right continuous and have left limits, let $\{x_t; t \geq 0\}$ denote the coordinate functions, $x_t(u) = u(t)$ and let $\mathcal{U}^0$ and $\mathcal{U}_t^0$ denote respectively $\sigma\{x_r; r \geq 0\}$ and $\sigma\{x_r; r \leq t\}$. We will use θ_t to denote the shift operator, $\theta_t u(s) = u(s + t)$, and rely on the context to distinguish this shift from the one associated with our original standard process. Let X also denote the mapping from Ω to U defined by

$$X(\omega)(t) = X_t(\omega).$$

Clearly X is measurable relative to $\mathcal{F}$ and $\mathcal{U}^0$ and so for each x we have a measure $\mathcal{P}^x$ on $\mathcal{U}^0$ characterized by the fact that the finite dimensional distributions under P^x of the process $\{X_t; t \geq 0\}$ are the same as those of $\{x_t; t \geq 0\}$ under $\mathcal{P}^x$. Consequently the x process is a time homogeneous Markov process with the same transition function as that of the X process. In fact if we define $\mathcal{U}$ to be the intersection over all x of the $\mathcal{P}^x$ completion of $\mathcal{U}^0$ and define $\mathcal{U}_t$ to be the intersection over all x of the $\mathcal{P}^x$ completion within $\mathcal{U}$ of $\mathcal{U}_{t+}^0$ then we claim that $x = (U, \mathcal{U}, \mathcal{U}_t, x_t, \theta_t, \mathcal{P}^x)$ is a standard process. Basically what has to be established to substantiate this claim is the right continuity and completeness properties of the σ-algebras, and the strong Markov and quasi-left-continuity properties. The fact that the x process has these properties can be thrown back on the corresponding properties of the X process as follows: we note, to start with, that the mapping X is measurable relative to $\mathcal{F}$ and $\mathcal{U}$. Indeed fix a point x in E and let Γ_0 be a set in $\mathcal{U}$. Then there are sets Γ_1 and Γ_2 in $\mathcal{U}^0$ with $\Gamma_0 \triangle \Gamma_1 \subset \Gamma_2$ and $\mathcal{P}^x(\Gamma_2) = 0$. If Γ_i' denotes $X^{-1}(\Gamma_i)$ then $\Gamma_0' \triangle \Gamma_1' \subset \Gamma_2'$ and

$P^x(\Gamma_2') = 0$. Since Γ_1' and Γ_2' are in $\mathcal{F}$ it follows that Γ_0' is in $\mathcal{F}$. The same argument, starting with the fact that X is measurable relative to $\mathcal{F}_t$ and $\mathcal{U}_{t+}^0$ shows that X is measurable relative to $\mathcal{F}_t$ and $\mathcal{U}_t$. Now let T be a $\{\mathcal{U}_t\}$ stopping time, and assume, to save a few words, that T is finite. Make the usual approximation

$$T_n = (k+1)/2^n \qquad \text{if} \qquad k/2^n \leq T < (k+1)/2^n.$$

If we define a function T_n' on Ω by

$$T_n'(\omega) = (k+1)/2^n \quad \text{if} \quad k/2^n \leq T(X(\omega)) < (k+1)/2^n$$

then each T_n' is a $\{\mathcal{F}_t\}$ stopping time, and as n increases, T_n' decreases to a $\{\mathcal{F}_t\}$ stopping time T' with $T' = T \circ X$. The fact that for each t, X is measurable with respect to $\mathcal{F}_t$ and $\mathcal{U}_t$ implies that if Γ is a set in $\mathcal{U}_T$ then $\Gamma' = X^{-1}(\Gamma)$ is in $\mathcal{F}_{T'}$, and furthermore it is obvious that $x_T \circ X = X_{T'}$. Now the x process is progressively measurable and so x_T is $\mathcal{U}_T$ measurable. If h is a positive $\mathcal{E}$ measurable function and t is positive then

$$P^x\big(h(X_{t+T'}); \Gamma'\big) = P^x\big(P^{X_{T'}}(h(X_t); \Gamma')\big)$$

by the strong Markov property of X. But clearly $h(x_{t+T}) \circ X = h(X_{t+T'})$ and we know already that $P^y\big(h(X_t)\big) = \mathcal{P}^y\big(h(x_t)\big)$ so the left side of the above equality is $\mathcal{P}^x\big(h(x_{t+T}); \Gamma\big)$ while the right side is $\mathcal{P}^x\big(\mathcal{P}^{x_T}(h(x_t); \Gamma)\big)$. This is the strong Markov property for the x process. A similar argument establishes quasi-left-continuity.

The process $x = (U, \mathcal{U}, \mathcal{U}_t, x_t, \theta_t, \mathcal{P}^x)$ is called the *canonical right continuous realization of* X. Usually one is given the original X and wishes to analyze some of its structure. The analysis may be simpler when done in terms of the canonical realization. And then one must interpret in terms of X whatever properties have been established.

9. Potential operators and resolvents.

Let $P(t, x, A); t > 0, x \in E, A \in \mathcal{E}$ be a sub-Markov transition function. We will assume that if f is continuous with compact support then $P_t f(x) (= \int P(t, x, dy) f(y))$ defines for each x a right-continuous function of t. This does hold in all cases of interest to us. Then $P_t f(x)$ is $\mathcal{E}$ measurable in x and right continuous in t for such f, and hence is measurable in (t, x) relative to $\mathcal{B}(0, \infty) \times \mathcal{E}$. By monotone class considerations the joint measurability assertion is valid if f is any element of $b\mathcal{E}$, the set of bounded $\mathcal{E}$ measurable functions. Thus the manipulations that are about to appear all will make sense. For α strictly positive the operator U^α defined by

$$(9.1) \qquad U^\alpha f(x) = \int_0^\infty e^{-\alpha t} P_t f(x) dt$$

defines a positive linear mapping from $b\mathcal{E}$ to itself and if $|f| \leq 1$ then $|U^\alpha f| \leq \alpha^{-1}$. Previously we have called U^α the α-potential operator and have used its interpretation in term of the underlying Markov-process when $P(t, x, A)$ is viewed as the transition function of a Markov process. The family $\{U^\alpha; \alpha > 0\}$ is called the *resolvent of the semi-group* $\{P_t; t > 0\}$. From the semi-group property $P_t P_s = P_{t+s}$ we obtain easily the *resolvent equation*

$$(9.2) \qquad U^\alpha - U^\beta = (\beta - \alpha) U^\alpha U^\beta, \quad \alpha, \beta > 0.$$

When α equals zero the expression (9.1) still makes sense for f positive; but the transformation $U(= U^0)$ is useful only in situations where $0 < Uf < \infty$ for many f (e.g. Brownian motion in R^d if $d \geq 3$, but not if $d = 1$ or 2). If f is in $b\mathcal{E}$ then $\alpha U^\alpha f$ is equal to $\int_0^\infty e^{-t} P_{t/\alpha} f dt$. Hence if f and the point x are such that $P_t f(x)$ approaches $f(x)$ as t approaches zero then $\alpha U^\alpha f(x)$ approaches $f(x)$ as α tends to infinity.

A useful fact is that a knowledge of the operator U^α for a single value of α determines the semigroup $\{P_t; t \geq 0\}$. Indeed note that

$$U^\alpha(U^\alpha f) = \int_0^\infty \int_0^\infty e^{-\alpha(s+t)} P_{s+t} f \, ds dt$$

$$= \int_0^\infty u e^{-\alpha u} P_u f \, du.$$

Repeating this step we obtain

$$(9.3) \qquad (U^\alpha)^{n+1} f = \int_0^\infty (t^n/n!) e^{-\alpha t} P_t f \, dt,$$

so for $\lambda < \alpha$

$$(9.4) \qquad \sum_0^\infty (-\lambda)^n (U^\alpha)^{n+1} f = \int_0^\infty e^{-(\lambda+\alpha)t} P_t f \, dt,$$

where the interchange of sum and integral is justified by the fact that $\lambda \|U^\alpha\| < 1$. Thus the Laplace transform of $P_t f$ is determined by U^α and hence so is $P_t f$ itself if f is continuous with compact support, since in that case $t \to P_t f$ is right continuous. This determines P_t since $P_t f(x)$ is for fixed x and t a measure in f.

In practice one encounters situations where rather than being given a transition function one has only the family $\{U^\alpha; \alpha > 0\}$ of positive linear operators usually defined only on a linear subset (but a measure theoretically large one) of $b\mathcal{E}$. Then one asks for the existence of a transition function such that (9.1) holds. This is the subject of the Hille-Yosida theorem. We will give a sketch of the version we will need in Chapter V. For a full discussion, which emphasizes the probabilistic interpretation of the various objects, the reader should consult volume II of Feller [F, 1]. Let $\mathcal{C}_0$ denote the set of continuous functions on E which vanish at ∞. We will suppose that each U^α is a positive linear operator from $\mathcal{C}_0$ to itself, that f in $\mathcal{C}_0$ and $0 \le f \le 1$ implies $0 \le U^\alpha f \le \alpha^{-1}$, that (9.2) holds and that $U^\alpha(\mathcal{C}_0)$, which is by (9.2) independent of α, is dense in the uniform norm in $\mathcal{C}_0$. It follows from standard reasoning (using the Riesz representation theorem applied to the positive linear functionals $f \to \alpha U^\alpha f(x)$) that there is a kernel $Q_\alpha(x, A)$, $x \in E$, $A \in \mathcal{E}$, which is a subprobability measure in A, $\mathcal{E}$ measurable in x and is such that $\alpha U^\alpha f(x) = \int Q_\alpha(x, dy) f(y)$. Let us suppose that in fact each $Q_\alpha(x, \cdot)$ is a probability measure, as it will be in case U^α arises from a transition function with $P(t, x, E)$ always equal to 1. Then for each strictly positive λ the formula

$$\lambda P_t f = \sum_{n=0}^\infty e^{-\lambda t} (\lambda t)^n / n! \; Q_\lambda^n f$$

defines a transition function for a standard process. This can be checked analytically but the probabilistic interpretation is more useful: starting at x in E the corresponding process remains there in the time interval $[0, S^\lambda)$ where S^λ is exponential with rate λ. At time S^λ it moves to a position X_{S^λ} independent of S^λ and distributed according to $Q_\lambda(x, \cdot)$. It proceeds then from the new position according to the original rules. The resolvent associated with the semigroup $\{_\lambda P_t; t > 0\}$ is

$$U_\lambda^\alpha = \left(\frac{1}{\alpha + \lambda}\right) \sum_0^\infty \left(\frac{\lambda^2}{\alpha + \lambda}\right)^n (U^\lambda)^n.$$

A little work with the resolvent equation yields the equality

$$U_\lambda^\alpha = (\alpha + \lambda)^{-1} I + \lambda^2/(\alpha + \lambda)^2 U^{\alpha\lambda/(\alpha+\lambda)}.$$

As λ approaches infinity this converges to U^α when applied to an element of $\mathcal{C}_0$ so there seems reason to hope that $_\lambda P_t$ will converge in a suitable manner to a limiting transition function whose resolvent is the given U^α. The proof of the Hille-Yosida theorem amounts to showing that indeed this does happen.

II Examples

In this chapter we will give examples of various kinds of Markov processes. Also we will construct Brownian motion and use it to construct some other processes that are particularly important in excursion theory.

1. Examples.

The most obvious way to define a particular process is to specify somehow its transition function. Usually this is what we will do. But in many cases some other description is more appropriate. For example Brownian motion can be described through its transition function as in I-3 or as a process with stationary independent increments or as a Gaussian process with a particular covariance function or as a diffusion process with the proper infinitesimal generator, the choice depending on the anticipated use. We will take a flexible viewpoint; and in particular will not force ourselves to describe every example in terms of the comprehensive set-up, $X = (\Omega, \cdots)$.

a) **Processes with stationary independent increments:** these are processes $\{X_t; t \geq 0\}$ with values in R^n having the properties (i) the distribution of $X_{t+h} - X_t$ does not depend on t, (ii) for any set $t_1 < \cdots < t_k$ of positive numbers, the random vectors $X_{t_0}, X_{t_1} - X_{t_0}, \cdots X_{t_k} - X_{t_{k-1}}$ are mutually independent. Obviously, for any such process the distribution of $X_1 - X_0$ will of necessity be an infinitely divisible distribution on R^n, call it μ_1. To avoid trivialities we will consider only such processes having the additional property that as $h \to 0$ the distribution of $X_{t+h} - X_t$ converges weakly to unit mass at 0. For processes with stationary increments this additional property is the same as the property that for each fixed t, X_s converges to X_t in probability as $s \to t$. We will say that a process with this property is *continuous in probability*. If $\{X_t; t \geq 0\}$ has stationary independent increments and is continuous in probability then μ_1, the distribution of $X_1 - X_0$, is embedded in a unique convolution semi-group $\{\mu_t; t \geq 0\}$ of probability measures on R^n which is continuous in the weak topology of measures; and of course μ_h is the distribution of $X_{t+h} - X_t$. It is obvious that $\{X_t; t \geq 0\}$ is a Markov process and that its transition function is given by

$$(1.1) \qquad\qquad P(t, x, A) = \mu_t(A - x)$$

with $t > 0$, $x \in R^n$, $A \in \mathcal{B}(R^n)$, and $A - x$ denoting $\{y - x | y \in A\}$.

Conversely suppose we are given a convolution semigroup $\{\mu_t; t \geq 0\}$ of probability measures on R^n. Then (1.1) defines a transition function on $(R^n, \mathcal{B}(R^n))$. For a bounded Borel function f we have

$$P_t f(x) = \int f(x + y)\mu_t(dy)$$

and so P_t maps $\mathcal{C}_0(R^n)$ into itself. If also the measures μ_t vary continuously in t, that is $\mu_h \to \varepsilon_0$ as $h \to 0$, then for each $f \in \mathcal{C}_0$, $P_t f \to f$ uniformly as $t \to 0$. Thus the hypothesis of I-6.1 are satisfied and so there is a standard process with this transition function. Whatever the initial distribution, the process with that initial distribution will have stationary independent increments and will be continuous in probability.

The most important special case by far is in R^1 when μ_t is the normal distribution with mean 0 and variance t. Then the transition function is the Brownian motion transition function from I-3.5. An apparently trivial example in R^1 is obtained by taking μ_t to be unit mass at t. The corresponding process is called "uniform motion to the right" because $P\{X_t = X_0 + t\} = 1$ for each t, (and the exceptional set of probability 0 is independent of t if the paths are right continuous). In spite of its simplicity this process has its uses, as we will see later in this chapter.

b) **Transformations:** Another class of examples arises by taking an already existing Markov process and applying to it a deterministic or random transformation which, because of some special feature of the original process and the transformation, leads one again to a Markov process. Consider the following general situation: $(E, \mathcal{E})$ and $(H, \mathcal{H})$ are measurable spaces, $\varphi : E \to H$ is measurable relative to $\mathcal{E}$ and $\mathcal{H}$ and $\varphi(E) = H$. Let $\mathcal{G} \subset \mathcal{E}$ be the σ-algebra generated by φ. Suppose $\{X_t, \mathcal{F}_t; t \geq 0\}$ is a time homogeneous Markov process over a probability space $(\Omega, \mathcal{F}, P)$ with state space E and transition function $P(t, x, A)$.

(1.2) Theorem. *If for each positive real valued $\mathcal{H}$ measurable function f, and each $t > 0$, $x \to P(t, x, f \circ \varphi)$ is $\mathcal{G}$ measurable then $\{\varphi(X_t), \mathcal{F}_t; t \geq 0\}$ is a time homogeneous Markov process with state space H.*

Proof. Given $y \in H$ pick $x \in E$ such that $\varphi(x) = y$ and let $Q(t, y, \cdot)$ be the distribution in H of φ relative to the measure $P(t, x, \cdot)$ on E. This means that $Q(t, y, f) = P(t, x, f \circ \varphi)$ for every positive $\mathcal{H}$ measurable f, and by the hypothesis the choice of x is irrelevant; that is $Q(t, y, f) = P(t, \varphi^{-1}(y), f \circ \varphi)$. One checks immediately that Q is a transition function. Given $0 \leq s < t$, $\Lambda \in \mathcal{F}_s$ and a positive $\mathcal{H}$ measurable function f we have

$$\int_\Lambda f(\varphi(X_t)) dP = \int_\Lambda P(t - s, X_s, f \circ \varphi) dP$$
$$= \int_\Lambda Q(t - s, \varphi(X_s), f) dP$$

so the assertion is valid. If the original process has the strong Markov property so will the new process. If E and H are topological spaces and φ is continuous then properties like sample function regularity and quasi-left-continuity will carry over, as well.

We used this approach in Chapter I to obtain reflecting Brownian motion. An important generalization is the following: let X be n-dimensional Brownian motion, that is $X_t = (X_t^1, \cdots, X_t^n)$ where the components are one dimensional Brownian motions and mutually independent, or, equivalently, the process has stationary independent increments in R^n and $X_t - X_0$ is distributed as an n-dimensional normal vector with means 0 and covariance matrix tI. Thus the transition function is $(2\pi t)^{-n/2} \exp(- \| x - y \|^2 /2t) dy$ with dy denoting lebesgue measure on R^n and $\| x - y \|$ the usual Euclidean distance. Let $\varphi(x) = \| x \|$, so that we are considering $\| X_t \|$, the radial part of n-dimensional Brownian motion. Obviously if f is a bounded Borel function on R^n which depends on x only through $\| x \|$ then $P(t, x, f)$ also is a function of $\| x \|$. Thus the process $\| X_t \|$, whose state space is $[0, \infty)$ is a time homogeneous Markov process. It is called the *n-dimensional Bessel process* (even though it is one dimensional). The cases $n = 1$ and $n = 3$ are the ones most important for excursion theory. For $n = 1$ we get reflecting Brownian motion. We obtained its transition density $p^+(t, x, y)$ in terms of the Gauss kernel, $p^+(t, x, y) = p(t, x, y) + p(t, -x, y)$, by following the change of variables in Theorem 1.2. When $n = 3$ one does the same thing, but obtaining an explicit formula requires first changing to spherical coordinates. The result turns out to be $P(t, x, dy) = b(t, x, y) dy, t > 0$, where for $x > 0$

$$(1.3) \qquad b(t, x, y) = \frac{y}{x}(p(t, x, y) - p(t, -x, y)),$$

$$b(t, 0, y) = \lim_{x \to 0} b(t, x, y) = \frac{2y^2}{\sqrt{2\pi t^3}} e^{-y^2/2t}.$$

An example of a transformation involving the path in a more complicated way is furnished by the process $X_t^+ = X_t - m_t$ from section 3 of Chapter I, where $\{X_t; t \geq 0\}$ is Brownian motion in R^1 and $m_t = \min(0, \inf_{s \leq t} X_s)$. This can be generalized: let $\{X_t; t \geq 0\}$ be a process in R^1 with stationary independent increments which is continuous in probability, and assume, as we may, that we have a version whose paths are right continuous and have left-hand limits. Define X_t^+ and m_t as above. Clearly m_t is measurable relative to $\sigma(X_r; r \leq t)$. Now the process $\{X_t^+; t \geq 0\}$ is a time homogeneous Markov process with state space $[0, \infty)$, and its transition function is, as it must be, simply $P^+(t, x, A) = P^x(X_t^+ \in A)$, where P^x denotes probabilities for the original process started at $x \geq 0$. The proof of this fact is exactly the proof given in I-3. It used only the stationary independent increments property. Suppose now the increments in our original process are symmetrically distributed. Then Theorem 1.2 applies, and the process $\{|X_t|; t \geq 0\}$ is also a Markov process. We will show in II-2 that when X is Brownian motion, $|X|$ and X^+ are probabilistically the same. But otherwise the two processes will be different. Usually it is X^+ which is called "X reflected at 0".

c) **Diffusion processes:** Time homogeneous Markov processes with continuous paths which also are strong Markov are called *diffusion processes*, or just *diffusions*. Thus Brownian motion (if we take the paths continuous) is a diffusion and so, under the same proviso is reflecting Brownian motion as well as the Bessel processes. Many facts about diffusion processes follow from general probabilistic potential theory; but there is a rich theory that is specific to the subject itself. One feature that appears constantly in the theory is that if $P(t, x, A)$ is the transition function of a diffusion and f is a sufficiently smooth function then $u(t, x) = P(t, x, f)$ satisfies a heat equation, $u_t = (\Delta u)_x$, where Δ is some sort of differential operator. In the case of Brownian motion in R^1 $\Delta f = f''/2$, and for reflecting Brownian Δ is the same with the "boundary condition" that to be in the domain of Δ a function must be smooth (of course) and satisfy $f'(0) = 0$. (This last, classically, is offered as an explanation for the term "reflecting".) The boundary conditions are very important, and provide genuine motivation for the introduction of excursion theory. Later we will consider Itô-McKean's classical 1962 paper [IM,1] as an illustration. The fundamental reference for diffusion processes is [IM,2].

d) **Special examples:** In excursion theory various rather special non-time-homogeneous Markov processes appear of which the most famous is called *Brownian excursion*. This is a process $\{X_t; 0 \le t \le 1\}$ with continuous paths, with $X_0 = X_1 = 0$, with $X_t > 0$ for all $t \in (0,1)$ and with absolute distributions and transition function, for $0 < s < t < 1$ given by

$$P(X_t \in dy) = \sqrt{2}y^2 \left(\pi t^3 (1-t)^3\right)^{-1/2} e^{-y^2/2t(1-t)} dy$$

(1.4)
$$P(s,x;t,dy) = p^-(t-s,x,y)\left(\frac{1-s}{1-t}\right)^{3/2} \frac{y}{x}\left(\frac{e^{-y^2/2(1-t)}}{e^{-x^2/2(1-s)}}\right) dy$$

where $p^-(t,x,y) = p(t,x,y) - p(t,-x,y)$, with $p(t,x,y)$ the Gauss kernel. One can check the Chapman-Kolmogorov equation and the fact that

$$\int P(X_s \in dx) P(s,x;t,dy) = P(X_t \in dy)$$

and conclude that a process with the desired finite dimensional distributions exists. Then one must make an additional argument that the paths can be taken to be continuous. An alternative argument for existence is the following:

(1.5) Theorem. *Let $\{Y_t; t \ge 0\}$ be the 3-dimensional Bessel process with continuous paths and with $Y_0 = 0$. Define $\{X_t; 0 \le t \le 1\}$ by*

$$X_t = (1-t)Y_{t/1-t} \qquad 0 \le t < 1$$
$$X_1 = 0.$$

Then $\{X_t; t \ge 0\}$ is Brownian excursion.

Proof. By the law of large numbers applied to the components of Brownian motion in R^3 it follows that with probability 1 $X_t \to 0$ as $t \to 1$; and X_t is continuous elsewhere because the Bessel process is continuous. Also the process is Markovian because the Bessel process is and the transformation at hand preserves the Markov property. Finally the absolute and joint distributions of (X_s, X_t) can be checked using the corresponding formulas for the Bessel process from (1.3) and making a linear change of variables.

e) **Linking:** In a number of situations one must construct a Markov process by piecing together processes that are defined only in finite, but random time intervals. We will need to consider the following instance of this: let X be a standard process, σ the time X hits a singleton $\{a\} \subset E$, β a strictly positive number and μ a probability measure on $\mathcal{E}$ with $\mu(\{a\}) = 0$. We want to construct a time-homogeneous Markov process $Y = (Y_t, \bar{P}^x)$ of the following description. Starting at a point x different from a the process Y behaves like X up until the time, σ, of hitting $\{a\}$ in the sense that $P^x(X_t \in A, t < \sigma) = \bar{P}^x(Y_t \in A, t < \sigma)$ for all $t \geq 0$ and $A \in \mathcal{E}$. Upon reaching the point a, Y stays at a for a length of time J which is exponentially distributed with rate β and independent of events up to time σ, and then it jumps to a point of $E - \{a\}$ chosen independently of all that has gone on before and distributed according to μ; and from there it proceeds according to the rules for X until the next time of hitting $\{a\}$ and so forth. As to the existence of such a process, it is quite clear that we can put measures $\bar{P}^x$ on the space of functions from $[0, \infty)$ to E which are right continuous and have left hand limits under which the coordinate process $\{Y_t; t \geq 0\}$ evolves probabilistically as we have described. What is more difficult is to give a proof of the Markovian character of Y, whose details aren't so complicated that they obscure the obvious validity of the assertion. A construction theorem containing our assertion as a very special case is given in [M,1]. But we hope the reader will accept this special case of linking without requiring formal justification, as we will have many occasions to use processes described this way.

(1.6) **Exercise:** Show that when X has stationary independent increments and right continuous paths the strong Markov property is equivalent to the statement that whenever T is a finite stopping time and $\{Y_t; t \geq 0\}$ is defined by

$$Y_t = X_{t+T} - X_T$$

then Y is distributed as X with initial position 0, and $\sigma\{Y_t; t \geq 0\}$ and $\mathcal{F}_T$ are independent. (Hint $P^x(X_t \in B) = P^0(X_t + x \in B)$.)

2. Brownian Motion.

a) **Existence:** We will give Lévy's famous construction of Brownian motion. The construction is elegant, involves only elementary methods, and allows one to compute conditional probabilities in a useful and unambiguous way.

To get started we will assume as we may that we have a probability space $(\Omega, \mathcal{F}, P)$ and a family $\{Y_q | q$ a dyadic rational in $[0,1]\}$ of random variables defined on Ω which are all normally distributed with mean 0 and variance 1 and are mutually independent. Let D_n denote the dyadics of rank n, that is $D_n = \{k/2^n; k = 0, \ldots, 2^n\}$. We will define a sequence $\{X_t^n; 0 \leq t \leq 1\}$ of stochastic processes over $(\Omega, \mathcal{F}, P)$ as follows:

$$
\begin{aligned}
X_t^0 &= tY_1 & & 0 \leq t \leq 1. \\
X_t^1 &= X_t^0 & & t \in D_0 \\
&= X_{1/2}^0 + 1/2 Y_{1/2} & & t \in D_1 - D_0 \\
&= \text{piecewise linear} & & t \in [0,1] - D_1
\end{aligned}
$$

and in general

$$
\begin{aligned}
X_t^{n+1} &= X_t^n & & t \in D_n \\
&= X_q^n + 2^{-(\frac{n}{2}+1)} Y_q & & t = q \in D_{n+1} - D_n \\
&= \text{piecewise linear} & & t \in [0,1] - D_{n+1}.
\end{aligned}
$$

Now for each n the finite dimensional distributions of the process X_t^n are jointly normal, and so questions of independence can be settled by computing covariances. All the means are 0, and an easy calculation and induction will convince the reader that for each n the random variables $X_{j+1/2^n}^n - X_{j/2^n}^n, j = 0, \ldots, 2^n - 1$ are mutually independent and each is normal with mean 0 and variance $1/2^n$. Hence the process $\{X_t^n; 0 \leq t \leq 1\}$, provided we view it only at times $t \in D_n$, has the properties required of Brownian motion. We will argue now that with probability one X_t^n converges uniformly in t as $n \to \infty$. Clearly the greatest value of $|X_t^n - X_t^{n+1}|$ will occur at a value of $t \in D_{n+1} - D_n$ and the magnitude there will be the absolute value of a normal random variable with mean 0 and variance $2^{-(n+2)}$. There are 2^n such dyadics and so we obtain for any $\epsilon > 0$

$$
P\{\max_t |X_t^{n+1} - X_t^n| > \epsilon\} \leq 2^n P\{|U| > \epsilon 2^{n/2+1}\}
$$

where U is normal with mean 0 and variance 1. Using the standard estimate $P\{|U| > x\} \leq x^{-1}e^{-x^2/2}$ and making the choice $\epsilon = 1/n^2$ we get

$$P\{\max_t |X_t^{n+1} - X_t^n| > n^{-2}\} \leq n^2 2^{n/2-1} e^{-2^{n+1}/n^4}.$$

The term on the right is the general term of a convergent series, so by the Borel-Cantelli lemma with probability one only finitely many of the events $\{\max_t |X_t^{n+1} - X_t^n| > n^{-2}\}$ will occur. Since Σn^{-2} converges the uniform convergence assertion is proved. Define the process $\{X_t; 0 \leq t \leq 1\}$ by $X_t = \lim_n X_t^n$ provided the limit exists uniformly, and $X_t \equiv 0$ otherwise. The paths of X_t are continuous since those of X_t^n are. If t is dyadic then X_t^n is constant from some value of n on, and it follows that if $0 < t_0 < \ldots < t_k \leq 1$ are dyadics then the increments $X_{t_0}, \ldots, X_{t_k} - X_{t_{k-1}}$ are independent and normal with the desired mean and variance. We can drop the requirement that the time points be dyadics because the paths are continuous and, as is well known, an almost everywhere limit of a sequence of jointly normal vectors is itself jointly normal and the means and covariances of the sequence converge to those of the limiting vector.

Thus we have constructed Brownian motion with time parameter running over $[0,1]$. To get a process with parameter interval $[0, \infty)$ we construct mutually independent copies, $\{_nX_t; 0 \leq t \leq 1\}$ for $n = 1, 2, \ldots$, of the process we have just constructed and extend the definition of our X_t by setting

$$
\begin{aligned}
X_t &= X_1 + {}_1X_{t-1} & 1 \leq t \leq 2 \\
&= X_2 + {}_2X_{t-2} & 2 \leq t \leq 3,
\end{aligned}
$$

and so forth. Alternatively one can obtain a Brownian motion $\{Y_t; t \geq 0\}$ from a Brownian motion $\{X_t; 0 \leq t \leq 1\}$ such that $X_0 = 0$ by setting $Y_t = (t+1)X_{1/t+1} - X_1$. Clearly the joint distributions of the Y process are normal, the means are 0, and upon checking one finds that covariance $(Y_s, Y_t) = \min(s, t)$. This does indeed prove that $\{Y_t; t \geq 0\}$ is Brownian motion, but the other procedure is more natural.

b) **Conditional Probabilities:** Consider again Brownian motion, only in the time interval $0 \leq t \leq 1$. The general definition of conditional probability tells us how to interpret properly the notion of conditional probabilities for $\{X_t; 0 \leq t \leq 1\}$ given the value of X_1. But here we can be more specific. Repeat the construction that we just made of $\{X_t; 0 \le t \le 1\}$ except that at the first stage replace X_1 with a fixed value y, and then proceed as before. The proof of uniform convergence is still valid and we get a process, let us call it $\{X_t^y; 0 \le t \le 1\}$, which has continuous paths and whose finite dimensional distributions are jointly normal and which satisfies $X_1^y = y$. For the probabilities this process induces on $C[0,1]$ the space of continuous paths on $[0,1]$, we write $_yP$. Write P for the probability distribution in $C[0,1]$ of our Brownian motion. We obtained Brownian motion by replacing y by a standard normal variable, independent of everything else. It follows that $P(\cdot) = \int_{-\infty}^{\infty} {}_yP(\cdot)n(y)dy$ where $n(y)$ is the standard normal density, that is the distribution of X_1. This means that $_yP$ does indeed give the conditional probabilities for the process X_t given $X_1 = y$. If $\{W_t; 0 \le t \le 1\}$ is the process $W_t = X_t - tX_1$ then W is called *tied down Brownian motion* or *the Brownian bridge*. It is called also "Brownian motion conditioned to be 0 at time 1." Now clearly the law of W is just $_0P$. In view of what we have just said this last name is perfectly reasonable and does not have to be justified by conditioning over events $\{|X_1| < \epsilon\}$ and taking limits as $\epsilon \to 0$.

The following fact is useful:

(2.1) Theorem. *The process $X_{1-t}^y - y$ is equivalent in distribution to the process X_t^{-y}.*

Proof. Clearly $X_t^y = X_t^0 + ty$. Also it is obvious that $\{X_t^0; 0 \le t \le 1\}$ and $\{X_{1-t}^0; 0 \le t \le 1\}$ are probabilistically the same. So, with the symbol $\cong$ denoting "equals in distribution" we have

$$X_{1-t}^y = X_{1-t}^0 + (1-t)y \cong X_t^0 + y - ty$$

or

$$X_{1-t}^y - y \cong X_t^0 - ty = X_t^{-y}.$$

We will use this to establish the claim made several times in Chapter I that the two processes we labelled "reflecting Brownian motion" are probabilistically the same.

(2.2) Theorem. *If $\{X_t; t \ge 0\}$ is Brownian motion then the processes $\{|X_t|; t \ge 0\}$ and $\{X_t - \min(0, \inf_{s \le t} X_s); t \ge 0\}$ have the same distribution.*

Proof: Both processes are known already to be diffusions so we need only check equality of the transition functions. Checking a Laplace transformed version will be enough; that is we must show that for any bounded Borel function f on $[0, \infty)$ and positive x the expression $E^x \int_0^\infty e^{-\lambda t} f(Z_t)\,dt$, with E^x denoting expectations for Brownian motion, yields the same thing regardless of which of the two processes we enter for Z_t. The two processes are identical (both are Brownian motion) up to the time, σ, at which Brownian motion reaches 0. So writing the integral as $\int_0^\sigma + \int_\sigma^\infty$ and using the strong Markov property we see that it is enough to consider only the case $x = 0$. That means we must show that for $b > 0$ and $r > 0$, $P^0(|X_r| > b) = P^0(X_r - \min_{s \le r} X_s > b)$. And according to the scaling from exercise (2.3) it will suffice to do this when $r = 1$. We have, in the notation of our conditional probability discussion, and using exercise I–4.7

$$
P^0\{|X_1| > b\} = 2P\{X_1 < -b\} = P\{\min_{s \le 1} X_s < -b\}
$$
$$
= \int_{-\infty}^\infty P(\min_{s \le 1} X_s^{-y} < -b) n(y)\,dy
$$
$$
= \int_{-\infty}^\infty P(\min_{s \le 1}(X_{1-s}^{-y} + y) < y - b) n(y)\,dy
$$
$$
= \int_{-\infty}^\infty P\{\min_{s \le 1} X_s^y - X_1^y < -b\} n(y)\,dy
$$
$$
= P^0\{X_1 - \min_{s \le 1} X_s > b\}
$$

as required.

Lévy's construction shows that we may find a version of Brownian motion all of where sample functions are continuous. And thus the same statement applies to reflecting Brownian motion and the Bessel processes. Henceforth we will assume that these versions have been chosen so that when we say "let X be Brownian motion" we are assuming that the paths are continuous.

(2.3) Exercise. Let $\{X_t; t \ge 0\}$ be Brownian motion starting at 0 and let c be a strictly positive constant. Show that the processes $\{X_t; t \ge 0\}$ and $\{c^{-1/2} X_{tc}; t \ge 0\}$ have the same finite dimensional distributions.

(2.4) Exercise. Let $\{X_t; 0 \le t \le 1\}$ be a stochastic process with values in a metric space with metric ρ. Suppose that for some $p > 0$ and $\alpha > 1$

$$E\{(\rho(X_s, X_t))^p\} \leq K|t - s|^\alpha \qquad 0 \leq s, t \leq 1$$

where K is a constant. Show that if $\theta < (\alpha - 1)/p$ then there is a process equal in law to X almost all of whose sample functions are Hölder continuous of order θ. Hints: (a) use Chebycheff's inequality to estimate $P\{\rho(X_{k/2^n}, X_{(k+1)/2^n}) \geq 2^{-n\theta}\}$; (b) conclude that for some ρ_n, with the series $\Sigma\rho_n$ convergent $P\{\rho(X_{k/2^n}, X_{(k+1)/2^n}) \geq 2^{-n\theta}$ for some $k < 2^n\} \leq \rho_n$; (c) use the Borel-Cantelli lemma on the events in (b) and then use the triangle inequality and the fact that if $|t - s| < 2^{-m}$ we can write

$$t = j/2^m + p_1/2^{m+1} + p_2/2^{m+2} + \cdots$$
$$s = i/2^m + q_1/2^{m+1} + q_2/2^{m+2} + \cdots$$

where the integers i and j differ by at most 1, and p's and q's are 0 or 1, and the sums are finite to conclude that with probability one when t is restricted to the dyadic rationals X_t is uniformly continuous; (d) define a process Y equal in law to X by $Y_t = \lim_{s \to t} X_s$ where s ranges only over the dyadic rationals and argue that Y has the desired path properties. (Note: this sort of result usually is called Kolmogorov's theorem).

(2.5) Exercise. When $\{X_t; 0 \leq t \leq 1\}$ is Brownian motion in R note the equality

$$E|X_t - X_s|^p = K_p|t - s|^{p/2}.$$

Use the previous exercise to conclude that for $\theta < 1/2$ almost all sample functions are Hölder continuous of order θ.

(2.6) Exercise. Let $\{X_t; t \geq 0\}$ be Brownian motion, $0 \leq r < s$ and $M_{rs} = \max\{X_t | r \leq t \leq s\}$.

    (a) Use I–4.7 to show that $P^x(M_{0t} = b) = 0$ for each x, t, b.

    (b) Then use I–4.6 with T fixed to conclude that if $[r, s]$ and $[u, v]$ do not overlap then $P^x(M_{rs} = M_{uv}) = 0$ for all x.

    (c) Letting r, s, u, v range over a countable dense set conclude that except for $\omega \in \Lambda$ where $P^x(\Lambda) = 0$ for all x if $X_t(\omega)$ has local maxima at t and r with $t \neq r$ then $X_t(\omega) \neq X_r(\omega)$.

3. Feller Brownian motions, and related examples.

A standard process (X_t, P^x) whose state space is $[0, \infty)$ is called a Feller Brownian motion if it behaves like ordinary Brownian motion up until the hitting time σ of the point 0; more precisely $P^x(X_t \in dy, t < \sigma) = p^-(t, x, y)dy$ where $p^-(t, x, y)$ is the density from I.7 for Brownian motion starting in $(0, \infty)$ and killed upon reaching $\{0\}$. Reflecting Brownian motion is a Feller Brownian motion. Another example is a process which starts as Brownian motion and upon reaching 0 stays there for a length of time having an exponential distribution, then jumps to a random point of $(0, \infty)$ chosen independently of all that has gone before, then proceeds from that point as a new Brownian motion, and so forth. The killed Brownian motion is called the *minimal process* and the Feller Brownian motions are then *extensions* of the minimal process. The problem of finding all the Feller Brownian motions can be generalized, either by allowing a different minimal process in $(0, \infty)$ or by replacing $(0, \infty)$ with a dense open subset V of a rather arbitrary state space E and seeking all standard processes with state space E which agree with a given minimal process on V until the time of hitting the boundary $E - V$. Later in the book we will use excursion theory to treat the extension problem when the boundary is a single point. For now we will focus on some special techniques that illustrate the general theory even though they apply only to Feller Brownian motions or to processes quite similar to them. We will start with a very simple minimal process.

(a) **Sawtooth processes:** Consider the situation in which the state space is $[0, \infty)$ and the minimal process is uniform motion to the left in $(0, \infty)$. That is if $X_0 = x$ with $x > 0$, then $X_t = x - t$ for $0 \leq t \leq x$ so that x is the time the process reaches the boundary point, 0. We want to find all, or at least the most interesting, of the standard processes $\{X_t\}$ having $[0, \infty)$ as state space and such that if $X_t = x > 0$ then $X_{t+s} = x - s$ for $s < x$. We will call any such process a *sawtooth process* in recognition of the appearance of the graph of the typical sample function. As usual we can obtain examples by the holding and jumping procedure: that is let β and μ denote respectively a strictly positive number and a probability measure on $\mathcal{B}(0, \infty)$. Then describe X by the requirement that when X_t first reaches 0 it holds there for a length of time having an exponential distribution with rate β and then independently jumps to a point distributed according to μ, then moves to the left at unit speed, etc.

To describe a more interesting class of examples we must first recall some analytic and probabilistic facts from the theory of processes with stationary independent increments: a process $\{T_t; t \geq 0\}$ with stationary independent increments is called a *subordinator* if $P(T_t - T_s \geq 0) = 1$ for all $s \leq t$. If η_r denotes the distribution of $T_{t+r} - T_t$ then $\{\eta_r; r \geq 0\}$ forms a continuous convolution semigroup of probability measure on $\mathcal{B}[0, \infty)$. The Laplace transform of the measures η_r will have the form

$$(3.1) \qquad \int_{[0,\infty)} e^{-\lambda u} \eta_r(du) = e^{-rg(\lambda)}$$

where $g(\lambda)$ has the special form

$$(3.2) \qquad g(\lambda) = \alpha\lambda + \int_{(0,\infty)} (1 - e^{-\lambda x})\nu(dx)$$

with $\alpha \geq 0$ and ν a measure on $\mathcal{B}(0, \infty)$ such that $\int_{(0,\infty)} x/(1+x)d\nu < \infty$. Conversely given any function g on $[0, \infty)$ having the representation (3.2) there is a convolution semigroup $\{\eta_r; r \geq 0\}$ and a subordinator $\{T_t; t \geq 0\}$ such that η_r is the distribution of $T_{t+r} - T_t$ and such that (3.1) holds. Furthermore we may assume that the sample functions of the subordinator are all right continuous (and hence non-decreasing) everywhere and that the strong Markov property holds. The increments will be strictly positive with probability one if and only if $\alpha > 0$ or $\|\nu\| = \infty$ and then we may, of course, assume that all the sample functions are strictly increasing. The probabilistic interpretation of ν is that for each $\varepsilon > 0$ and $t > 0$ the number of points $s \leq t$ such that $T_s - T_{s-} > \varepsilon$ has a Poisson distribution with parameter $t\nu(\varepsilon, \infty)$. It follows that if $\nu(0, \infty)$ is finite then the time, τ, at which the first path discontinuity occurs has an exponential distribution with rate $\|\nu\|$, and that the size, $T_\tau - T_{\tau-}$, of the jump has $\nu/\|\nu\|$ for probability distribution, and these two random variables are independent. Hence we have the following approximation: given $\varepsilon > 0$ and $n \geq 1$ let $\tau_{n1}, \tau_{n2} \cdots$ be the successive times at which the path T_t has a discontinuity whose size is in the interval $(\varepsilon/2^n, \varepsilon/2^{n-1}]$ and let $J_{nk} = T_{\tau_{nk}} - T_{\tau_{nk}-}$. If we set

$$T_t^\varepsilon = \Sigma J_{nk}$$

where the sum is over all n and k such that $\tau_{nk} \leq t$, then the process $\{\bar{T}_t^\epsilon; t \geq 0\}$ defined by $\bar{T}_t^\epsilon = T_t - T_t^\epsilon$, that is the one obtained by removing all jumps of size $\leq \epsilon$, is a subordinator with analytic data (3.2) given by $\bar{\alpha} = \alpha$ and $\bar{\nu} = \nu|_{(\epsilon,\infty)}$. We have

$$ET_t^\epsilon \leq t \sum_{n=1}^{\infty} \frac{\epsilon}{2^{n-1}} \nu(\epsilon/2^n, \epsilon/2^{n-1}]$$

$$\leq 2t \sum_{n=1}^{\infty} \frac{\epsilon}{2^n} \nu(\epsilon/2^n, \epsilon/2^{n-1}]$$

$$\leq 2t \int_{(0,\epsilon]} x\nu(dx).$$

This last quantity approaches 0 with ϵ. It follows easily that with probability one as $\epsilon \to 0$ $\bar{T}_t^\epsilon \to T_t$ uniformly on compact t sets.

Now let Ω denote the set of all functions $w : [0,\infty) \to [0,\infty)$ which are right continuous strictly increasing and unbounded. For $t \geq 0$ and $w \in \Omega$ set

$$T(t, w) = w(t)$$

$$T^{-1}(t, w) = \inf\{s \geq 0 | T(s, w) > t\}.$$

$$X_t(w) = T(T^{-1}(t, w), w) - t.$$

Note that for each w, $T^{-1}(t, w)$ is a continuous increasing function of t, and that if $T(0, w) = x > 0$ then $T^{-1}(t, w) = 0$ for $0 \leq t \leq x$, and so $X_t(w) = x - t$ for $0 \leq t \leq x$. Define σ algebras $\mathcal{G}_t$ and $\mathcal{F}$ on Ω by $\mathcal{G}_t = \sigma\{T(s); s \leq t\}, \mathcal{F} = \sigma\{T(s); s \geq 0\}$. Then $\{T^{-1}(t) \leq s\} = \{T(s) \geq t\}$ and so $T^{-1}(t)$ is a $\{\mathcal{G}_t\}$ stopping time. The process $\{T(t), \mathcal{G}_t; t \geq 0\}$ is progressively measurable and so X_t is $\mathcal{G}_{T^{-1}(t)}$ measurable and hence is a random variable over $(\Omega, \mathcal{F})$. Define a family $\{\theta_t; t \geq 0\}$ of mappings from Ω to Ω by

$$\theta_t w(\cdot) = w(\cdot + T^{-1}(t, w)) - w(T^{-1}(t, w)) + X_t(w).$$

A routine, if uninspiring, examination of the definitions will convince the reader that $X_t \circ \theta_r = X_{t+r}$ for all positive t and r. Suppose we are given a function g as in (3.2) with either $\alpha > 0$ or $\|\nu\| = \infty$ and the corresponding convolution semi-group $\{\eta_r\}$ of probability measures on $\mathcal{B}(0,\infty)$. For each $x > 0$ let P^x denote the unique measure on $\mathcal{F}$ under which the coordinate process $\{T(t); t \geq 0\}$ is a subordinator with $P^x(T(0) = x) = 1$ and the increment $T(t) - T(s)$ distributed according to η_{t-s} for $s < t$.

(3.3) Theorem. *The process (X_t, P^x) is a sawtooth process.*

Remark: In case ν from (3.2) is a finite measure the process $\{X_t\}$ is just the holding and jumping process described earlier with β equal to $\alpha^{-1}\|\nu\|$ and $\mu = \nu/\|\nu\|$. The more interesting case where $\|\nu\| = \infty$ cannot be described in such a step-by-step manner.

Proof: We will content ourselves with verifying the simple Markov property. We will leave it to the reader to verify that the transformations $P_t g(x) = P^x(g(X_t))$ map $C_0([0, \infty)$ to itself and on C_0 converges strongly to the identity as $t \to 0$. Obviously the paths are right continuous so by I-5.1 and the discussion in I-6 the strong Markov property and quasi left continuity hold and the σ-algebras $\sigma\{X_s; x \leq t\}$ can be extended so as to meet the other requirements of a standard process. If $X_t(w) = x > 0$ then $T(0, \theta_t w) = x$ and so $X_{t+r}(w) = X_r(\theta_t w) = x - r$ for $0 \leq r \leq x$, so $\{X_t; t \geq 0\}$ will indeed be a sawtooth process. For checking the Markov property let t and r be positive, $A \in \mathcal{B}[0, \infty)$ and $\Gamma \in \sigma\{X_s; s \leq t\}$. Then $\Gamma \in \mathcal{G}_{T^{-1}(t)}$ and $X_{r+t}(w)$ is simply $X_r(\theta_t w)$. By the form of the strong Markov property valid for processes with stationary independent increments (see (1.6)) applied to the subordinator $\{T(t); t \geq 0\}$ for any x the path $\theta_t w(\cdot)$ is, relative to P^x, distributed as that of the original subordinator started at $X_t(w)$ and independent of any event in $\mathcal{G}_{T^{-1}(t)}$. It follows that

$$P^x(X_{r+t} \in A; \Gamma) = P^x\big(P^{X_t}(X_r \in A); \Gamma\big).$$

and so $\{X_t, \mathcal{G}_{T^{-1}(t)}; t \geq 0\}$ is a Markov process.

(b) **Local time:** For a situation such as the one at hand, where the probabilistic behavior of the process is prescribed except when it is at 0, the measure theoretic description of the process in terms of its excursions away from 0 requires a suitable way of measuring the time that X_t spends at the boundary point 0. What is desired is a real valued stochastic process $\{\ell(t); t \geq 0\}$ over $(\Omega, \mathcal{F})$ with the properties: (1) for each t $\ell(t)$ is measurable relative to $\sigma\{X_s; s \leq t\}$, (2) except for $w \in \Delta$, with $P^x(\Delta) = 0$ for all x, the function $t \to \ell(t, w)$ is non-decreasing and continuous, satisfies $\ell(0, w) = 0$, and has the additivity property $\ell(t+s, w) = \ell(s, w) + \ell(t, \theta_s w)$ for all t and s, and (3) except for $w \in \Delta$ as above the measure $d_t \ell(t, w)$ has as support the closure of $\{t | X_t(w) = 0\}$. (See (3.25).) Properties (1) and (2) define what we will later on call a *continuous additive functional* of the process and property (3) is an expression, appropriate for the situation at hand, of the fact that $\ell(t)$ grows at exactly those points where $X_t = 0$. The general theory of additive functionals guarantees that a local time exists and is unique up to multiplication by a constant. But in the present situation we need not rely on that and can describe a version of $\{\ell(t); t \geq 0\}$ explicitly. Namely: consider $\{T^{-1}(t); t \geq 0\}$. For every w the function $t \to T^{-1}(t, w)$ satisfies the smoothness requirements of property (2). Checking that it satisfies also the additivity property is straightforward. As to property (3), let, for the moment, $T(t)$ denote any right continuous strictly increasing unbounded function from $[0, \infty)$ to $[0, \infty)$ and let T^{-1} denote its inverse. Let $\mathcal{R}$ denote $\{t | T^{-1}(t + \varepsilon) - T^{-1}(t) > 0 \text{ for all } \varepsilon > 0\}$ and $\mathcal{L}$ denote $\{t > 0 | T^{-1}(t) - T^{-1}(t-\varepsilon) > 0 \text{ for all } \varepsilon > 0\}$. One shows easily that the three sets $\mathcal{R}, T[0, \infty)$, and $\{t | T(T^{-1}(t)) = t\}$ are the same. The closed set $\mathcal{R} \cup \mathcal{L}$ is contained in the closure of $\mathcal{R}$ since T^{-1} is continuous. This shows that T^{-1} satisfies the growth requirements of (3) so we need only show that for every positive t, $T^{-1}(t)$ satisfies the measurability requirements of (1). To do this suppose first that in the representation (3.2) the measure ν has finite mass, so that necessarily $\alpha > 0$. Let $\ell(t, w) = \int_0^t I_{\{0\}}(X_s(w)) ds$, that is the lebesgue measure of $\{s \leq t | X_s = 0\}$. Clearly $\ell(t) \in \sigma\{X_s; s \leq t\}$, and also it is clear that $T^{-1}(t) = \alpha^{-1} \ell(t)$ so the measurability of $T^{-1}(t)$ is established in this case. In the remaining case $\|\nu\| = \infty$ and then the following more satisfactory approach is available. For ε and t positive let $N_\varepsilon(t)$ denote the number of points $s \leq t$ such that $T(s) - T(s-)$ strictly exceeds ε. Then $N_\varepsilon(t)$ has the Poisson distribution with parameter $t\nu(\varepsilon, \infty)$. Since $\nu(\varepsilon, \infty) \to \infty$ as $\varepsilon \to 0$ the ratio $N_\varepsilon(t)/\nu(\varepsilon, \infty)$ converges in probability to t as $\varepsilon \to 0$. We can take a sequence $\varepsilon_n \to 0$ such that the convergence take place with probability one for all rational t and hence for all t since $N_\varepsilon(t)$ is non decreasing in t. Then with probability one $N_{\varepsilon_n}(T^{-1}(t))/\nu(\varepsilon_n, \infty)$ will converge to $T^{-1}(t)$. Now the reader will show easily that for each $\varepsilon > 0$ the mapping $r \to T(r-)$ sets up a one-to-one correspondence between those points $r \leq T^{-1}(t)$ such that $T(r) - T(r-) > \varepsilon$ and those points $s \leq t$ such that $X_s > \varepsilon$ but s is a left limit of points s' with $X_{s'} = 0$. Quite clearly the number of such points is an integer valued random variable measurable relative to $\sigma\{X_s; s \leq t\}$. Thus $N_\varepsilon(T^{-1}(t))$ is measurable relative to the same σ-algebra, and this establishes the required measurability for $T^{-1}(t)$.

We will conclude the discussion of this example by characterizing the transition function of the sawtooth process $X_t = T(T^{-1}(t)) - t$. The operators $P_t f(x)$ are of course uniquely determined by the potential operators

$$(3.4) \qquad U^\beta f(x) = E^x \int_0^\infty e^{-\beta t} f(X_t) dt.$$

Let $\sigma = \inf\{t | X_t = 0\}$ and set

$$V^\beta f(x) = E^x \int_0^\sigma e^{-\beta t} f(X_t) dt$$

so that for the trivial minimal process at hand

$$V^\beta f(x) = \int_0^x e^{-\beta t} f(x - t) dt.$$

Writing the integral on the right of (3.4) as $\int_0^\sigma + \int_\sigma^\infty$ and using the Markov property we have

$$U^\beta f(x) = V^\beta f(x) + E^x e^{-\beta \sigma} U^\beta f(0)$$
$$= V^\beta f(x) + e^{-\beta x} U^\beta f(0),$$

so we need only calculate $U^\beta f(0)$. Suppose the subordinator $\{T(t); t \geq 0\}$ is such that the Lévy measure ν in the representation (3.2) is finite. Assuming that $X_0 = 0$ let J be the time X_t leaves 0 so that J is exponentially distributed with rate $\alpha^{-1}\|\nu\|$ and X_J is independent of J and with $\nu/\|\nu\|$ as distribution. Given a positive Borel function f on $[0, \infty)$ we have

$$U^\beta f(0) = E^0 \int_0^J + \int_J^{J + \sigma \circ \theta_J} + \int_{J + \sigma \circ \theta_J}^\infty e^{-\beta t} f(X_t) dt$$
$$= f(0)\beta^{-1} E^0 (1 - e^{-\beta J})$$
$$+ E^0 e^{-\beta J} E^0 V^\beta f(X_J)$$
$$+ E^0 e^{-\beta J} E^{X_J} (e^{-\beta J}) U^\beta f(0).$$

This can be solved for $U^\beta f(0)$ and all the remaining terms are calculated easily. The result is

$$(3.5) \qquad U^\beta f(0) = \left(\alpha f(0) + V^\beta f(\nu)\right)/\left(\beta\alpha + \beta V^\beta 1(\nu)\right)$$

with, of course $V^\beta f(\nu)$ standing for $\int V^\beta f(x)\nu(dx)$. Coming to the general case, recall the approximation $\bar{T}^\varepsilon(t)$ of $T(t)$ obtained by deleting the jumps of size less than ε. With probability one the path of the process $\{\varepsilon t + \bar{T}^\varepsilon(t); t \geq 0\}$ converges, uniformly on compact t sets, to the path $T(t)$. The reader will verify easily that as a consequence if $T^\varepsilon(t)$ denotes the sawtooth process based on $\{\varepsilon t + \bar{T}^\varepsilon(t); t \geq 0\}$ then if the path $\varepsilon t + \bar{T}^\varepsilon(t)$ converges to the path $T(t)$ uniformly on compact sets, X_t^ε converges to X_t except perhaps at those points t which are points of left increase of T^{-1} but not points of right increase. That set is countable and so with probability one X_t^ε converges to X_t for almost all (lebesgue measure) t. So if we set

$$U_\varepsilon^\beta f(x) = E^x \int_0^\infty e^{-\beta t} f(X_t^\varepsilon)\,dt$$

then as $\epsilon \to 0$ $U_\varepsilon^\beta f(0) \to U^\beta f(0)$ for all bounded continuous f. If the analytic description of the subordinator $\{T_t; t \geq 0\}$ is given by (3.2) then the description for $\{\varepsilon t + \bar{T}_t^\varepsilon; t \geq 0\}$ is obtained by replacing α by $\alpha + \varepsilon$ and ν by its restriction to (ε, ∞). Applying (3.5) to this latter case to calculate $U_\varepsilon^\beta f(0)$ and then letting $\varepsilon \to 0$ we find that (3.5) describes $U^\beta f(0)$ in the general case.

(c). **Feller Brownian motions:** The task here is to find all the standard processes with infinite lifetime and state space $[0, \infty)$ which behave like ordinary Brownian motion when the path is in $(0, \infty)$, as defined more precisely at the beginning of II-3. Solving the problem in this or any similar setting involves two steps. The first is to assume that such a process is given and then derive analytic data that characterizes uniquely the transition function of the process. This program goes back to Ventcel [V, 1]. It has been carried out in considerable generality in the fundamental paper of Motoo [Mo, 1], but even that paper contains hypotheses that rule out some interesting and elementary special cases. The second step is to establish the existence of an extension of the minimal process corresponding to given analytic data. This construction step has been carried out successfully only in situations such as the present one in which the terminal set ($\{0\}$ in the present case) consists of a finite number of points. No doubt the proper method for doing the construction in this case is to use Itô's excursion method [I, 1]. But we will treat the present example of Feller Brownian motion by the special methods of [IM, 1] since that approach is interesting in its own right.

Let us first of all find all the possible Feller Brownian motions. We need some notation and a few calculations. We will let $\{V^\lambda; \lambda > 0\}$ denote the resolvent operators for Brownian motion, $\{X_t; t \geq 0\}$, starting at $x > 0$ and killed at σ, the time the path reaches 0. Thus for a positive or bounded Borel function f on $[0, \infty)$

$$V^\lambda f(x) = E^x \int_0^\sigma e^{-\lambda t} f(X_t) dt$$
$$= \int_0^\infty v^\lambda(x, y) f(y) dy$$

with $v^\lambda(x, y) = \int_0^\infty e^{-\lambda t} p^-(t, x, y) dy$. An easy calculation, using the Gauss kernel, will show that

$$(3.6) \qquad v^\lambda(x, y) = \frac{1}{\sqrt{2\lambda}}(e^{-\sqrt{2\lambda}|x-y|} - e^{-\sqrt{2\lambda}|x+y|}),$$

with $x > 0, y > 0$. In particular

$$\lambda V^\lambda 1(x) = E^x(1 - e^{-\lambda\sigma}) = 1 - e^{-x\sqrt{2\lambda}}.$$

For later use notice that $V^1 1$ is a strictly positive continuous function on $(0, \infty)$ which approaches 1 as $x \to \infty$ and is asymptotic to a multiple of x as $x \to 0$. We will let $\{R^\lambda; \lambda > 0\}$ denote the resolvent operators for reflecting Brownian motion on $[0, \infty)$. Then

$$R^\lambda f(x) = \int_0^\infty e^{-\lambda t} f(|X_t|) dt$$
$$= \int_0^\infty r^\lambda(x, y) f(y) dy$$

and, from the discussion in section 3 of Chapter I

$$r^\lambda(x, y) = \frac{1}{\sqrt{2\lambda}}(e^{-\sqrt{2\lambda}|x-y|} + e^{-\sqrt{2\lambda}|x+y|}), \qquad x \geq 0, y > 0.$$

If f is a bounded Borel function on $(0,\infty)$ then one shows easily that as $x \to 0$, $x^{-1}V^\lambda f(x) \to \sqrt{2\lambda}R^\lambda f(0)$. From this we get a simple but useful fact:

(3.7) if f is a bounded Borel function on $(0,\infty)$ then $\lim\limits_{x\to 0} V^\lambda f(x)/V^\lambda 1(x) = \lambda R^\lambda f(0)$.

Let $\{X_t; t \geq 0\}$ be a Feller Brownian motion and let $\{U^\alpha; \alpha > 0\}$ denote the potential (or resolvent) operators, $U^\alpha f(x) = E^x \int_0^\infty e^{-\alpha t} f(X_t)dt$ defined at least for Borel functions f on $[0,\infty)$ which are positive or bounded. The operators V^λ and R^λ are those from (3.7). If $\sigma = \inf\{t > 0 | X_t = 0\}$ then, since the process is Brownian motion prior to σ, we have

$$U^\alpha f(x) = E^x \int_0^\sigma e^{-\alpha t} f(X_t)dt + E^x e^{-\alpha\sigma} E^{X_\sigma} \int_0^\infty e^{-\alpha t} f(X_t)dt$$
$$= V^\alpha f(x) + e^{-x\sqrt{2\alpha}}U^\alpha f(0).$$

In particular $U^\alpha f(0)$ determines $U^\alpha f(x)$ for all x. According to I-9 a knowledge of the operator U^α for a fixed positive α determines the family $\{U^\alpha; \alpha > 0\}$ and hence determines the transition function. Thus, we will consider the following as an adequate description of all the possible Feller Brownian motions.

(3.8) Theorem. *There are positive numbers p and q and a measure η on $\mathcal{B}(0,\infty)$ such that for every positive Borel function f on $[0,\infty)$*

$$(3.9) \qquad U^1 f(0) = pf(0) + qR^1 f(0) + \int_0^\infty V^1 f d\eta.$$

We have $p + q + \int_0^\infty (1 - e^{-x\sqrt{2}})\eta(dx) = 1$. The numbers and measure are determined uniquely by the operator $U^1 f(0)$. If $p = q = 0$ then $\|\eta\| = \infty$.

Proof: Given $\varepsilon > 0$ let $\sigma_\varepsilon = \inf\{t | X_t \geq \varepsilon\}$ and let $\bar\sigma_\varepsilon = \sigma_\varepsilon + \sigma \circ \theta_{\sigma_\varepsilon}$. Then for any positive Borel function f write $\int_0^\infty e^{-t} f(X_t)dt$ as $\int_0^{\sigma_\varepsilon} + \int_{\sigma_\varepsilon}^{\bar\sigma_\varepsilon} + \int_{\bar\sigma_\varepsilon}^\infty$. Applying the strong Markov property in the usual way we have

$$U^1 f(0) = E^0 \int_0^{\sigma_\varepsilon} e^{-t} f(X_t)dt + E^0 e^{-\sigma_\varepsilon} V^1 f(X_{\sigma_\varepsilon})$$
$$+ E^0 \left(e^{-\bar\sigma_\varepsilon} U^1 f(X_{\bar\sigma_\varepsilon})\right).$$

Since $X_{\bar{\sigma}_\epsilon} = 0$ if $\bar{\sigma}_\epsilon < \infty$ the last term is $E^0(e^{-\bar{\sigma}_\epsilon})U^1 f(0)$ and so subtracting and dividing we obtain

$$(3.10) \qquad U^1 f(0) = \theta_\epsilon^{-1} E^0 \int_0^{\sigma_\epsilon} e^{-t} f(X_t)\,dt + \theta_\epsilon^{-1} E^0\big(e^{-\sigma_\epsilon} V^1 f(X_{\sigma_\epsilon})\big),$$

with $\theta_\epsilon = E^0(1-e^{-\bar{\sigma}_\epsilon})$. Denote by η_ϵ the measure $A \to \theta_\epsilon^{-1} E^0 e^{-\sigma_\epsilon} I_A(X_{\sigma_\epsilon})$. Since $\sigma_\epsilon \leq \bar{\sigma}_\epsilon$ it follows that if $p_\epsilon = \theta_\epsilon^{-1} E^0(1 - e^{-\sigma_\epsilon})$ then $p_\epsilon \leq 1$. Also since $\eta_\epsilon(V^1 1) \leq 1$ the masses $\eta_\epsilon([\delta, \infty)$ are for each $\delta > 0$ bounded over ϵ and so letting $\epsilon \to 0$ through a subsequence we may assume $p_\epsilon \to p$ and the measures η_ϵ converge to a measure η in the sense that $\int g\,d\eta_\epsilon \to \int g\,d\eta$ for every continuous g with compact support in $(0, \infty)$. If k is any positive number and f is the indicator of $[k, \infty)$ then $V^1 f(x) \to 1$ as $x \to \infty$ whereas $U^1 f(0)$ is small if k is large. It follows that in the convergence of η_ϵ to η no mass escapes to ∞, and so $\int g\,d\eta_\epsilon \to \int g\,d\eta$ for g bounded and continuous provided g vanishes near 0. The latter condition cannot be removed: by passing to another subsequence we may assume $\lim_{\epsilon \to 0} \eta_\epsilon(V^1 1)$ exists: call it q'. But we can say only that $\eta(V^1 1) \leq q'$. Define q to be $q' - \eta(V^1 1)$. Obviously what we can assert is that $\eta_\epsilon(g) \to \eta(g)$ for all bounded continuous functions g such that $g(x)/V^1 1(x) \to 0$ as $x \to 0$. Now let f in (3.10) be a bounded continuous function on $[0, \infty)$. We write the first term on the right as

$$p_\epsilon f(0) + \theta_\epsilon^{-1} E^0 \int_0^{\sigma_\epsilon} (f - f(0))(X_t)e^{-t}\,dt.$$

Obviously as $\epsilon \to 0$ the second summand approaches 0 since it is in absolute value less than $\sup_{x < \epsilon}|f(x) - f(0)|$. Write the second term on the right of (3.10) as

$$\eta_\epsilon\big(V^1 f - R^1 f(0)V^1 1\big) + R^1 f(0)\eta_\epsilon(V^1 1).$$

The integrand in the first term is a bounded continuous function which, by (3.7), vanishes at 0 faster than does $V^1 1$. So as $\epsilon \to 0$ this entire expression approaches

$$\eta\big(V^1 f - R^1 f(0)V^1 1\big) + q' R^1 f(0)$$

and so the representation (3.9) is established for bounded continuous f and hence for all bounded Borel f since each side of the equation defines a measure in f. The second sentence in (3.8) follows from the fact that

$U^1 1(0) = 1$. As to the uniqueness of the components in (3.9), if f is the indicator of $\{0\}$ then $U^1 f(0) = p$ as the other two terms are 0, and hence p is determined. As to the other components let f_λ denote the function $e^{-\lambda y}$ restricted to $(0, \infty)$. Inserting f_λ for f in (3.9) and doing a little calculation based on the formulas for the kernels v_λ and r_λ yields, for $\lambda \neq 2$,

$$U^1 f_\lambda(0) = (\lambda^2 - 2)^{-1} \{ (\lambda - \sqrt{2}) q\sqrt{2} + \int_0^\infty e^{-x\sqrt{2}}(1 - e^{-x(\lambda - \sqrt{2})}) \eta(dx) \}.$$

Thus the operator U^1 determines, for each positive β, the expression

$$\beta q\sqrt{2} + \int_0^\infty e^{-x\sqrt{2}}(1 - e^{-\beta x}) \eta(dx).$$

This is an expression of the sort appearing in (3.2) with $q\sqrt{2}$ in the role of α and $e^{-x\sqrt{2}}\eta(dx)$ in the role of $\nu(dx)$. We will take as known from the theory of subordinators — or leave as an easy Laplace transform exercise — the fact that in (3.2) $g(\lambda)$ determines α and ν. Consequently in our case q and η are uniquely determined. A consequence of the uniqueness statement in (3.8) is that in defining p, q and η as limits it was not necessary to pass to subsequences. We still must establish the assertion that if $p = q = 0$ then $\|\eta\| = \infty$. If (3.9) holds with $p = q = 0$ and a finite measure η then upon starting our Feller Brownian motion with initial distribution η we obtain

$$\int U^1 f(x)\eta(dx) = \int V^1 f d\eta + \left(\int e^{-x\sqrt{2}}\eta(dx) \right) U^1 f(0)$$

or $\eta U^1 f = k U^1 f(0)$ for some constant k and all positive Borel f. If we take a bounded Borel function g and a positive number δ and then take f to be $\delta^{-1}(g - e^{-\delta}P_\delta g)$ we get $U^1 f(x) = \delta^{-1} E^x \int_0^\delta e^{-r} g(X_r) dr$. If g also is continuous this converges boundedly to $g(x)$ for all x as $\delta \to 0$. Consequently a finite measure μ is determined by the expressions $\mu U^1 f$. It follows then that η is a multiple of unit mass at 0 and this, obviously, is impossible.

So all the conclusions of (3.8) are established, and we can turn to the construction problem. What we will show is that given positive numbers p and q and a measure η such that $p + q + \int(1 - e^{-x\sqrt{2}})\eta(dx) = 1$ and $\| \eta \| = \infty$ if $p = q = 0$ there is indeed a Feller Brownian motion satisfying (3.9) with the given parameters. The construction involves some interesting probabilistic techniques which will be used elsewhere, but describing the procedure and carrying out the details does take some time.

The most obvious example of a Feller Brownian motion is reflecting Brownian motion. It will be used to construct the others, and for this the notion of local time will be vital. The definition is that given in the discussion of sawtooth processes: that is if $\{Y_t; t \geq 0\}$ denotes reflecting Brownian motion then *local time at* $\{0\}$ for Y is a stochastic process $\{\ell(t); t \geq 0\}$ satisfying the measurability, additivity, continuity and growth properties (1), (2) and (3) from paragraph (b). Of course the shift operator θ_t is understood to be the one appropriate to the process Y. For our purposes it will be enough to establish the existence of local time for a particular way of constructing Y: specifically as in I-3 let $\{X_t; t \geq 0\}$ with measures P^x be Brownian motion with continuous paths; let

$$m_t(\omega) = \min\big(0, \min_{s \leq t} X_s(\omega)\big)$$

and

$$Y_t(\omega) = X_t(\omega) - m_t(\omega).$$

Then we showed in I-3 that Y relative to the measures P^x for $x \geq 0$ is a standard process and in II-2 we identified it as reflecting Brownian motion. Assuming we have taken the space Ω for our original Brownian motion to be the space of continuous real valued functions ω on $[0, \infty)$ and $X_t(\omega) = \omega(t)$ then $\theta_t \omega(r) = \omega(t+r) - m_t(\omega)$ defines the shift for Y, that is $Y_r \circ \theta_t = Y_{t+r}$. We will show now that if we define $\{\ell(t); t \geq 0\}$ by

$$\ell(t, \omega) = -m_t(\omega)$$

then $\{\ell(t); t \geq 0\}$ is local time at $\{0\}$. Quite clearly for all ω, $t \to \ell(t, \omega)$ defines a continuous non-decreasing function with $\ell(0, \omega) = 0$. The additivity property $\ell(t + s) = \ell(s) + \ell(t) \circ \theta_s$ is easy to check. Coming to the growth properties, if $Y_s(\omega) > 0$ throughout an interval (a, b) then $X_s(\omega) > m_a(\omega)$ for all $s \in (a, b)$ and so $m_a(\omega) = m_b(\omega)$. Conversely suppose a and b are rationals and $T = \inf\{s \geq a | Y_s = 0\}$. Then T is a stopping time for the Y process and hence also for X. On the set $T < \infty$ we have $X_T = m_a$. Now a Brownian motion starting at 0 immediately takes on negative values and so by the strong Markov property for X, $P^x(X_{T+r} < X_T$ for some

$r < \varepsilon) = 1$ for all $\varepsilon > 0$. Hence $P^x(T < b, m_b = m_a) = 0$. It follows that with probability one relative to each P^x the support of $d_t \ell(t)$ and $\{t|Y_t = 0\}$ are the same. So what remains is verifying that m_t can be recovered from a knowledge of $\{Y_s; s \le t\}$ at least once we eliminate a set of paths having probability zero. In the probabilistic argument which follows we will take P^0 as basic measure and leave to the reader the incorporation of the other P^x measures. Fix $\varepsilon > 0$ and let $N_\varepsilon(t)$ denote the number of times the path $s \to Y_s$ crosses from strictly above ε down to 0 as s ranges over the interval $[0, t]$. More precisely let

$$T_0 = \inf\{s \ge 0|Y_s > \varepsilon\}$$
$$\sigma = \inf\{s \ge 0|Y_s = 0\}$$

and

$$J_{n+1} = T_n + \sigma \circ \theta_{T_n}, T_n = J_n + T_0 \circ \theta_{J_n}$$

for $n = 1, 2, \ldots$. Thus $J_1, J_2, \ldots$ are the times at which the successful crossings from ε to 0 are completed; and $N_\varepsilon(t) = n$ if $J_n \le t < J_{n+1}$. Let

$$R_t = \inf\{s| - m_s > t\} = \inf\{s|X_s < -t\}$$

and set

$$Z_t^\varepsilon = N_\varepsilon(R_t).$$

Now R_t is a stopping time for the process X and $Y_{R_t} = 0$. It is obvious that $N_\varepsilon(R_{t+s}) - N_\varepsilon(R_t)$ is simply the value $N_\varepsilon(R_s)$ that one obtains when $\{X_{r+R_t} - X_{R_t}; r \ge 0\}$ is taken as the basic Brownian motion, so from the strong Markov property for X it is clear that the process $\{Z_t^\varepsilon; t \ge 0\}$ has stationary independent increments. Clearly $Z_0^\varepsilon = 0$, the paths $t \to Z_t^\varepsilon$ are right continuous and the values of the increments $Z_{t+s}^\varepsilon - Z_t^\varepsilon$ are non-negative integers. In fact almost surely P^0 the discontinuities in $t \to Z_t^\varepsilon$ are all jumps of height one. To justify this assertion we need consider only the first jump: let $\tau = \inf\{t \ge 0|Z_t^\varepsilon > 0\}$ and let $W = Z_\tau^\varepsilon$. If W exceeds 2 then $N_\varepsilon(R_\tau) \ge 2$ and $N_\varepsilon(R_s) = 0$ for all $s < \tau$. Thus for all $s < \tau$ $R_s < J_1$ while $J_2 \le R_\tau$. It follows that the path of the Brownian motion $\{X_{r+J_1} - X_{J_1}; r \ge 0\}$ is positive throughout an initial interval and this is an event of probability 0. Thus $\{Z_t^\varepsilon; t \ge 0\}$ is a Poisson process. Let λ_ε denote its rate; that is $E^0(Z_t^\varepsilon) = t\lambda_\varepsilon$. Then $P^0(Z_1^\varepsilon \ge 1) = 1 - e^{-\lambda_\varepsilon}$. The event $\{Z_1^\varepsilon \geq 1\}$ surely will occur if the original Brownian motion reaches ε before it reaches -1 and so $\lambda_\varepsilon \to \infty$ as $\varepsilon \to 0$. Hence $\lambda_\varepsilon^{-1} Z_t^\varepsilon$ converges in P^0 probability to t as $\varepsilon \to 0$. Let $\varepsilon \to 0$ through a subsequence chosen so that the convergence $\lambda_\varepsilon^{-1} Z_t^\varepsilon \to t$ takes place with P^0 probability one for every rational t. Since the limit function is continuous and the approximating functions are monotone the convergence will be uniform on compact t sets. Let us restrict ourselves to a Brownian motion path and a sequence of ε such that $\lambda_\varepsilon^{-1} Z_\varepsilon(t) \to t$ uniformly on compacts. If s is a positive number and $t = -m_s$ then $R_{t-} \leq s \leq R_t$ and so

$$\lambda_\varepsilon^{-1} Z^\varepsilon(t-) \leq \lambda_\varepsilon^{-1} N_\varepsilon(R_t-) \leq \lambda_\varepsilon^{-1} N_\varepsilon(s)$$
$$\leq \lambda_\varepsilon^{-1} N_\varepsilon(R_t) = \lambda_\varepsilon^{-1} Z_\varepsilon(t).$$

Thus

$$-m_s = \lim_{\varepsilon \to 0} \lambda_\varepsilon^{-1} N_\varepsilon(s).$$

Clearly $N_\varepsilon(s)$ is measurable relative to $\sigma\{Y_r; r \leq s\}$ so the identification of $\{-m_t; t \geq 0\}$ as local time is complete.

We will break up the construction of the general Feller Brownian motion into two stages: To set things up properly let $\{x(t); t \geq 0\}$ be Brownian motion and as usual set

$$m_t = \min\big(0, \min_{s \leq t} x(s)\big).$$

Let $\{T(t); t \geq 0\}$ be a subordinator, independent of the Brownian motion with $T(0) = 0$ and such that in the representation (3.2) either $\alpha > 0$ or $\| \nu \| = \infty$, so that we may assume the sample functions of the subordinator are strictly increasing. Of course we assume the paths of the Brownian motion are continuous and those of the subordinator are right continuous. For x positive, P^x will denote probabilities computed under the assumption that $x(0) = x$ and $T(0) = 0$. One can set things up more precisely by taking a basic probability space consisting of the product of the space of continuous function on $[0, \infty)$ with the space of right continuous strictly increasing function on $[0, \infty)$. We will leave such details to the reader. The inverse of the subordinator will be defined as in the discussion of sawtooth processes.

(3.11) Theorem. *The process $\{X(t); t \geq 0\}$ defined by*

(3.12) $$X(t) = T\big(T^{-1}(-m_t)\big) + x(t)$$

is a Feller Brownian motion. In the representation (3.9) we have

$$p = 0$$

(3.13)
$$q = \alpha\sqrt{2}/\Delta$$

$$\eta = \nu/\Delta,$$

with $\Delta = \alpha\sqrt{2} + \int_0^\infty (1 - e^{-x\sqrt{2}})\nu(dx)$.

Proof. First of all note that if we write (3.12) as

$$T\big(T^{-1}(-m_t)\big) + m_t + x(t) - m_t$$

then the representation looks like the representation of a sawtooth process, but run according to the local time, $-m_t$, for the reflecting Brownian
motion and with that reflecting Brownian motion added on.

For a preliminary orientation and use later on we will verify (3.13)
assuming the Markovian character of X has been established already. As-
sume first of all that in the representation (3.2) we have $\| \nu \| < \infty$ (so
that necessarily $\alpha > 0$). If we refer to the remark following (3.3) we see
that in this case $T^{-1}(t) = t/\alpha$ up until the time, J, at which the sawtooth process $T\big(T^{-1}(t)\big) - t$ has its first jump. Thus until the time, let
us call it K, when $-m_t$ reaches J, we have $T\big(T^{-1}(-m_t)\big) = -m_t$; and so
for $0 \le t \le K, X(t) = x(t) - m_t$, that is, it equals a reflecting Brownian
motion. Then independent of what has gone before, the process X has a
jump equal in size to the jump of T at $T^{-1}(-m_K)$. The time J is exponen-
tially distributed with rate $\alpha^{-1} \| \nu \|$ and the jump height is distributed
according to $\nu/ \| \nu \|$. From there X proceeds again as a Brownian motion
whose initial position is indeed the height of that first jump, since necessarily the time K is an increase point of $t \to -m_t$ so that $x(K) - m_K = 0$.
Given a positive Borel function f on $[0,\infty)$, we compute the potential operator $U^1 f(0) = E^0 \int_0^\infty e^{-t} f(X(t)) dt$ in the usual way; that is writing
$\int_0^\infty = \int_0^K + \int_K^{K+\sigma\circ\theta_K} + \int_{K+\sigma\circ\theta_K}^\infty$ and using the strong Markov property we
obtain

$$U^1 f(0) = E^0 \int_0^K e^{-t} f(x(t) - m_t) dt$$

(3.14)
$$+ E^0 e^{-K} V^1 f(\nu/ \| \nu \|)$$

$$+ E^0 e^{-K} E^{\nu/\|\nu\|} e^{-\sigma} U^1 f(0).$$

(Note: we have not introduced shift operators for the process X but clearly $K + \sigma \circ \theta_K$ is to denote the first return to 0 after time K.) If we write the first integral on the right of (3.14) as $\int_0^\infty - \int_K^\infty$ and use the strong Markov property and the fact that $x(K) - m_K = 0$ we see that it may be written as $E^0(1 - e^{-K})R^1 f(0)$, with R^1 being, as usual, the potential operator for the reflecting Brownian motion $x(t) - m_t$. We have $E^\nu/\|\nu\| e^{-\sigma} = \int_0^\infty e^{-x\sqrt{2}} \nu(dx)/\|\nu\|$ and so (3.14) can be solved for $U^1(0)$ once we evaluate $E^0 e^{-K}$. To do this write $E^0(e^{-K})$ as $P^0(K < e_1)$ where e_1 denotes an exponentially distributed random variable of rate 1 and independent of all the processes under consideration. Since K is the time when $-m_t$ reaches the level J we have $P^0(K < e_1) = P^0(-m_{e_1} > J)$. This last expression is the probability that a Brownian motion starting at J reaches 0 before time e_1, that is the expected value of $P^J(\sigma < e_1) = E^J e^{-\sigma} = e^{-J\sqrt{2}}$. Inserting the fact that J has an exponential distribution of rate $\|\nu\|/\alpha$ and taking expected value we obtain $E^0 e^{-K} = \|\nu\|/(\alpha\sqrt{2} + \|\nu\|)$. From this a little algebra yields (3.13) under the additional assumptions that ν is finite and $\alpha > 0$. We can remove the restriction $\|\nu\| < \infty$ in exactly the same way as we did in obtaining (3.5) for sawtooth processes: specifically let $\bar{T}^\varepsilon$ denote the approximation to the subordinator T obtained by removing the jumps of size less than ε and adding a linear term εt. If X^ε denotes the Feller Brownian motion based on $\bar{T}^\varepsilon$ then what we have just done implies that $E^0 \int_0^\infty e^{-t} f(X^\varepsilon(t)) dt$ is given by the right side of (3.9) with α in (3.13) replaced by $\alpha + \varepsilon$ and ν in (3.13) replaced by its restriction to (ε, ∞). We noted following (3.5) that as $\varepsilon \to 0, \bar{T}^\varepsilon(\bar{T}^\varepsilon)^{-1}(t)$ converges to $T(T^{-1}(t))$ except perhaps at those points t lying in a countable set dependent on the T process. The conclusion of I-4.7 implies that for each t, m_t has a continuous distribution, and it follows that for each t, with P^0 probability 1, $X^\varepsilon(t)$ converges to $X(t)$. Now the validity of (3.13) in the general case is obvious.

Next we have to deal with the Markov (and strong Markov) properties of the presumed Feller Brownian motion defined by (3.12). Once we have done this we can observe that if the initial position of $x(t)$ is $x > 0$ then $m_t = 0$ until the time when $x(t)$ reaches 0, and so since $T^{-1}(0) = 0$ and $T(0) = 0$ we will have $X(t) = x(t)$ in that interval; that is $X(t)$ indeed will be a Feller Brownian motion. For s positive let q_s denote $T^{-1}(-m_s)$.

Define $\mathcal{H}_s$ by saying that Δ is in $\mathcal{H}_s$ if for every $r \geq 0, \Delta \cap \{q_s \leq r\}$ is in $\sigma\{T(t); t \leq r, x(t); t \leq s\}$. The reader will verify easily that for each s, $q_s \in \mathcal{H}_s$, $\mathcal{H}_s$ is a σ-algebra, and $\mathcal{H}_r \subset \mathcal{H}_s$ if $r \leq s$. Next we argue that for each s $X(s)$ is $\mathcal{H}_s$ measurable. Obviously in doing so we need consider only $T(q_s)$ since $x(s)$ is $\mathcal{H}_s$ measurable. Let f be a continuous function on $[0, \infty)$ and let r be a positive number. Then $f(T(q_s))I_{\{q_s \leq r\}}$ is the limit as $n \to \infty$ of the expression

$$\sum_{k < k_0} f\big(T(k/2^n)\big) I_{\frac{k-1}{2^n} \leq q_s < \frac{k}{2^n}} + f(T(r)) I_{\frac{k_0}{2^n} \leq q_s \leq r}$$

where k_0 is defined by $k_0/2^n \leq r < (k_0 + 1)/2^n$. Obviously each summand is measurable relative to $\sigma\{T(t); t \leq r, x(t); t \leq s\}$ and so $T(q_s) \in \mathcal{H}_s$. Now fix s. Let $\bar{x}$ denote the process $\bar{x}(t) = x(t + s) - x(s)$ and let T^+ denote the process $T^+(t) = T(t + q_s) - T(q_s)$. Let Γ_1 and Γ_2 denote Borel sets in the space of right continuous functions on $[0, \infty)$ and let Δ be a set in $\mathcal{H}_s$. If $\mathcal{H}$ denotes $\sigma\{T(t); t \geq 0, x(t); t \leq s\}$ then the event $\{T^+ \in \Gamma_2, \Delta\}$ is in $\mathcal{H}$, whereas the Brownian motion $\bar{x}$ is independent of $\mathcal{H}$. Furthermore the process $\{T(t); t \geq 0\}$ is Markovian relative to the σ-algebras $\sigma\{T(r); r \leq t, x(r); r \leq s\}_t$ and q_s is a stopping time relative to this family, with Δ a set in the pre-q_s σ-algebra. So applying the first independence assertion to $\bar{x}$ and then the strong Markov property of the T process at q_s we obtain

$$(3.15) \qquad P^x(\bar{x} \in \Gamma_1, T^+ \in \Gamma_2, \Delta) = P^0(x \in \Gamma_1)P^0(T \in \Gamma_2)P^x(\Delta).$$

In words (3.15) says that relative to any P^x the processes $\bar{x}$ and T^+ are probabilistically the same as the original Brownian motion and subordinator relative to P^0 and that the pair is independent of $\mathcal{H}_s$. In particular we have

$$(3.16) \quad P^x\big(X(s) + \bar{x} \in \Gamma_1, T^+ \in \Gamma_2; \Delta\big) = P^x\big(P^{X(s)}(x \in \Gamma_1, T \in \Gamma_2, \Delta)\big)$$

since $X(s) + \bar{x}$ is a Brownian motion with initial position $X(s)$ and is independent of T^+. It is now clear that to verify the Markov property of $\{X(t); t \geq 0\}$ all we need do is check that for each positive t, $X(s + t)$ is in fact what we obtain by evaluating at t the process defined by (3.12) but with all the quantities involved based on the Brownian motion $X(s) + \bar{x}$ and the subordinator T^+. We will use the superscript $+$ to denote quantities defined in terms of these processes; so, for example

$$X^+(t) = X(s) + \bar{x}(t)$$
$$m_t^+ = \min\big(0, \min_{u \leq t} x^+(u)\big)$$
$$\sigma^+ = \inf\{t | x^+(t) = 0\}$$
$$q_t^+ = (T^+)^{(-1)}(-m_t^+)$$

and so forth. Let $\Delta = X(s) - \big(x(s) - m(s)\big) = T(q_s) + m_s$. If $t \leq \sigma^+$ then $m_{t+s} \geq m_s - \Delta$, or $T(q_s) \geq -m_{t+s}$ so that $q_{t+s} = q_s$. Then we have

$$\begin{aligned} X(t+s) &= T(q_{t+s}) + x(t+s) \\ &= T(q_s) + x(s) + \bar{x}(t) \\ &= x^+(t). \end{aligned}$$

But in this case $m^+(t) = 0$ so $T^+(q_t^+) = 0$ and $X(t+s) = X^+(t)$ as required. Suppose now $t > \sigma^+$. Then $m_{t+s} = m_t^+ + x(s) - X(s)$, and we have

$$\begin{aligned} q_t^+ &= \inf\{r | T(r + q_s) - T(q_s) > -m_t^+\} \\ &= \inf\{r | T(r + q_s) > T(q_s) - m_{t+s} + x(s) - X(s)\} \\ &= -q_s + q_{t+s}. \end{aligned}$$

And so

$$\begin{aligned} T^+(q_t^+) + x^+(t) &= T(q_t^+ + q_s) - T(q_s) + X(s) + x(t+s) - x(s) \\ &= T(q_{t+s}) + x(t+s) = X(t+s) \end{aligned}$$

as required.

We still have to verify the strong Markov property for $\{X(t); t \geq 0\}$. With enough attention to σ-algebras this can be thrown back on the strong Markov properties of the original Brownian motion and subordinator. But it is less involved simply to invoke I-4.4 and then check that for each positive s and bounded continuous f on $[0, \infty)$, $E^x f\big(X(s)\big)$ varies continuously with x. The validity of this last assertion follows almost immediately from the fact that we can evaluate integrals relative to P^x by using the measure P^0 and replacing $x(s)$ by $x(s) + x$. To be specific, the distribution of $-m_s + x$ relative to P^0 is the same as the distribution of $-m_s$ relative to P^x. If $\{x_n\}$ is a sequence of positive numbers converging to x then

$-m_s + x_n \to -m_s + x$ and $T^{-1}(-m_s + x_n) \to T^{-1}(-m_s + x)$ since T^{-1} is continuous. If x_n decreases to x then by the right continuity of T we have

$$(3.17) \qquad T\big(T^{-1}(-m_s + x_n)\big) \to T\big(T^{-1}(-m_s + x)\big).$$

If x_n increases to x then (3.17) fails only if $-m_s + x$ is a point which is not in the range of T, but which is left limit of such points. The fact that m_s has a continuous distribution and that m_s and T are independent implies that the P^0 probability of this is 0, so regardless of how x_n approaches x (3.17) holds with probability one. The continuity in x of $E^x f(X(s))$ is now obvious.

Next we have to deal with the question of how to obtain Feller Brownian motions for which the representation (3.7) yields values of p other than 0. These will be obtained by making a random time change in the processes we have just constructed, and for this first of all we must investigate the notion of local time at $\{0\}$ for these processes. Let $\{X(t); t \geq 0\}$ be the Feller Brownian motion we have just constructed, based on the Brownian motion $\{x(t); t \geq 0\}$ its minimum process $\{m_t; t \geq 0\}$ and an independent subordinator $\{T(t); t \geq 0\}$. What we want is a stochastic process $\{\ell(t); t \geq 0\}$ with continuous paths satisfying the same properties relative to $\{X(t); t \geq 0\}$ as were satisfied by $\{T^{-1}(t); t \geq 0\}$ relative to the sawtooth process $T(T^{-1}(t)) - t$ or by $\{-m_t; t \geq 0\}$ relative to the reflecting Brownian motion $x(t) - m_t$. That is for each t, $\ell(t)$ should be measurable relative to $\sigma\{X(r); r \leq t\}$, $\ell(t)$ should grow exactly on $\{t \mid X(t) = 0\}$ and $\{\ell(t); t \geq 0\}$ should satisfy an additivity property: since we have not introduced a shift operator for the X process, the appropriate expression of the additivity property is that for each t and s

$$\ell(t + s) = \ell(s) + \ell_s^+(t)$$

where $\ell_s^+(t)$ is to denote a quantity defined exactly as $\ell(t)$ but in terms of the process $X_s^+(r) = X(r + s); r \geq 0$. We will show now that such a local time process is obtained by taking

$$\ell(t) = T^{-1}(-m_t).$$

Obviously $\ell(t)$ is continuous in t. Pending the establishment of the measurability property the additivity already has been shown: it is simply the relationship $q_t^+ + q_s = q_{t+s}$ derived in the course of checking the Markov property of X. As to the growth properties since $T^{-1}(-m_r)$ is continuous in r we need to show only that except for a set of paths of probability 0 and a countable set of r values which can depend on the path, $\{r \mid X(r) = 0\}$ coincides with the set of points of right increase of the function $r \to T^{-1}(-m_r)$.

Firstly $T\big(T^{-1}(t)\big)$ exceeds t, and so if $X(r) = 0$ then $T\big(T^{-1}(-m_r)\big) = -m_r$ and $x(r) - m_r = 0$. The first equality implies that $-m_r$ is a point of right increase of T^{-1}. Now except for a set of probability 0 all but countably many r such that $x_r - m_r = 0$ are points of right decrease of the minimum function $t \to m_t$ and so if r is also one of these then $T^{-1}(-m_{r+\varepsilon}) - T^{-1}(-m_r) > 0$ for all strictly positive ε. Conversely if for each $\varepsilon > 0$ $T^{-1}(-m_{r+\varepsilon}) > T^{-1}(-m_r)$ then r is a decrease point of the minimum function and so $x_r - m_r = 0$. Also $-m_r$ is a right increase point of T^{-1} and so $T\big(T^{-1}(-m_r)\big) = -m_r$, and so $X(r) = 0$. Thus we are left with the establishing the measurability fact that with probability one $T^{-1}(-m_r)$ can be recovered from the path $X(t)$ as t ranges over $[0, r]$.

Once again as in the discussion of sawtooth processes we have to consider separately the cases $\| \nu \| = \infty$ and $\| \nu \| < \infty$ in the representation (3.2) relative to the subordinator T. Suppose first $\| \nu \| = \infty$. Let $N_\varepsilon(t)$ denote the number of points $s \leq t$ such that $T(s) - T(s-) > \varepsilon$. We argued that there is a sequence $\varepsilon_n \to 0$ such that with probability one for all t $N_{\varepsilon_n}\big(T^{-1}(t)\big)/\nu(\varepsilon_n, \infty)$ converges to $T^{-1}(t)$. Hence $N_{\varepsilon_n}\big(T^{-1}(-m_r)\big)/\nu(\varepsilon_n, \infty))$ will converge to $T^{-1}(-m_r)$ for all r. The reader will have no trouble seeing that the number of points s less than $T^{-1}(-m_r)$ such that $T(s) - T(s-) > \varepsilon_n$ is the same as the number of points t less than r such that $X(t) - X(t-) > \varepsilon_n$. Obviously this latter random integer is measurable relative to $\sigma\{X(t); t \leq r\}$ and so the measurability of $T^{-1}(-m_r)$ is established in this case. In the case $\| \nu \| < \infty$, so that $\alpha > 0$, we use the "downcrossing" argument that established m_r as a measurable function of $\{x(t) - m_t; t \leq r\}$. Specifically we noted that if $N_\varepsilon(t)$ now denotes the number of times that the path of $x(r) - m(r)$ crosses from above ε down to 0 as r ranges over $[0, t]$ then as $\varepsilon_n \to 0$ through a properly chosen sequence $\lambda_{\varepsilon_n}^{-1} N_{\varepsilon_n}(t) \to -m_t$ for all t with probability one. In the case at hand $T^{-1}(t) = \alpha^{-1} t$ throughout the interval $0 \leq t \leq J_1$, where J_1 is the time at which the sawtooth process takes its first jump. So that if

$-m_r < J$ then $T^{-1}(-m_r)$ is simply α^{-1} times $\lim_n \lambda_{\varepsilon_n}^{-1} N'_{\varepsilon_n}(r)$ where $N'_\varepsilon(r)$ denotes the number of times in $[0,r]$ that the X process crosses from above ε down to 0. (In this interval $N'_\varepsilon(r) = N_\varepsilon(r)$.) A little bookkeeping will convince the reader that this relationship holds for all r, and since the latter quantity is calculated from $X(t)$ for $0 \le t \le r$, the desired measurability assertion follows.

Now on to the final construction: let $\{X(t); t \ge 0\}$ denote the process we have just constructed, based on the "parameters" α and ν and use $\{\ell(t); t \ge 0\}$ to denote its local time at $\{0\}$, so that $\ell(t) = T^{-1}(-m_t)$. Let β be a positive number and consider the additive functional

$$A(t) = t + \beta\ell(t).$$

Obviously without exceptional points in the probability space $t \to A(t)$ defines a continuous strictly increasing function. If we let τ denote the inverse of A,

$$\tau(t) = \inf\{s | A(s) > t\},$$

then for each t, $\tau(t) \le s$ if and only if $A(s) \ge t$ and so for each t, $\tau(t)$ is a stopping time for the process X. The additivity property of A yields the equality

$$(3.18) \qquad\qquad \tau(t+s) = \tau(t) + \tau(s) \circ \theta_{\tau_t}$$

where the symbol $\tau(s) \circ \theta_{\tau_t}$ is to be interpreted as meaning a quantity defined exactly as $\tau(s)$ but based on the process $\{X(\tau(t) + r); r \ge 0\}$. The reader should have no difficulty concluding that (3.18) along with the strong Markov property for X implies that the process $\bar{X}$ defined by

$$(3.19) \qquad\qquad \bar{X}(t) = X(\tau(t)) \qquad\qquad t \ge 0$$

is a Markov process. The regularity details and the fact that $\bar{X}$ is also a standard process are discussed thoroughly in V-2.11 of [BG,1]; we will take that material as known, but it is not needed for understanding the rest of our presentation.

(3.20) Theorem. *The process $\{\bar{X}(t); t \ge 0\}$ is a Feller Brownian motion. In the representation (3.9) we have*

$$p = \Delta^{-1}\beta$$
$$(3.21) \qquad\qquad q = \Delta^{-1}\alpha\sqrt{2}$$
$$\eta = \Delta^{-1}\nu$$

with

$$\Delta = \beta + \alpha\sqrt{2} + \int (1 - e^{-x\sqrt{2}})\nu(dx).$$

Proof: We are taking as already established the fact that $\bar{X}$ is a standard process. Since $\ell(t) = 0$ for $t \leq \sigma$, the time when X first reaches 0, we have in that interval $\tau(t) = t$, and hence $\bar{X}(t) = X(t)$; and so $\bar{X}$ is indeed a Feller Brownian motion. It remains then only to verify the quantities in (3.21). Assume, to start off with that $\| \nu \| < \infty$ so that $\alpha > 0$. Let K have the same meaning as in (3.14) and define $\bar{K}$ by $\tau(\bar{K}) = K$ so that at time $\bar{K}$ the process $\bar{X}$ jumps from 0 to a point distributed according to $\nu / \| \nu \|$ and independent of developments up to that time. Letting $\bar{U}^1 f$ denote the 1 potential of f relative to $\bar{X}$ we have as usual

$$
\begin{aligned}
\bar{U}^1 f(0) = E^0 &\int_0^{\bar{K}} e^{-t} f(\bar{X}(t))\, dt \\
&+ E^0 e^{-\bar{K}} V^1 f(\nu / \| \nu \|) \\
&+ E^0 e^{-\bar{K}} E^{\nu/\|\nu\|}(e^{-\sigma}) \bar{U}^1 f(0).
\end{aligned}
$$

(3.22)

Obviously we intend to evaluate the first summand on the right, and $E^0 e^{-\bar{K}}$, and then solve for $\bar{U}^1 f(0)$. The second evaluation is the simpler one: as noted in the proof of (3.11), K is the time $x(r)$ reaches a level $-J$, where J is a random variable having an exponential distribution with rate $\| \nu \| / \alpha$ and independent of the basic Brownian motion. Thus $\bar{K} = K - (\beta/\alpha)m_K = K + (\beta/\alpha)J$ and so

$$
\begin{aligned}
E^0 e^{-\bar{K}} &= E^0 e^{-(\beta/\alpha)J} E^J e^{-\sigma} \\
&= E^0 \left(e^{-((\beta/\alpha)+\sqrt{2})J} \right) \\
&= \| \nu \| / (\| \nu \| + \beta + \alpha\sqrt{2}).
\end{aligned}
$$

To evaluate the first summand note that if $t < \bar{K}$ then $t = \tau(r)$ with $r < K$, and for such a value of r

$$
\begin{aligned}
X(r) &= T(T^{-1}(-m_r)) + m_r + x(r) - m_r \\
&= x_r - m_r.
\end{aligned}
$$

Thus in that interval $\bar{X}$ behaves as the Feller Brownian motion one obtains by running $x(t) - m_t$ according to the inverse of $t - (\beta/\alpha)m_t$, since $T^{-1}(r) =$

$\alpha^{-1}r$ for $r < K$. Thus if we let $W^1 f$ denote the potential operator for that Feller Brownian motion and use the fact that at time $\bar{K}$ that process is at 0 we have

$$(3.23) \qquad E^0 \int_0^{\bar{K}} e^{-t} f(\bar{X}(t)) dt = E^0 \left(\int_0^\infty - \int_{\bar{K}}^\infty \right)$$

$$= E^0 (1 - e^{-\bar{K}}) W^1 f(0).$$

Now we must evaluate $W^1 f(0)$. The Feller Brownian motion on which it is based has continuous paths, and a glance at the proof of (3.8) shows that this implies that in (3.9) the measure η is absent. So we have $W^1 f(0) = p' f(0) + q' R^1 f(0)$ for some choice of p' and q', and $p' + q' = 1$. Let f be the indicator of $(0, \infty)$, let $|x|(t)$ denote the reflecting Brownian motion $x(t) - m(t)$ and let $\tau'(t)$ denote the inverse of its additive functional $t - (\beta/\alpha)m_t$. Then since $f(0) = 0$ and $R^1 f(0) = 1$ we have

$$q' = E^0 \int_0^\infty e^{-t} f\left(|x|(\tau(t)) \right) dt$$

$$(3.24) \qquad = E^0 \int_0^\infty e^{-\left(u - (\beta/\alpha)m_u\right)} f(|x|(u)) d\left(u - (\beta/\alpha)m_u\right)$$

$$= E^0 \int_0^\infty e^{-\left(u - (\beta/\alpha)m_u\right)} du,$$

where we have dropped the measure $-dm_u$ as it puts all its mass on $\{u| \ |x|(u) = 0\}$ and also used the fact that for each u $P^0 \left(|x|(u) > 0 \right) = 1$.

The last integral in (3.24) is easy to evaluate: we have shown, I-4.7, that

$$P^0(-m_u > x) = 2P^0 \left(x(u) > x \right) = 2P^0 \left(x(1) > xu^{-\frac{1}{2}} \right)$$

and so $-m_u$ has for density $\sqrt{2/\pi} e^{-x^2/2u} u^{-1/2}$. So after a change in order of integration we have

$$\int_0^\infty e^{-u} E^0 e^{(\beta/\alpha)m_u} du$$

$$= \int_0^\infty e^{-(\beta/\alpha)x} dx \int_0^\infty \frac{2}{\sqrt{2\pi u}} e^{-u} e^{-x^2/2u} du.$$

The inner integral is $r^1(0, x) = \sqrt{2} e^{-x\sqrt{2}}$ from the discussion following (3.6) and so

$$q' = \frac{\alpha \sqrt{2}}{\alpha \sqrt{2} + \beta}$$

$$p' = \frac{\beta}{\alpha \sqrt{2} + \beta}.$$

If we insert these calculations back into (3.23) and then into (3.22) and then solve for $\overline{U}^1 f(0)$ we will obtain (3.21). Finally the assumption that $\|\nu\|$ is finite is removed by the same method as we used in discussing sawtooth processes, that is by replacing the subordinator $T(t)$ with $\overline{T}^\varepsilon(t) + \varepsilon t$, to which our calculations do apply, and then letting ε approach 0. We will leave it to the reader to supply the necessary continuity arguments.

To all intents and purposes Theorems (3.8) and (3.20) give a complete description of all Feller Brownian motions. That is (3.8) associates with any Feller Brownian motion "parameters" p, q, and η, gives the additional conditions these parameters must satisfy and shows how the parameters determine the probability law of the process. Conversely suppose we are given p, q and η, with $p + q + \int_0^\infty (1 - e^{-x\sqrt{2}})\eta(dx) = 1$ and suppose for the moment that if $\|\eta\| < \infty$ then $q > 0$. Then (3.20) says that a Feller Brownian motion with these parameters exists; we take $\beta = p$, $\alpha = q/\sqrt{2}$ and $\nu = \eta$. There remains only the case where $p > 0, q = 0$ and $\|\eta\| < \infty$. The reader should have no trouble checking that the corresponding Feller Brownian motion is one of the holding and jumping type, that is starting at 0 it remains there for an initial interval of time whose length, J, is exponential of rate $\|\eta\|/p$; and then at time J it jumps to a position $X(J)$ independent of J and distributed according to $\eta/\|\eta\|$, proceeds from there as a Brownian motion, and so forth.

(3.25) Remark. Strictly speaking condition (1) of paragraph (b) in the definition of local time is not correct—we must allow $\ell(t)$ to be measurable relative to some completion of $\sigma\{X_r; r \le t\}$. The correct definition, along with a description of the proper completion, is given in section 3 of Chapter III. However the basic idea is that $\ell(t)$ should depend only on the process up to time t and not on some essentially larger σ-algebra $\mathcal{F}_t$ even if the process $\{X_t, \mathcal{F}_t; t \ge 0\}$ is Markovian.

(3.26) Remark. To shorten an already overlong presentation, in our treatment of Feller Brownian motions we left out the case of processes with finite lifetime—that is processes which at some random time jump (necessarily from the origin) to a cemetary state Δ and then remain there forever. (We have made a similar omission in Chapter VII.) But these processes are easily described: if $\{X_t; t \ge 0\}$ is one of our Feller Brownian motions, with local time $\{\ell(t); t \geq 0\}$ and p, q and η as in II–3.9, and if we define a stopping time τ by $\tau = \inf\{t | \ell(t) = S^\lambda\}$ where S^λ has an exponential distribution with rate λ and is independent of X, then the process $\overline{X}$ defined by

$$\overline{X}_t = X_t \qquad t < \tau$$
$$= \Delta \qquad t \geq \tau$$

is a Feller Brownian motion with finite lifetime τ. Since $X_\tau = 0$ (τ is an increase point of $\ell(t)$) an obvious calculation shows that $E^0 \int_0^\infty e^{-t} f(\overline{X}_t) dt$ is equal to $\delta E^0 \int_0^\infty e^{-t} f(X_t) dt$ with δ equal to $E^0(1-e^{-\tau})$. (Recall that functions f on $[0, \infty)$ are extended to Δ by setting $f(\Delta) = 0$.) Clearly δ is equal to $E^0 \int_0^\infty e^{-\lambda \ell(t)} e^{-t} dt$ and so any δ with $0 < \delta \leq 1$ is possible. Thus the potential operator for $\overline{X}$ is of the form II–3.9, but $p+q+\int(1-e^{-x\sqrt{2}})\eta(dx) < 1$ is possible also. The reader should have no trouble showing that this is the most general form. For an example where the finite lifetime case is used for a computation see [W,3].

III Point Processes of Excursions

In this chapter we will develop the theory of Poisson point processes, and give Itô's theory of the analysis of a Markov process in terms of its excursions away from a fixed point in the state space.

1. Additive processes.

A real valued stochastic process $\{X_t; t \geq 0\}$ is called an *additive process* if $X_0 = 0$ with probability one, and if for all finite sequences $0 \leq t_0 < \cdots < t_n$ the increments $X_{t_1} - X_{t_0}, \cdots, X_{t_n} - X_{t_{n-1}}$ are mutually independent. Except for the normalization $X_0 = 0$ this is a generalization of what we called a Lévy process, the generalization being that we have dropped the assumption that the distribution of $X_b - X_a$ depends only on $b - a$. Usually one makes the additional requirement that $t \to X_t$ is continuous in probability, that is $X_{t_n} \to X_t$ in probability whenever t_n is a sequence converging to t. If we assume this then it follows that the increments $X_b - X_a$ have infinitely divisible distributions. A fact somewhat more difficult to prove is that if $\{X_t; t \geq 0\}$ is additive and continuous in probability then there is an additive process $\{Y_t; t \geq 0\}$, defined on the same probability space, all of whose sample functions are right continuous and have everywhere left-hand limits, and is such that for each t, $X_t = Y_t$ with probability one. In other words under the continuity in probability assumption we may as well assume that the sample function of X_t have these regularity properties to start with. We will not need this fact because the additive processes we will encounter arise in a specific way that makes the regularity of the sample functions automatic.

Let $(\Omega, \mathcal{F}, P)$ be a probability space. Let $\{\mathcal{F}_{st}; 0 \leq s < t < \infty\}$ be a family of sub σ-algebras of $\mathcal{F}$ having these properties: (1) $\mathcal{F}_{st} = \sigma\{\mathcal{F}_{su}, \mathcal{F}_{ut}\}$ whenever $s < u < t$, and (2) whenever $0 \leq t_0 < \cdots < t_n$, the σ-algebras $\mathcal{F}_{t_0, t_1}, \cdots, \mathcal{F}_{t_{n-1}, t_n}$ are mutually independent. Then we will say that *the family* $\{\mathcal{F}_{st}; 0 \leq s < t < \infty\}$ *is additive*. Obviously if $\{X_t; t \geq 0\}$ is an additive process and we set $\mathcal{F}_{st} = \sigma\{X_r - X_u; s \leq u < r \leq t\}$ then we obtain an additive family. Conversely suppose $\{\mathcal{F}_{st}\}$ is an additive family and $\{X_t; t \geq 0\}$ is a real valued process over $(\Omega, \mathcal{F}, P)$ with $X_0 = 0$ and is such that $\sigma\{X_t - X_s\} \subset \mathcal{F}_{st}$ whenever $0 \leq s < t$. Then clearly the process is additive and the additive family it generates is coarser than the one we started with. We will express the condition "$\sigma\{X_t - X_s\} \subset \mathcal{F}_{st}$ for $s < t$"

by saying "$\{X_t\}$ *is adapted to* $\{\mathcal{F}_{st}\}$".

The deepest properties of infinitely divisible distributions and additive processes follow most naturally from Itô's stochastic integral representation of additive processes. We will need the following lemma, which is the basic step in Itô's development. See [I,2] for a complete discussion of the entire representation.

(1.1) Lemma. *Suppose over a complete probability space* $(\Omega, \mathcal{F}, P)$ *we have an additive family* $\{\mathcal{F}_{st}\}$ *and two processes* $\{X_t\}$ *and* $\{Y_t\}$ *both adapted to* $\{\mathcal{F}_{st}\}$ *and both continuous in probability. Suppose also that almost all the sample functions of both processes are right continuous and have left hand limits, and that the increments* $Y_t - Y_s, s \leq t$ *are almost surely non-negative integers. Finally, suppose that*

$$P\{X_t \neq X_{t-} \text{ and } Y_t \neq Y_{t-} \text{ for some } t\} = 0.$$

Then $\sigma\{X_t; t \geq 0\}$ *and* $\sigma\{Y_t; t \geq 0\}$ *are independent.*

Remarks: Probably the statement is best remembered as: "two additive processes (one being integer valued) adapted to the same family and without a jump time in common are independent." As to measurability, if $t \to x_t$ and $t \to y_t$ are functions right continuous and with left limits then there will be a point t with $x_t \neq x_{t-}$ and $y_t \neq y_{t-}$ if and only if there are strictly positive integers m and k such that for every strictly positive integer n there are rationals $p_n < q_n < p_n + 1/n < m$ with $|x_{p_n} - x_{q_n}| > 1/k$ and $|y_{p_n} - y_{q_n}| > 1/k$. Thus the condition about common discontinuity times involves only countably many $\mathcal{F}$ sets so its probability is defined.

Proof of (1.1). We will show first that for any positive t, X_t and Y_t are independent. Pick a positive integer n to be held fixed for a few moments. For $1 \leq k \leq n$ set $x_k = X_{kt/n} - X_{(k-1)t/n}$ and define y_k similarly in terms of $\{Y_t\}$. Set $x = \Sigma x_k$ with the understanding that the sum is only over those k for which $y_k = 0$. We will show first that for each real number r and non-negative integer p the two expressions

$$E(e^{irx}, Y_t = p)P(Y_t = 0);$$

(1.2)

$$E(e^{irx}, Y_t = 0)P(Y_t = p)$$

differ by a term of the form pM_n where M_n approaches 0 as $n \to \infty$. To do this let $\varepsilon = (\varepsilon_1, \cdots, \varepsilon_n)$ denote a set of non-negative integers adding up to p and set $z(\varepsilon) = \{k | \varepsilon_k = 0\}$, $n(\varepsilon) = \{k | \varepsilon_k > 0\}$. The exponential e^{irx} is a product, some of whose factors are of the form e^{irx_k} while the rest are 1. The values are determined by conditions on non-overlapping increments of the X and Y processes and so

$$E(e^{irx}; y_1 = \varepsilon_1, \cdots, y_n = \varepsilon_n)P(Y_t = 0)$$

(1.3)

$$= \prod_{z(\varepsilon)} E(e^{irx_k}; y_k = 0) \prod_{n(\varepsilon)} P(y_k = \varepsilon_k) \prod_k P(y_k = 0).$$

If we write the last product as $\prod_{z(\varepsilon)} \cdot \prod_{n(\varepsilon)}$ and switch the terms $\prod_{n(\varepsilon)} P(y_k = \varepsilon_k)$ and $\prod_{n(\varepsilon)} P(y_k = 0)$ then the right side of (1.3) reads

$$(1.4) \qquad \prod_{z(\varepsilon)} E(e^{irx_k}; y_k = 0) \prod_{n(\varepsilon)} P(y_k = 0) \prod_k P(y_k = \varepsilon_k).$$

The first of the two expressions in (1.2) is obtained by summing the left side of (1.3), and hence the expression in (1.4), over all possible choices of $(\varepsilon_1, \cdots, \varepsilon_n)$ as above. The other expression in (1.2) is a sum over the same choices of $(\varepsilon_1, \cdots, \varepsilon_n)$ of terms $\prod_k P(y_k = \varepsilon_k)$ multiplied by a common factor $E(e^{irx}; Y_t = 0)$, which can be written also as $\prod_k E(e^{irx_k}; y_k = 0)$. Any product such as this last one can be written, for any fixed choice of $\varepsilon = (\varepsilon_1, \cdots, \varepsilon_n)$, as a product over $z(\varepsilon)$ times a product over $n(\varepsilon)$, and using this one concludes immediately that the absolute difference of the two expressions in (1.2) is no greater than

$$\sum_\varepsilon \left| \prod_{n(\varepsilon)} P(y_k = 0) - \prod_{n(\varepsilon)} E(e^{irx_k}; y_k = 0) \right| \prod_k P(y_k = \varepsilon_k),$$

where once again the sum is over all n-tuples $\varepsilon = (\varepsilon_1, \cdots, \varepsilon_n)$ of non-negative integers adding up to p. For complex numbers $\alpha_1, \cdots, \alpha_j$ and $\beta_1, \cdots, \beta_j$ of modulus less than 1 we have the inequality $|\Pi\alpha_i - \Pi\beta_i| \leq \sum |\alpha_i - \beta_i|$ and so the last displayed expression does not exceed

$$\sum_\varepsilon \left(\sum_{n(\varepsilon)} |P(y_k = 0) - E(e^{irx_k}; y_k = 0)| \right) \prod_k P(y_k = \varepsilon_k)$$

$$\leq \sum_\varepsilon \left(\sum_{n(\varepsilon)} E|1 - e^{irx_k}| \right) \prod_k P(y_k = \varepsilon_k)$$

$$\leq p \max_k E|1 - e^{irx_k}|.$$

Reintroduce the parameter n and recall that $x = \Sigma x_k$ depends on n, let us now call it $x(n)$, and that $\max_k E|1 - e^{irx_k}|$ also depends on n, call it M_n. We will now leave as exercises for the reader to check first of all that because of the continuity in probability of $t \to X_t$ we have $M_n \to 0$ as $n \to \infty$ and secondly that because of the absence of common discontinuities, $x(n)$ converges to X_t with probability one as $n \to \infty$. The conclusion from applying these facts to the expressions in (1.2) is that

$$E(e^{irX_t}; Y_t = 0)P(Y_t = p)$$

(1.5)

$$= E(e^{irX_t}; Y_t = p)P(Y_t = 0).$$

Let us sum (1.5) over all $p = 0, 1, \cdots$ to obtain

$$E(e^{irX_t}, Y_t = 0) = E(e^{irX_t})P(Y_t = 0)$$

and putting this in the left side of (1.5), we get

$$E(e^{irX_t})P(Y_t = p)P(Y_t = 0) = E(e^{irX_t}, Y_t = p)P(Y_t = 0).$$

Using the continuity in probability of $t \to Y_t$ we argue easily that $P(Y_t = 0)$ is for all t strictly positive. Thus

$$E(e^{irX_t}, Y_t = p) = E(e^{irX_t})P(Y_t = p)$$

for all r and p, and so X_t and Y_t are independent. In essence this completes the entire proof: if $\{Z_t\}$ is an additive process then so is $\{W_t\}$ defined by $W_t = Z_{t+s} - Z_s$ so we have already shown that $X_t - X_s$ and $Y_t - Y_s$ are independent for any choice of $s < t$. If $\{\mathcal{A}_i; 1 \le i \le n\}$ and $\{\mathcal{B}_i; 1 \le i \le n\}$ are σ-algebras such that for each i, $\mathcal{A}_i$ and $\mathcal{B}_i$ are independent and such that the family $\{\sigma(\mathcal{A}_i, \mathcal{B}_i); 1 \le i \le n\}$ is independent then $\sigma(\mathcal{A}_1 \cup \cdots \cup \mathcal{A}_n)$ and $\sigma(\mathcal{B}_1 \cup \cdots \cup \mathcal{B}_n)$ are independent. What we have done, along with this and a monotone class argument completes the proof of lemma (1.1).

One consequence of lemma (1.1) is the following fact whose proof we leave as an exercise for the reader.

(1.6) Suppose that $\{Z_t; t \ge 0\}$ is an additive process, continuous in probability and all of whose sample functions are right continuous and non-decreasing. For $i = 1, 2, \cdots, n$ let $Z_i(t)$ denote the number of points $r \le t$ such that $Z_r - Z_{r-} = i$. Then the processes $\{Z_i(t); t \ge 0\}$, $i = 1, \cdots, n$, are additive and they are mutually independent.

2. Poisson Point Processes

We will give (repeating verbatim) the theory of Poisson point processes as given by Itô in [I, 1].

(a) Let U be a set and $\mathcal{U}$ a σ-algebra of subsets of U. Let T denote the open positive real axis $(0, \infty)$ and let $\mathcal{B}$ denote the σ-algebra in $T \times U$ generated by $\mathcal{B}(T) \times \mathcal{U}$, where as usual $\mathcal{B}(T)$ denotes the Borel sets of T. We say that an object p is a *point function from T to U* if there is a countable (possibly empty) subset D_p of T and for each $t \in D_p$ a point $p(t)$ of U; that is p is just a function from a countable subset, called the *domain of p*, of T to U. Let Π denote the set of all point functions from T to U. If E is any subset of $T \times U$ and $p \in \Pi$ then $N(E, p)$ denotes the number of points $t \in D_p$ such that $(t, p(t))$ is an element of E. Obviously $N(E, p)$ is a positive integer or $+\infty$. We will make Π into a measurable space $(\Pi, \mathcal{P})$ by taking $\mathcal{P}$ to be the smallest σ-algebra with respect to which the functions $p \to N(E, p)$ are measurable for each $E \in \mathcal{B}$. If E is a subset of $T \times U$ we define $p|E$, *the restriction of p to E*, to be the point function p' whose domain, $D_{p'}$, is $\{t \in D_p | (t, p(t)) \in E\}$ and which is defined by $p'(t) = p(t)$ for $t \in D_{p'}$. Note that for $D \subset T \times U$ we have $N(D, p') = N(D \cap E, p)$. When $A \subset T$ we define the *domain restriction $p|_d A$ of p to A* by $p|_d A = p | A \times U$ and when $V \subset U$ we define the *range restriction $p|_r V$ of p to V* by $p|_r V = p | T \times V$. For $t > 0$ we define the *stopped point function $\alpha_t p$* by $\alpha_t p = p|_d (0, t]$ and we define the *shifted point function $\theta_t p$* by saying $D_{\theta_t p} = \{s > 0 | t + s \in D_p\}$ and $\theta_t p(s) = p(s + t)$ for $s \in D_{\theta_t p}$. If $\{p_n\}_{n=1}^{\infty}$ is a sequence of point functions whose domains, D_{p_n}, are pairwise disjoint then we define $p = \Sigma p_n$ by $D_p = \underset{n}{\cup} D p_n$ and $p(t) = p_n(t)$ if $t \in D_{p_n}$.

(b) Let $(\Omega, \mathcal{F}, P)$ be a probability space. A function Y from Ω to Π is called a *point process* if it is measurable relative to $\mathcal{F}$ and $\mathcal{P}$. Usually one denotes by Y_ω the point function associated with the point $\omega \in \Omega$, and then $Y_\omega(t)$ is defined as a point of U if $t \in D_{Y_\omega}$. The measurability hypothesis says that if $E \in \mathcal{B}$ then $N(E, Y_\omega)$, as a function of ω, is a random variable over $(\Omega, \mathcal{F})$. The process Y is called *discrete* whenever for all positive t, $P(N((0, t] \times U, Y) < \infty) = 1$; it is called *$\sigma$-discrete* if for each t there is a sequence $\{U_n(t)\}_{n=1}^{\infty}$ of $\mathcal{U}$ sets such that $U = \underset{n}{\cup} U_n(t)$ and for each n, $P(N((0, t] \times U_n(t), Y) < \infty) = 1$. Note that if Y is a point process and if $E \in \mathcal{B}$ then the restriction process Y^E defined by $Y_\omega^E = Y_\omega | E$ is also a point process. The notation $\alpha_t Y$ of course denotes the stopped process, $(\alpha_t Y)_\omega = \alpha_t(Y_\omega)$; and the shifted process $\theta_t Y$ is defined similarly. A point process is called a Poisson *point process* if it is σ-discrete and if for each t, the distribution in $(\Pi, \mathcal{P})$ of the point process $\theta_t Y$ is the same as the distribution of Y and the processes $\alpha_t Y$ and $\theta_t Y$ are independent. The use of the label "Poisson" is justified by the following theorem.

(2.1) Theorem. *If Y is a Poisson point process then for every $E \in \mathcal{B}, N(E, Y)$ has a Poisson distribution; and if $E_1, \cdots, E_k$ are pairwise disjoint elements of $\mathcal{B}$ then the random variables $N(E_1, Y), \cdots, N(E_k, Y)$ are mutually independent.*

Proof: (Note that we are interpreting a variable taking the value ∞ with probability 1 as having a Poisson distribution with mean ∞.) Because Y and $\theta_t Y$ have the same distribution it follows that if $N\big((0, t] \times V, Y\big)$ is finite with probability one then so is $N\big((0, kt] \times V, Y\big)$ for all k. Thus we may assume that there is a sequence $\{U_n\}_{n=1}^{\infty}$ of $\mathcal{U}$ sets not depending on t such that $P\big(N\big((0, t] \times U_n, Y\big) < \infty \text{ for all } t\big) = 1$ and such that $U_1 \subset U_2 \subset \cdots$ and $U = \bigcup_n U_n$. If E is any subset of $T \times U$ then $N(E, Y)$ is the limit of the increasing sequence $N\big(E \cap (0, n] \times U_n, Y\big)$, and since the Poisson distribution and independence properties are preserved under limits we may carry out the proof of (2.1) under the assumption that for some n E and $E_1, \cdots, E_k$ are subsets of $(0, n] \times U_n$. Let $Z_t = N\big((0, t] \times U_n, Y\big)$. Then the process Z_t has stationary independent increments, and if we restrict ourselves to the probability one set where $Z_t < \infty$ for all t, its sample functions are right continuous and increase only by jumps of height one. It follows immediately that Z_t is a Poisson process, and in particular the process is continuous in probability. Let E be an element of $\mathcal{B}$. For $0 \leq s < t$ let $\mathcal{F}_{st}$ denote the σ-algebra generated by the random variables $N\big((a, b] \times V, Y\big)$, where $a < b$ range over $[s, t]$ and V over $\mathcal{U}$. The fact that $\alpha_t Y$ and $\theta_t Y$ are independent implies that the family $\{\mathcal{F}_{st}\}$ is additive. Let

$$X_t = N\big(E \cap (0, t] \times U_n, Y\big)$$
$$Y_t = N\big(E^c \cap (0, t] \times U_n, Y\big).$$

These processes are adapted to $\{\mathcal{F}_{st}\}$ and $X_t + Y_t = Z_t$. They are nondecreasing in t, integer valued and, being dominated by the Z process, are

continuous in probability. They have no discontinuity in common and so
by (1.1) for each t, X_t and Y_t are independent. Since their sum has a
Poisson distribution it follows from Raikov's classical theorem that X_t has
a Poisson distribution. As already noted this is enough to establish the first
assertion of (2.1). For the second assertion let $Z_t^i = N\big(E_i \cap (0,t] \times U_n, Y\big)$
for $i = 1, 2, \cdots, k$ and let $Z_t^* = \Sigma i Z_t^i$. The process $\{Z_t^*\}$ is adapted to $\{\mathcal{F}_{st}\}$
and hence is additive. Its sample functions are right continuous and non-
decreasing and since $|Z_t^* - Z_s^*| \le k|Z_t - Z_s|$ the process $\{Z_t^*\}$ is continuous
in probability. Now Z_t^i is the number of jumps of Z_t^* of height i in the time
interval $(0,t]$. So by (1.6) the random variables $Z_t^1, \cdots, Z_t^k$ are independent,
and this completes the proof.

If Y is a Poisson point process consider the measure m on $\mathcal{B}$ defined
by

$$m(E) = P\big(N(E,Y)\big),$$

that is the expected number of points $t \in D_{Y_\omega}$ such that $\big(t, Y_\omega(t)\big)$ lies in
E. The fact that the process is σ-discrete and invariant under θ_t implies
that m is σ-finite and shift invariant in t; and so m is in fact a product
measure $dm = dt \times dn$ where dt denotes lebesgue measure on $\mathcal{B}(0, \infty)$ and
n is a σ-finite measure on $\mathcal{U}$. The measure n is called the *characteristic
measure* of the Poisson point process Y. Two Poisson point processes with
the same characteristic measure have the same distribution, that is they in-
duce the same measure on $(\Pi, \mathcal{P})$. This follows from the fact that the joint
distributions of the n-tuples $\big(N(E_1, Y), \cdots, N(E_n, Y)\big)$ are determined by
those in which $E_1, \cdots, E_n$ are pairwise disjoint, and by (2.1) any such joint
distribution is determined by the measure m, which, in turn, is determined
by n. What will be more important for us than the existence of a characteristic measure for any Poisson point process is the fact that given a
σ-finite meaure on $\mathcal{U}$ it is in fact the characteristic measure of a Poisson
point process.

(2.2) Theorem. *If n is a σ-finite measure on $\mathcal{U}$ there is a Poisson point
process whose characteristic measure is n.*

Proof: Suppose first of all that $n(U) < \infty$. Let $\beta = n(U)$ and let ν be
the probability measure on $\mathcal{U}$ defined by $\nu(E) = n(E)/\beta$. Over some probability space $(\Omega, \mathcal{F}, P)$ let $\tau_1, \tau_2, \cdots$ be independent random variables each

having the exponential distribution with rate β, let $Y_1, Y_2, \cdots$ be independent $(U, \mathcal{U})$ random variables each having the distribution ν and suppose also that the τ's and Y's are independent. Now we assert that if we define a point process Y over $(\Omega, \mathcal{F}, P)$ by taking

$$D_{Y_\omega} = \{\tau_1(\omega), \tau_1(\omega) + \tau_2(\omega), \cdots\}$$

$$Y_\omega(t) = Y_n(\omega) \quad t = \tau_1(\omega) + \cdots + \tau_n(\omega)$$

then we obtain a Poisson point process with n as characteristic measure. The argument for this is about the same as the one which equates the two definitions of a Poisson process, one as a Lévy process with Poisson increments and the other as the accumulated number of events when inter-event times are independent and exponentially distributed, and we will leave this to the reader. To pass to the general case write $U = \overset{\infty}{\underset{k=1}{\cup}} U_k$ where the sets $U_k \in \mathcal{U}$ are pairwise disjoint with $n(U_k)$ finite for each k, and let Y^k be the Poisson point process with characteristic measure $n_k(E) = n(E \cap U_k)$, the family $\{Y^k ; k \geq 1\}$ being supposed independent. The probability that any two of the processes have a domain point in common is seen easily to be 0. Thus we can form the sum process $Y = \underset{k}{\sum} Y^k$. We will leave it to the reader to show that Y is a Poisson point process with n for characteristic measure.

(2.3) Exercise. Let Y be a Poisson point process with characteristic measure n and let φ be a non-negative function on $T \times U$ which is $\mathcal{B}$ measurable. Show that then

$$E \exp\left(- \sum_{t \in D_Y} \varphi(t, Y_t)\right)$$

$$= \exp\left\{ \int_{T \times U} (e^{-\varphi(t,u)} - 1) dt \, n(du) \right\}.$$

This is called the *exponential formula for Poisson measures*. First consider the case where φ is the indicator of $[a, b] \times V$ with V in $\mathcal{U}$. Then apply (2.1) to the case where φ is a linear combination of such indicators with the sets being pairwise disjoint. Complete the argument using general measure-theoretic considerations.

For use in Chapter VI we will need a generalization of these concepts. Let $(V, \mathcal{V})$ be a measurable space and let μ be a σ-finite measure on $\mathcal{V}$.

A Poisson *random measure with intensity measure* μ is a probability space $(\Omega, \mathcal{F}, P)$ and for each ω in Ω a measure Y_ω on $\mathcal{V}$ with the following properties:

(a) $\omega \to Y_\omega(A)$ is for each A in $\mathcal{V}$ measurable relative to $\mathcal{F}$ and the Borel sets of $[0, \infty]$,

(b) for each A in $\mathcal{V}$, $Y(A)$ has the Poisson distribution with mean $\mu(A)$,

(c) if $A_1, \ldots, A_n$ are in $\mathcal{V}$ and pairwise disjoint then $Y(A_1), \ldots, Y(A_n)$ are mutually independent.

One shows easily that these conditions determine the joint distribution of $(Y(A_1), \ldots, Y(A_n))$ for any choice of $A_1, \ldots, A_n$ from $\mathcal{V}$ and thus the law of the stochastic process $\{Y(A); A \in \mathcal{V}\}$.

(2.4) Exercise. Show that if μ is finite such a random measure can be constructed by taking an independent family $\{N, Z_1, Z_2, \ldots\}$, where N has the Poisson distribution with mean $\mu(V)$ and the Z_i are $(V, \mathcal{V})$ random variables with distribution $\mu/\mu(V)$, and then setting

$$Y(A) = \sum_{k=1}^{N} I_A(Z_k), \qquad (\sum_{1}^{0} = 0).$$

When μ is σ-finite write $V = \bigcup B_n$, with $\mu(B_n)$ finite and $\{B_n\}$ disjoint. Take independent Poisson random measures Y_n such that Y_n has intensity measure μ restricted to B_n and show that $Y = \sum Y_n$ is a Poisson random measure with μ as intensity measure.

(2.5) Exercise. Show that the exponential formula from (2.3) now reads

$$E(\exp - \int \varphi dY)$$

$$= \exp\{\int (e^{-\varphi} - 1)d\mu\}$$

for a positive $\mathcal{V}$ measurable function φ.

Note that if Y is a Poisson point process we regard it as a Poisson random measure by taking it to be the measure on $\mathcal{B}(T) \times \mathcal{U}$ that puts unit mass at the point $(t, Y_\omega(t))$ for each t in D_{Y_ω}. (A measure η on a measurable space $(X, \mathcal{X})$ is called unit mass at x in X if $\eta(A) = I_A(x)$ for

A in $\mathcal{X}$.) Then the intensity measure is $dt \times dn$ where n is the characteristic measure. The restriction of Y to a measurable subset E of $T \times U$ also is a Poisson random measure with intensity measure $dt \times dn$ restricted to E. In particular $\alpha_t Y$ has as intensity measure the product with n of lebesgue measure restricted to $[0, t]$. But $\alpha_t Y$ is not a Poisson point process.

(2.6) Exercise. Show that if Y is a Poisson point process and T is a finite stopping time for the filtration $\{\alpha_t Y; t \geq 0\}$ then $\theta_T Y$ is equal in law to Y and $\theta_T Y$ and $\alpha_t Y$ are independent.

3. Poisson Point Processes of Excursions

Let $X = (\Omega, \mathcal{F}, \mathcal{F}_t, X_t, \theta_t, P^x)$ be a standard process with state space $(E, \mathcal{E})$, and let b be a point of E. We wish to describe the process X in terms of $\{t | X_t = b\}$ and the pieces of the path, or "excursions away from b", that are defined over the complement of that t set. Itô observed in [I, 1] that if $\{t | X_t = b\}$ is parametrized in terms of the local time at b then the excursions away from b can be regarded as the "points" in a Poisson point process, whose characteristic measure describes many interesting features of the original process. And he noted that conversely this approach gives a way of constructing processes such as Feller Brownian motion, which previously had been constructed only by clever *ad hoc* methods or whose actual existence was somewhat in doubt. We will present his ideas. But first we must deal with local time and additive functionals in more generality than we did in Chapter II.

(a) **Additive functionals:** Throughout we will assume that the σ-algebras $\mathcal{F}_t$ are the proper completions of the σ-algebras $\mathcal{F}_t^0 = \sigma\{X_s; s \leq t\}$ generated by the random variables of the process; that is Λ is in $\mathcal{F}_t$ if and only if for each probability measure μ on $\mathcal{E}$ there is a set Γ in $\mathcal{F}_t^0$ and a set Δ in $\sigma\{X_s; s \geq 0\}$ such that $\Lambda \Delta \Gamma \subset \Delta$ and $P^\mu(\Delta) = 0$. As noted previously this, in conjunction with the other hypotheses of a standard process, makes the filtration $\{\mathcal{F}_t\}$ right continuous. It is the definition that gives a useful filtration, but one that preserves the fact that sets in $\mathcal{F}_t$ are essentially defined by conditions on the process up to time t. A family $\{A_t; t \geq 0\}$ of functions from Ω to $[0, \infty]$ is called an *additive functional of X* provided (a) $t \to A_t(\omega)$ is non-decreasing, right continuous and satisfies $A_0(\omega) = 0$ except for $\omega \in \Lambda$ where $P^\mu(\Lambda) = 0$ for all μ; (b) for each t, $A_t \in \mathcal{F}_t$, and (c) for each t and s, $A_{t+s} = A_t + A_s \circ \theta_t$ almost surely P^μ for each μ. The most obvious example is the "classical" additive functional one obtains by taking a positive bounded $\mathcal{E}$ measurable function on E and setting

$$A_t(\omega) = \int_0^t f(X_s(\omega))\,ds.$$

A less obvious example arose in Chapter II when we took X to be reflecting Brownian motion obtained from a Brownian motion $\{x_t\}$ and its minimum function m_t as $X_t = x_t - m_t$ and claimed that $\{-m_t; t \geq 0\}$ is an additive functional of the reflecting Brownian motion. With the shift operators θ_t defined as we did the additivity condition was obvious as were the continuity and monotoneity properties. But the measurability assertion (b) hardly appeared even to be true, and its verification required a clever probability argument and hence the adjunction to $\sigma\{X_s; s \leq t\}$ of sets of probability zero. The entire subject of additive functionals is discussed thoroughly in [BG, 1]. At this point we will sketch only as much of the theory as is required to obtain local time; and we will postpone giving a more nearly complete development until it is needed. First we need the notion of the potential of an additive functional and this requires that we introduce the notion of excessive function.

(b) **Excessive functions:** If φ is a positive universally measurable function on E and λ is a positive number we say that φ is λ-excessive (for the process X) if

$$e^{-\lambda t} P_t \varphi \leq \varphi, \qquad \text{and}$$

(3.1)

$$e^{-\lambda t} P_t \varphi \to \varphi \qquad \text{as} \qquad t \to 0.$$

When $\lambda = 0$ we say simply, "excessive". The property of being λ-excessive depends only on the semi-group of transition operators, but most important properties of excessive functions are obtained by using the probabilistic structure of the process X. One basic property that involves the process is the following.

(3.2) Theorem. *If φ is λ-excessive then almost surely relative to each P^μ the mapping $t \to \varphi(X_t)$ is right continuous and has left hand limits on $[0, \infty)$.*

We refer the reader to [BG, 1] for a proof. In view of our limited objectives possibly we could dodge the use of (3.2). But it is such a fundamental condition in all of probabilistic potential theory that to avoid its use would be misleading.

Positive constant functions obviously are λ-excessive, and so are sums of λ-excessive functions as well as limits of increasing sequences of λ-excessive functions. If φ_1 and φ_2 are λ-excessive and $\varphi = \min(\varphi_1, \varphi_2)$ then obviously φ satisfies the inequality in (3.1). If we use Theorem (3.2) then we conclude that φ is in fact λ-excessive, but intervention of the process at this point seems to be essential. If f is a positive universally measurable function and φ is its λ-potential,

$$\varphi = U^\lambda f = \int_0^\infty e^{-\lambda s} P_s f ds,$$

then

$$e^{-\lambda t} P_t \varphi = \int_t^\infty e^{-\lambda s} P_s f ds$$

and so obviously φ is λ-excessive. More generally suppose $\{A_t; t \geq 0\}$ is an additive functional, and for a fixed positive λ set

$$\varphi(x) = E^x \int_0^\infty e^{-\lambda t} dA_t.$$

The function φ is called the λ-*potential of the additive functional A*.

Using the additivity property of A we obtain

$$(3.3) \qquad \varphi(x) = E^x \int_{[0,t]} e^{-\lambda s} dA_s + E^x e^{-\lambda t} E^{X_t} \int_0^\infty e^{-\lambda s} dA_s.$$

The second term on the right of (3.3) is simply $e^{-\lambda t} P_t \varphi(x)$ so φ satisfies the inequality in (3.1). If for each x the expression $E^x \int_0^t e^{-\lambda s} dA_s$ is finite for some strictly positive t, for example if φ is everywhere finite, then the first term on the right of (3.3) decreases to zero as t does, and so φ is λ-excessive. We obtain the case of a λ-potential, $\varphi = U^\lambda f$, by taking A to be the classical additive functional $A_t = \int_0^t f(X_s) ds$; of course some condition must be imposed on f to ensure right continuity in t. Suppose φ is λ excessive and finite and that at a point $x \in E$ we have

$$e^{-\lambda t} P_t \varphi(x) \to 0 \qquad\qquad t \to \infty.$$

Then

$$U^\lambda \big(n(\varphi - e^{-\lambda/n} P_{1/n} \varphi) \big)(x)$$
$$= n \int_0^{1/n} e^{-\lambda s} P_s \varphi(x) ds + \lim_{r \to \infty} n \int_r^{r+1/n} e^{-\lambda s} P_s \varphi(x) ds.$$

The limit on the right of this display is 0, and as $n \to \infty$ the first term increases to $\varphi(x)$. If φ satisfies this condition for all x, for example if φ is bounded and λ is strictly positive, one concludes that φ is the limit of an increasing sequence of λ-potentials of positive functions. Now suppose φ is λ-excessive and that φ is known to be the limit of an increasing sequence $\{U^\lambda f_n; n = 1, 2, \cdots\}$ of λ-potentials of positive functions f_n. Given a stopping time T and a universally measurable positive or bounded function h introduce the notation

$$(3.4) \qquad P_T^\lambda h(x) = E^x\left(e^{-\lambda T}h(X_T); T < \infty\right).$$

Then $P_T^\lambda U^\lambda f(x) = E^x \int_T^\infty e^{-\lambda s} f(X_s)ds$ and so $P_T^\lambda U^\lambda f_n \leq U^\lambda f_n$; and by monotone convergence

$$P_T^\lambda \varphi \leq \varphi.$$

By a fairly straightforward argument (see Chapter II of [BG, 1]) one can eliminate the hypothesis that φ can be approximated by potentials—that is the inequality holds for any λ-excessive function. We will take this as having been established.

Coming to the λ-excessive function of most concern to us, fix a point $b \in E$ and let σ denote the time of hitting $\{b\}$; that is $\sigma = \inf\{t > 0 | X_t = b\}$. In discussing excursions away from b we are interested only in the case where b is regular for itself, that is $P^b(\sigma = 0) = 1$ and so we will make this assumption throughout the discussion of this example. Let φ be the function

$$(3.5) \qquad \varphi(x) = E^x e^{-\lambda \sigma},$$

where when $\lambda = 0$ we interpret this as $\varphi(x) = P^x(\sigma < \infty)$. For t positive let us rewrite the right side of (3.5) as $E^x(e^{-\lambda \sigma}; \sigma > t) + E^x(e^{-\lambda \sigma}; \sigma \leq t)$. The expression $t + \sigma \circ \theta_t$ is equal to σ if $\sigma > t$ and otherwise it exceeds σ and so

$$\varphi(x) \geq E^x(e^{-\lambda(t+\sigma\circ\theta_t)}) = e^{-\lambda t} P_t \varphi(x).$$

And, as t decreases to 0, $t + \sigma \circ \theta_t$ decreases to σ almost surely P^x, because if $x \neq b$ then $P^x(\sigma \leq t)$ decreases to 0, whereas $P^b(\sigma = 0) = 1$ and for every $\delta > 0$, $P^b(X_r = b$ for some $r \in (t, \delta))$ increases to 1 as t decreases to 0 and so $t + \sigma \circ \theta_t$ decreases to 0 almost surely relative to P^b. Hence φ satisfies both the conditions of (3.1), that is φ is λ-excessive. Also $\varphi(b) = 1$ since $P^b(\sigma = 0) = 1$. If h is any universally measurable function on E then

$$(3.6) \qquad P_\sigma^\lambda h(x) = \varphi(x)h(b)$$

since $P^x(X_\sigma \neq b, \sigma < \infty) = 0$. Finally, in addition to being λ-excessive φ is also what is called *uniformly λ-excessive*; that is $e^{-\lambda t}P_t\varphi$ converges to φ uniformly as $t \to 0$. To see this suppose $\varepsilon > 0$ and pick t so small that $e^{-\lambda t}P_t\varphi(b) \geq 1 - \varepsilon$. Then applying (3.6) with h being $\varepsilon + e^{-\lambda t}P_t\varphi(x)$ we have for any x

$$\varepsilon + e^{-\lambda t}P_t\varphi(x) \geq P_\sigma^\lambda h(x) = \varphi(x)h(b) \geq \varphi(x)$$

as asserted. According to the general theory of additive functionals, as set forth in Theorem (3.7) below, any bounded uniformly λ-excessive function, such as φ, is the λ-potential of a unique continuous additive functional L, that is

$$\varphi(x) = E^x \int_0^\infty e^{-\lambda t}dL_t.$$

We will outline the proof and also establish the fact that L increases exactly on $\{t | X_t = b\}$ so the name "local time at b" for L is justified.

(3.7) Theorem. *Let λ be positive. If f is a bounded uniformly λ-excessive function and $e^{-\lambda t}P_t f \to 0$ as $t \to \infty$ then there is an additive functional L such that f is the λ-potential of L. The additive functional is continuous, that is (a) $t \to L_t(\omega)$ is continuous except for $\omega \in \Lambda$ with $P^x(\Lambda) = 0$ for all x, and it can be assumed to be perfect, that is one may assume (b) the relation $L_{t+s}(\omega) = L_t(\omega) + L_s(\theta_t\omega)$ holds for all s and t except for $\omega \in \Lambda$ with $P^x(\Lambda) = 0$ for all x. Finally any continuous additive functional $\{l_t; t \geq 0\}$ whose λ-potential is f is equal to L in the sense that $l_t(\omega) = L_t(\omega)$ for all t, except for $\omega \in \Lambda$ where $P^x(\Lambda) = 0$ for all x.*

Proof: We will write $P_t^\lambda g$ for the operator $e^{-\lambda t}P_t g$. Let $g_n = n(f - P_{1/n}^\lambda f)$ and $f_n = U^\lambda g_n$ so that as we have noted already

$$f_n = n \int_0^{1/n} P_t^\lambda f\, dt$$

and hence f_n increases with n and approaches f uniformly as $n \to \infty$. Let $A_n(t)$ denote the continuous additive functional

$$A_n(t) = \int_0^t g_n(X_s)ds$$

so that f_n is the λ potential of $\{A_n(t); t \geq 0\}$. Set

$$C_n(t) = \int_0^t e^{-\lambda s}dA_n(s).$$

If we apply the additive functional property of A_n we obtain $C_n(\infty) = C_n(t) + e^{-\lambda t} C_n(\infty) \circ \theta_t$, and from this we get that for each x

$$(3.8) \qquad E^x(C_n(\infty)|\mathcal{F}_t) = C_n(t) + e^{-\lambda t} f_n(X_t),$$

that is the right side of (3.8) is, relative to any of the measures P^x, a positive martingale. If we apply Doob's martingale version of Kolmogorov's inequality we conclude that for any $\delta > 0$

$$
\begin{aligned}
P^x(\sup_t |e_n(t) - e_m(t)| \geq \delta) & \\
(3.9) \qquad &\leq \delta^{-2} E^x \left(e_n(\infty) - e_m(\infty) \right)^2 \\
&= \delta^{-2} E^x \left(C_n(\infty) - C_m(\infty) \right)^2
\end{aligned}
$$

where $e_n(t)$ denotes the right side of (3.8) and the supremum in (3.9) is taken as t ranges over any fixed countable subset of $[0, \infty)$. If we take n and m large enough that $\| f_n - f_m \|$ is less than δ then $|e_n(t) - e_m(t)|$ will exceed δ if $|C_n(t) - C_m(t)|$ exceeds 2δ so we can write

$$
\begin{aligned}
(3.10) \qquad P^x(\sup_t |C_n(t) - C_m(t)| > 2\delta) & \\
&\leq \delta^{-2} E^x \left(C_n(\infty) - C_m(\infty) \right)^2
\end{aligned}
$$

and now the supremum is over all t since $C_n(t)$ is a continuous function of t. To estimate the expectation on the right side of (3.10) we note that it is equal to

$$
\begin{aligned}
E^x \Big(\int_0^\infty \int_0^\infty & e^{-\lambda t} e^{-\lambda s} (g_n - g_m)(X_t)(g_n - g_m)(X_s)\, dt\, ds \Big) \\
(3.11) \qquad = 2E^x & \int_0^\infty \int_t^\infty \Delta_{n,m}\, ds\, dt \\
= 2E^x & \int_0^\infty e^{-\lambda t}(g_n - g_m)(X_t) e^{-\lambda t}(f_n - f_m)(X_t)\, dt
\end{aligned}
$$

where $\Delta_{n,m}$ denotes the integrand in the first line of (3.11), and where we get from the second line in (3.11) to the third by applying the Markov property and the fact that $f_j = U^\lambda g_j$. Quite clearly the last line in (3.11) is less than $\| f_n - f_m \| \, (2U^\lambda(g_n + g_m))$ and this is not greater than $4 \| f_n - f_m \| \, \| f \|$. Now the proof of the existence part of (3.7) is practically complete: that is since $\| f_n - f_m \|$ tends to 0 as n and m

become large we can find a subsequence $n_1 < n_2 < \cdots$ so that for each x and k

$$P^x(\sup_t |C_{n_k}(t) - C_{n_{k+1}}(t)| > k^{-2})$$
$$\leq k^{-2}.$$

The Borel Cantilli lemma will imply that the set on which $\lim_{k \to \infty} C_{n_k}(t)$ fails to converge uniformly in t will have P^x measure 0 for each x; and, since

$$A_n(t) = \int_0^t e^{\lambda s} dC_n(s),$$

on the complement of that exceptional set the limit as $k \to \infty$ of $A_{n_k}(t)$ exists uniformly on compact t sets. We define

$$L_t = \lim_{k \to \infty} A_{n_k}(t)$$

on the set in Ω where the limit on the right exists uniformly on compacts, and $L_t \equiv 0$ on the complement. The reader should have little difficulty in verifying that L_t is a perfect continuous additive functional whose λ-potential is f.

Coming to the uniqueness, let A_t and B_t be additive functionals both of which have f as λ-potential. Then exactly the same argument that led to (3.9) allows us to conclude that for each x and positive δ

$$\begin{aligned}
(3.12) \qquad & P^x(\sup_t |\int_0^t e^{-\lambda s} dA_s - \int_0^t e^{-\lambda s} dB_s| \geq \delta) \\
& \leq \delta^{-2} E^x(\int_0^\infty e^{-\lambda s} dA_s - \int_0^\infty e^{-\lambda s} dB_s)^2,
\end{aligned}$$

and so the uniqueness will be established once we show that, if A and B also are *continuous* additive functionals then the expected value in the right side of (3.12) is 0.

To compute this expected value we must evaluate three terms of the form

$$(3.13) \qquad E^x \int_0^\infty e^{-\lambda t} da_t \int_0^\infty e^{-\lambda s} db_s$$

where a_t and b_t are taken from the pair A_t, B_t with repetitions allowed. Since the additive functionals are continuous, the diagonal, $\{(s,t)|s = t\}$, carries no mass relative to $da_t db_s$ and so each of the expressions in (3.13) is a sum of two expressions, one of the form

$$(3.14) \qquad E^x \int_0^\infty e^{-\lambda t} da_t \int_{(t,\infty)} e^{-\lambda s} db_s,$$

and one in which a_t and b_t change places. We will argue in a moment that an application of the Markov property and the fact that both additive functionals have f as λ-potential will show that (3.14) may be calculated as

$$(3.15) \qquad E^x \int_0^\infty e^{-2\lambda t} f(X_t) da_t$$

with a_t having the same meaning as it does in (3.14). From this the fact that the right side of (3.12) vanishes is obvious. To show that (3.14) and (3.15) are the same let $\beta_n(t)$ denote the function whose value is $\int_{(k/2^n,\infty)} e^{-\lambda s} db_s$ if t lies in the interval $[(k-1)/2^n, k/2^n)$. By monotone convergence, (3.14) is the limit as $n \to \infty$ of

$$(3.16) \qquad E^x \int_0^\infty e^{-\lambda t} \beta_n(t) da_t.$$

Using the Markov property we may write (3.16) as

$$E^x \int_0^\infty \sum_k I_k e^{-\lambda t} e^{-\lambda k/2^n} f(X_{k/2^n}) da_t.$$

where I_k is the indicator of $\{t | (k-1)/2^n \le t < k/2^n\}$. Since, by (3.2), $t \to f(X_t)$ is right continuous, the integrand in this expression approaches $e^{-2\lambda t} f(X_t)$ and so the entire expression in (3.16) converges to that in (3.15), using bounded convergence and the facts that f is bounded and that the λ-potential $E^x \int_0^\infty e^{-\lambda t} da_t$ is finite. This completes the proof of uniqueness.

(c) **Local time.** Returning to the situation in subsection (b) suppose λ is strictly positive and consider the λ-excessive function φ from (3.5). Theorem (3.7) applies, and so φ is the λ-potential of a unique continuous additive functional L. Of course the definition of L depends on the choice of λ, but we will see shortly that changing λ changes L only through multiplication by a constant. For definiteness we will take $\lambda = 1$ and call the continuous additive functional $\{L_t; t \ge 0\}$ such that

$$E^x e^{-\sigma} = E^x \int_0^\infty e^{-t} dL_t$$

the *local time at $\{b\}$ for the process X.*

We will now justify the name "local time" by showing that except for a set of probability 0 the set of points t where $X_t = b$ and those where L_t grows are the same. Let

$$L^I = L^I(\omega) = \{t | L_{t-\varepsilon}(\omega) < L_{t+\varepsilon}(\omega) \quad \text{for all } \varepsilon > 0\}$$

and

$$L^R = L^R(\omega) = \{t | L_t(\omega) < L_{t+\varepsilon}(\omega) \quad \text{for all } \varepsilon > 0\}.$$

Thus L^I and L^R are respectively the set of points of increase and of right increase of the function $t \to L_t$. We noted already in Chapter II that L^I is the closure of L^R and that the difference of the two sets is countable. Now define a stopping time R by

$$R = \inf\{t | L_t > 0\}.$$

Then $L_R = 0$, and the computation

$$E^x \int_0^\infty e^{-t} dL_t = \varphi(x) = P_\sigma^1 \varphi(x)$$

$$= E^x \int_\sigma^\infty e^{-t} dL_t$$

shows that almost surely relative to P^x we have $L_\sigma = 0$, that is $\sigma \leq R$. In fact almost surely $\sigma = R$. To see this we need show only that $P^b(R > 0) = 0$. First of all $P^b(R < \infty) > 0$ or else we would have $\varphi(b) = 0$. Also $L_{R+R\circ\theta_R} = (L_R)\circ\theta_R = 0$ and so $R\circ\theta_R = 0$ on $\{R < \infty\}$. Thus $\sigma\circ\theta_R = 0$ on $\{R < \infty\}$ and so $X_R = b$ on $\{R < \infty\}$. Finally then we have

$$0 = P^b(R\circ\theta_R > 0, R < \infty) = P^b(P^b(R > 0); R < \infty)$$

and so $P^b(R > 0) = 0$. Having established this let Λ denote the set in Ω on which $\sigma\circ\theta_r$ equals $R\circ\theta_r$ for every rational r. We have concluded already that $P^x(\Lambda^c) = 0$ for all x. Suppose $\omega \in \Lambda$ and $t \in L^R(\omega)$. If $s > t$ then there is a rational r with $t < r < s$ such that $L_s(\omega) - L_r(\omega)$ is strictly positive. Thus $r + R\circ\theta_r$ is strictly less than s, and hence so is $r + \sigma\circ\theta_r$ and so there is a point $t' \in (r, s)$ such that $X_{t'}(\omega) = b$. Since s is arbitrary and the paths are right continuous it follows that $X_t(\omega) = b$.

That is $L^R(\omega) \subset \{t|X_t(\omega) = b\}$ unless $\omega \in \Lambda^c$. Suppose now that $\omega \in \Lambda$ and $X_t(\omega) = b$. If t is strictly positive and r is a rational with $r < t$ then $r + \sigma \circ \theta_r \leq t$ and so $r + R \circ \theta_r \leq t$. Clearly $r + R \circ \theta_r$ is in L^R and so $[r, t] \cap L^R(\omega)$ is not empty. It follows, since r is arbitrary, that $\{t > 0|X_t(\omega) = b\}$ is contained in the closure of $L^R(\omega)$ if $\omega \in \Lambda$. The restriction $t > 0$ is easy to eliminate. What we have shown is that almost surely

$$(3.17) \qquad\qquad L^R \subset \{t|X_t = b\} \subset L^I.$$

Finally the measure $dL_t(\omega)$ puts all its mass on $L^I(\omega)$ and hence all its mass on $L^R(\omega)$ as the difference of the two sets is countable and $dL_t(\omega)$ puts no mass at points. Thus if f is any positive $\mathcal{E}$ measurable function then almost surely the equation

$$\int_0^t f(X_s)dL_s = f(b)L_t$$

holds as an identity in t. This relationship and (3.17) justify the term "local time at b" for the continuous additive functional L.

Recall that in II-3 we gave constructions of local time in some special cases without appeal to general theory. If ℓ denotes any local time at $\{b\}$ and $\psi(x)$ is defined as $E^x \int_0^\infty e^{-t}d\ell_t$ then $\psi(x) = \psi(b)E^x e^{-\sigma}$ and so by the uniqueness theorem $\ell = \psi(b)L$ where L is the local time we have just constructed. For example in the representation $X_t = x_t - m_t$ for reflecting Brownian motion (x is Brownian motion and m its minimum function) we showed that $-m_t$ serves as a local time at $\{0\}$ for X. From the calculations following II-3.23 it follows easily that $-E^x \int_0^\infty e^{-t}dm_t = (1/\sqrt{2})$ and so for reflecting Brownian motion our "general theory" local time at $\{0\}$ is $-m_t$ multiplied by $\sqrt{2}$. There are other ways of obtaining local time, all important for one situation or another. We refer the reader to sections 1 and 2 of Chapter VI, which are accessible once Chapter IV is completed, for discussion and application of these.

(d) **Point processes of excursions.** Let X be a standard process, let b be a fixed point of the state space E and let σ denote the time of hitting $\{b\}$. Let $M = M(\omega)$ denote the closure of $\{t \geq 0|X_t(\omega) = b\}$. The complement in $(0, \infty)$ of M consists of maximal open intervals; if (r, s) is one of these intervals then $s = r + \sigma \circ \theta_r(\omega)$ and the path u defined by

$$u(t) = X_{r+t}(\omega) \qquad\qquad 0 \leq t < s - r$$

is right continuous in $[0, s-r)$ and has left limits in $(0, s-r)$, and it satisfies $u(t) \neq b$ for all t in $(0, s-r)$. Either of the possibilities $u(0) = b$ or $u(0) \neq b$ may occur. It is customary to call paths obtained this way *excursions of* X *away from* b and to refer to the corresponding half open interval $[r, s)$ as an *excursion interval*. (If $X_0 \neq b$ the interval $[0, \sigma)$ is not regarded as an excursion interval.) It is useful to have the path u defined on all of $[0, \infty)$. We will do this by setting $u(t) = b$ if $t \geq s - r$.

More formally let U denote the set of all functions from $[0, \infty)$ to E which are continuous on the right and have left limits. Make a measurable space $(U, \mathcal{U}^0)$ by taking $\mathcal{U}^0$ to be the σ-algebra generated by the coordinate functions. For $u \in U$ set

$$\sigma_b(u) = \inf\{t > 0 \,|\, u(t) = b\}.$$

Let X' denote the function from Ω to U defined by

$$X'(\omega)(t) = X_t(\omega) \qquad\qquad t < \sigma(\omega)$$
$$= b \qquad\qquad t \geq \sigma(\omega).$$

Then, for example, we have $\sigma = \sigma_b \circ X'$; and the excursion of X corresponding to the excursion interval $[r, s)$ (and extended to all of $[0, \infty)$ by leaving it at b once it arrives there) can be written as $X' \circ \theta_r$. The question of whether or not σ_b is $\mathcal{U}^0$ measurable has not been discussed, but the fact that $\sigma = \sigma_b \circ X'$ and that σ is already known to be $\mathcal{F}$-measurable implies that sets such as

$$\{\omega \,|\, \sigma_b(X'(\omega)) > \delta\}$$

are in $\mathcal{F}$. Similarly the discussion in section 7 of Chapter I of the killed process implies that X' is measurable relative to $\mathcal{F}$ and $\mathcal{U}^0$.

For the rest of this chapter we will suppose that the point b is regular for itself, that is $P^b(\sigma = 0) = 1$. Let $\{L_t; t \geq 0\}$ denote local time at b normalized as in paragraph (c). Also we will suppose that $\{b\}$ is *recurrent* in the sense that $P^x(\sigma < \infty) = 1$ for all x in E. The recurrence hypothesis will guarantee that for all x $P^x(L_t \to \infty$ as $t \to \infty) = 1$, as the reader will verify easily. This hypothesis eliminates some important examples; we will indicate in the last paragraph of this section how to modify our presentation to include these. In the definitions that follow we will consider only sample points for which the inclusions $L^R \subset \{t | X_t = b\} \subset L^I$ among the points of increase and right increase of L and the b-values of X_t hold. According to paragraph (c) the exceptional set in Ω has P^μ measure 0 for all μ. For $t \geq 0$ set

$$\beta_t = \beta_t(\omega) = \inf\{s | L_s(\omega) > t\},$$
$$\beta_t^- = \lim_{s \to t, s < t} \beta_s \qquad t > 0$$
$$\beta_0^- = 0.$$

Then $\{\beta_t; t \geq 0\}$ is the right continuous inverse of L, β_t^- is equal to $\inf\{s | L_s \geq t\}$, and $\beta_t - \beta_t^-$ is equal to $\sigma \circ \theta_{\beta_t^-}$. Let $G = G(\omega)$ denote the set of strictly positive left hand end points of the excursion intervals for the sample function $t \to X_t(\omega)$. A point s is in G if and only if $s = \beta_t^-$ for some $t > 0$ such that $\beta_t - \beta_t^- > 0$, and then the corresponding excursion, extended to all of $[0, \infty)$, is the path

$$X' \circ \theta_{\beta_t^-},$$

that is

$$u(r) = X_{r+\beta_t^-} \qquad 0 \leq r < \beta_t - \beta_t^-$$
$$= b \qquad r \geq \beta_t - \beta_t^-.$$

We will take P^b as the basic probability measure on the sample space, and will define over $(\Omega, \mathcal{F}, P^b)$ a point process Y with values in U by setting, in the notation of section 2,

$$D_{Y_\omega} = \{t > 0 | \beta_t(\omega) - \beta_t^-(\omega) > 0\}$$

and associating with each t in D_{Y_ω} the path

$$X' \circ \theta_{\beta_t^-}(\omega).$$

(3.18) Theorem. Y *is a Poisson point process.*

Proof: First we must attend to some finiteness and measurability issues. For $\delta > 0$ set

$$\Gamma_\delta = \{u | \sigma_b(u) > \delta\}.$$

Every path $Y_\omega(t)$ lies in Γ_δ for some strictly positive δ. Fix such a number δ and a positive t and let N denote the number of paths $Y(s)$ with $s \le t$ and $Y(s)$ in Γ_δ. Then N is just the number of values $s \le t$ such that $\beta_s - \beta_s^-$ strictly exceeds δ, and so

$$N \le \delta^{-1}\beta_t < \infty.$$

It follows that $\{N\big((0,t] \times \Gamma_\delta, Y\big) < \infty\}$ is an $\mathcal{F}$ set of P^b measure 1 and this lets us conclude that Y is σ-discrete once the rest of the measurability is established. Clearly for the measurability it will suffice to show that $N\big((0,t] \times V, Y\big)$ is $\mathcal{F}$ measurable whenever t is finite and V is a subset of Γ_δ for some $\delta > 0$. Given such a V let τ denote the smallest value of t such that $\beta_t - \beta_t^- > \delta$. Each random variable β_t is a stopping time relative to the filtration $\{\mathcal{F}_t\}$ and one sees without difficulty that τ is a stopping time relative to the filtration $\{\mathcal{F}_{\beta_t}\}$. For the present all that is important is that τ is $\mathcal{F}$-measurable. The stochastic process $\{\beta_t^-; t \ge 0\}$ over $(\Omega, \mathcal{F}, P^b)$ has sample functions which are left continuous and hence it is jointly measurable in (t,ω) relative to $\mathcal{B}[0,\infty) \times \mathcal{F}$; and so the random variable $T = \beta_\tau^-$ is $\mathcal{F}$ measurable. More generally if τ_n denotes the nth smallest value of t such that $\beta_t - \beta_t^- > \delta$ and $T_n = \beta_{\tau_n}^-$ then τ_n and T_n are $\mathcal{F}$-measurable. The random quantity $N\big((0,t] \times V, Y\big)$ is simply the number of integers n such that $\tau_n \le t$ and $X' \circ \theta_{T_n}$ is in V. Now if R denotes any non-negative $\mathcal{F}$ measurable function the mapping from Ω to U defined by $X' \circ \theta_R$ is measurable relative to $\mathcal{F}$ and $\mathcal{U}^0$. Indeed we need only show that for each positive t the E-valued random point

$$
\begin{array}{ll}
X_{t+R} & \quad t < \sigma \circ \theta_R \\[1ex]
b & \quad t \ge \sigma \circ \theta_R
\end{array}
$$

is measurable relative to $\mathcal{F}$ and $\mathcal{E}$. We will leave this as an exercise for the reader. The conclusion is that each set of the form $X' \circ \theta_{T_n} \in V$ is an $\mathcal{F}$ set. And now it follows that Y is a point process and is σ-discrete. As to the Poisson property, if t is positive then one checks easily that for all positive s we have

$$\beta_{s+t} = \beta_t + \beta_s \circ \theta_{\beta_t}.$$

It follows that a value of s such that $s+t \in D_{Y_\omega}$ and the corresponding excursion are exactly those associated with the point $\theta_{\beta_t(\omega)}(\omega)$ by the original point process Y, that is,

$$\theta_t Y = Y \circ \theta_{\beta_t}$$

where the usage of θ_t on the left side denotes the shifted point process defined in section 2 and should not be confused with the shift operators for X. Also it is clear that the stopped process $\alpha_t Y$ is measurable relative to $\mathcal{F}_{\beta_t}$. We apply the strong Markov property of our original Markov process at the stopping time β_t. The value β_t is a point in L^R and hence $X_{\beta_t} = b$ so that the conditional independence of $\mathcal{F}_{\beta_t}$ and $Y \circ \theta_{\beta_t}$ becomes full independence and the distribution of $Y \circ \theta_{\beta_t}$ is that of Y relative to P^b, that is the two quantities have the same distribution. Thus Y is indeed a Poisson point process. It is called a *point process of excursions* or the *point process of excursions of X away from b.*

We note that in some situations the name *excursion process* is used for the point process W that one obtains by taking

$$D_{W_\omega} = D_{Y_\omega}$$

and associating with t in D_{W_ω} the path

$$X \circ \theta_{\beta_t^-}(\omega)$$

that is the "excursion" is the entire path $r \to X_{r+\beta_t^-}$ rather than the "stopped" path, that is to say the one held fast at the point b once it returns there. The argument of (3.18) shows that W is a point process; but it is not a Poisson point process because $\alpha_t W$ is not $\mathcal{F}_{\beta_t}$ measurable. We will find it essential to consider both processes.

(e) **The characteristic measure.** In finding the characteristic measure for Y we will assume that our process X is itself the canonical right continuous realization introduced in I-8. Basically this means that Ω is the space U of right continuous left limit functions, the X_t are the coordinate variables with θ_t being the usual shift operator and $\mathcal{F}$ and $\mathcal{F}_t$ are the appropriate completions of the σ algebras generated by the appropriate collection of coordinate variables. This assumption saves us having to deal with some notational problems that have only nuisance value. Recall the notation $G = G(\omega)$ for the strictly positive left ends of the excursion intervals. Except for the null set on which at least one of the inclusions of

(3.17) fails, G is just the set of strictly positive numbers which are points of increase, but not of right increase for the local time, L_t.

Let f be a positive bounded function on $\Omega (= U)$ which is measurable relative to $\mathcal{F}^0$, the σ-algebra generated by the coordinate functions. Define a positive function, A_f on Ω by

$$A_f = \Sigma_{s \in G} e^{-s} \{(1 - e^{-\sigma})f\} \circ \theta_s.$$

First of all we note that A_f is $\mathcal{F}$ measurable. Indeed let G_n^ε denote the nth smallest left end point in G such that the length of the associated excursion interval strictly exceeds ε. The reader should be able to argue without difficulty that $G_n^\varepsilon + \varepsilon$ is a stopping time. Every s in G such that ε is strictly less than $\sigma \circ \theta_s$ is equal to G_n^ε for some n (depending on ε). From this and the fact that $\sigma = \sigma \circ \theta_\varepsilon + \varepsilon$ if σ strictly exceeds ε it is clear that

$$(3.19) \qquad A_f = \lim_{\varepsilon \to 0} \Sigma_n e^{-G_n^\varepsilon} f \circ \theta_{G_n^\varepsilon} (1 - e^{-\sigma}) \circ \theta_{G_n^\varepsilon + \varepsilon}$$

and that the approximating sum increases as ε decreases. The terms $e^{-G_n^\varepsilon}$ and $f \circ \theta_{G_n^\varepsilon}$ are $\mathcal{F}$ measurable since G_n^ε is $\mathcal{F}$ measurable and f is $\mathcal{F}^0$ measurable. And the term $(1 - e^{-\sigma}) \circ \theta_{G_n^\varepsilon + \varepsilon}$ is $\mathcal{F}$ measurable because $G_n^\varepsilon + \varepsilon$ is a stopping time, so that the arguments from the end of I-6 are applicable. This shows that A_f is $\mathcal{F}$ measurable.

Define a function ψ_f on E by

$$\psi_f(x) = E^x A_f.$$

Then ψ_f is universally measurable. If T is a stopping time we have

$$(3.20) \qquad \begin{aligned} &E^x e^{-T} \psi_f(X_T) \\ &= E^x \Sigma_{s \in G \circ \theta_T} e^{-(s+T)} \{(1 - e^{-\sigma})f\} \circ \theta_{s+T} \end{aligned}$$

We have noted already that for any fixed t the values $s + t$ with s in $G \circ \theta_t$ are just those values s in G which exceed t strictly. Thus the right side of (3.20) is

$$(3.21) \qquad E^x \Sigma_{s \in G, s > T} e^{-s} \{(1 - e^{-\sigma})f\} \circ \theta_s.$$

Let us take $T = \sigma$ in (3.21). Obviously no value of s strictly less than σ can be in G and neither is σ in G since σ is a stopping time, $X_\sigma = b$, and b is regular for itself. It follows that

$$(3.22) \qquad \psi_f(x) = E^x e^{-\sigma} \psi_f(X_\sigma) = \psi_f(b) E^x e^{-\sigma}.$$

Obviously $\psi_f(b)$ defines a measure in f, let us call it P_1. Then (3.22) may be written as

$$(3.23) \qquad E^x A_f = P_1(f)E^x \int_0^\infty e^{-t} dL_t$$

where as usual L denotes local time at $\{b\}$. Let τ denote the mapping, $\tau(X) = X'$ from Ω to itself where X' is defined by

$$\begin{aligned} X'_t(\omega) &= X_t(\omega) & t < \sigma(\omega) \\ &= b & t \geq \sigma(\omega). \end{aligned}$$

(3.24) Theorem. *The measure P_1 has total mass less than 1. The P_1 measure of $\{\sigma = 0\}$ is equal to 0. Let P_2 be the measure defined by*

$$P_2(f) = P_1\left[f/(1 - e^{-\sigma})\right].$$

The measure P_3 defined as the distribution relative to P_2 of τ is the characteristic measure of the point process Y.

Proof: First of all if f equals 1 then

$$P_1(1) = E^b \int_0^\infty e^{-s} I(X_s) ds$$

where I is the indicator of $E - \{b\}$ so the inequality $P_1(1) \leq 1$ is established.

Next we must give a measurability argument since P_1 is defined as a measure on $\mathcal{F}^0$ whereas σ is known only to be $\mathcal{F}$ measurable. Let $\{V_n\}$ be a decreasing sequence of open sets with compact closure whose intersection is $\{b\}$. Each σ_{V_n} is $\mathcal{F}^0$ measurable since V_n is open, and for each positive δ the quantity $\delta + \sigma_{V_n} \circ \theta_\delta$ is $\mathcal{F}^0$ measurable. Let

$$\bar{\sigma} = \lim_{\delta \to 0} \lim_{n \to \infty} \delta + \sigma_{V_n} \circ \theta_\delta.$$

Then $\bar{\sigma}$ is $\mathcal{F}^0$ measurable $\bar{\sigma} \leq \sigma$, and the discussion in section 5 of Chapter I shows that $P^\mu(\bar{\sigma} < \sigma) = 0$ for each initial distribution μ on $\mathcal{E}$. Thus for $0 < \varepsilon < \delta$ and any n we can find a set Γ in $\mathcal{F}^0$ containing $\{\bar{\sigma} < \sigma\}$ and such that

$$\begin{aligned} P^b(I_\Gamma \circ \theta_{G_n^\varepsilon + \delta} &> 0) \\ &= P^b\left(P^{X_{G_n^\varepsilon + \delta}}(\Gamma)\right) = 0, \end{aligned}$$

where the first equality uses the fact that $G_n^\varepsilon + \delta$ is a stopping time. Clearly we may assume that Γ is independent of n, ε and δ if the latter two are restricted to be rational. Now

$$P_1(I_\Gamma \circ \theta_\delta) = \lim_{\varepsilon \to 0} E^b(\sum_n e^{-G_n^\varepsilon}\{(1 - e^{-\sigma}) \circ \theta_{G_n^\varepsilon}\} I_\Gamma \circ \theta_{G_n^\varepsilon + \delta}$$

and so $P_1(I_\Gamma \circ \theta_\delta) = 0$ for all strictly positive rational δ. It is obvious that if $\bar{\sigma} < \sigma$ then $I_\Gamma \circ \theta_\delta$ is strictly positive for some strictly positive rational δ. These are $\mathcal{F}^0$ events of P_1 measure 0. And so σ differs from $\bar{\sigma}$ only on an $\mathcal{F}^0$ set of P_1 measure 0. Coming to the second assertion of the theorem, let $\overline{\mathcal{F}}$ denote the completion relative to P_1 of $\mathcal{F}^0$ and regard P_1 as extended to $\overline{\mathcal{F}}$. If f is $\overline{\mathcal{F}}$ measurable and positive and $f_1 \le f \le f_2$ where f_1 and f_2 are $\mathcal{F}^0$ measurable and $P_1(f_2 - f_1) = 0$ then $A_{f_1} \le A_f \le A_{f_2}$ and so A_f is $\overline{\mathcal{F}}$ measurable and $P_1(f)$ is still satisfies (3.23). If f is the indicator of $\{\sigma = 0\}$ then clearly A_f is identically 0, so that $P_1(\sigma = 0) = 0$ as required. Coming to the description of P_2 note first that while P_2 may have infinite mass it satisfies

$$P_2(1 - e^{-\sigma}) = P_1(1) \le 1$$

and $P_2(\sigma = 0) = 0$ so that P_2 is σ-finite. Formula (3.23) now reads

$$(3.25) \qquad E^x \Sigma_{s \in G} e^{-s} f \circ \theta_s = P_2(f) E^x \int_0^\infty e^{-t} dL_t.$$

Denote the left side of (3.25) by $\rho_f(x)$, so that

$$\rho_f(x) = P_2(f) E^x e^{-\sigma}.$$

If T is a stopping time and g is positive and $\mathcal{F}_T$ measurable then the argument that connected (3.20) and (3.21) shows that

$$\begin{aligned}
E^x g e^{-T} \rho_f(X_T) &= E^x \Sigma_{s \in G, s > T} g e^{-s} f \circ \theta_s \\
&= P_2(f) E^x (g E^{X_T} \int_0^\infty e^{-t} dL_t) \\
&= P_2(f) E^x (\int_T^\infty e^{-t} g \, dL_t).
\end{aligned}$$

This can be written more compactly as

$$(3.26) \qquad E^x \Sigma_{s \in G} Z_s f \circ \theta_s = P_2(f) E^x \int_0^\infty Z_s dL_s$$

where the process $\{Z_s; s \geq 0\}$ is defined by

$$Z_s = e^{-s}g \qquad\qquad s > T$$
$$= 0 \qquad\qquad s \leq T.$$

Let R be an $\{\mathcal{F}_t\}$ stopping time and c and d be positive numbers with $0 \leq c \leq d$. Obviously for any process $\{Z_s\}$ of the form

$$Z_s = ge^{-s}I_{(R\wedge c, R\wedge d]}(s)$$

where g is bounded, positive and $\mathcal{F}_{R\wedge c}$ measurable the equality in (3.26) holds because such a process is the difference of two for which the equality has been established. Hence if n is fixed and $R_k = R \wedge k/2^n$, the equality holds for

$$Z_s^n = \sum_{k=0}^{\infty} e^{R_k}e^{-s}I_{(R_k, R_{k+1}]}(s).$$

As n increases to ∞, Z_s^n increases to $I_{(0,R]}(s)$ and so (3.26) holds for $Z_s = I_{(0,R]}(s)$ as well.

With this in mind, in (3.26) let f be the indicator of a set Γ in $\mathcal{F}^0$, take $x = b$ and $Z_s = I_{s \leq \beta_1}$. The right side of (3.26) is just $P_2(\Gamma)E^b L_{\beta_1} = P_2(\Gamma)$. On the left the sum whose expectation we are taking consists exactly of terms $f \circ \theta_s$ with $s = \beta_r^-$ for an r with $\beta_r - \beta_r^- > 0$ and $s \leq \beta_1^-$, that is $r \leq 1$. That is exactly the number of points W_t in Γ corresponding to points t in D_W with $t \leq 1$; thus $P_2(\Gamma)$ is the expected number of such points.

The fact that P_3 is the characteristic measure for Y follows immediately: that is, if f is $\mathcal{F}^0$ measurable and positive then $f \circ \tau$ is $\overline{\mathcal{F}}$ measurable and $P_3(f) = P_2(f \circ \tau)$. Thus (3.26) applied to $f \circ \tau$ reads

$$(3.27) \qquad E^x \Sigma_{s \in G} Z_s(f \circ \tau) \circ \theta_s = P_3(f)E^x \int Z_s dL_s.$$

Once again taking $Z_s = I_{s \leq \beta_1}$, $x = b$, and $f = I_\Gamma$ with Γ in $\mathcal{F}^0$ we obtain $P_3(\Gamma)$ on the right while on the left we have the expected number of paths Y_t in Γ with t in D_Y and t less than 1.

This completes the proof of (3.24). Note that the domain of P_3 can be extended to the σ-algebra $\mathcal{F}'$ of all those sets Γ such that $\tau^{-1}(\Gamma)$ is in $\overline{\mathcal{F}}$. The equality (3.27) with f being $\mathcal{F}'$ measurable remains valid. We will assume without further mention that this extension has been made. The function σ is $\mathcal{F}'$ measurable because $\sigma = \sigma \circ \tau$.

The characteristic measure P_3 is referred to usually as the *excursion measure*. We will use this term from now on.

Only notational changes are needed to include situations where X is perhaps something other than the canonical realization. In (3.27) one simply replaces $f \circ \tau$ with $f \circ X'$. Remember though that the measures P_2 and P_3 still are defined on the relevant σ-algebras in path space U.

More common notation for the measure P_3 is $\hat{P}$. We will use this notation from now on. Many proofs use only (3.27) with an appropriate choice of Z and f. Often we call (3.27) the *excursion* formula.

(f) Markov properties of the excursion measure. The key to making computations with the excursion measure is the fact that with this measure on U the coordinate process is a Markov process on the *open* time interval $(0, \infty)$ whose transition data is that of the original process X stopped at σ. Since the measure $\hat{P}$ has infinite mass some slight alteration in the basic definitions is required. Also the fact that $\hat{P}(X_0 = b)$ might be strictly positive indicates that the time parameter 0 cannot be included. We will now develop this. In doing so there is no loss in assuming that the original process X is the canonical right continuous realization since for the present purposes it is only an auxiliary device for making computations involving the measure $\hat{P}$. As usual P^x denotes probabilities for our original process starting at x. We will let $\bar{P}^x$ denote the corresponding probabilities for the process X stopped at time σ that is

$$\bar{P}^x(\Gamma) = P^x(X' \in \Gamma)$$

where
$$X'_t = X_t \qquad t < \sigma$$
$$ = b \qquad t \geq \sigma$$

and Γ is a set in $\mathcal{F}^0$.

(3.28) Theorem. *If T is an $\{\mathcal{F}^0_{t+}\}$ stopping time with $T > 0$, g is a positive function which is $\mathcal{F}^0_{T+}$ measurable and f is a positive $\mathcal{F}^0$ measurable function then*

$$P_2(g \cdot f \circ \theta_T) = P_2\big(g P^{X_T}(f)\big)$$

and

$$\hat{P}(g \cdot f \circ \tau \circ \theta_T; T < \sigma)$$
$$= \hat{P}(g\bar{P}^{X_T}(f); T < \sigma).$$

Proof. Suppose G is a positive $\mathcal{F}$ measurable random variable which need not be a stopping time but is such that $G + \delta$ is an $\{\mathcal{F}_t\}$ stopping time; and suppose T is an $\{\mathcal{F}_{t+}^0\}$ stopping time such that $T \geq \delta$, δ being a strictly positive number. Then $G + T \circ \theta_G$ is an $\{\mathcal{F}_t\}$ stopping time. Indeed

$$\{G + T \circ \theta_G < t\} = \cup\{G \leq q, \theta_G^{-1}\{T < t - q\}$$

where q ranges over rationals strictly less than $t - \delta$. We have $\{G \leq q\} = \{G + \delta \leq q + \delta\} \in \mathcal{F}_{q+\delta} \subset \mathcal{F}_t$. The set $\{T < t - q\}$ is in $\mathcal{F}_{t-q}^0$ and the σ-algebra consisting of the trace on $\{G \leq q\}$ of the θ_G inverse images of such sets is generated by sets of the form

$$\{X_{r+G} \in B\} \cap \{G \leq q\}$$

with $B \in \mathcal{E}$ and $r + q \leq t$. Any such set is in $\mathcal{F}_t$, and so the assertion is established. A similar argument shows that if g is $\mathcal{F}_{T+}^0$ measurable then $g \circ \theta_G$ is measurable relative $\mathcal{F}_{G+T\circ\theta_G}$. Now let δ be a strictly positive number and suppose that $T(\omega) \geq \delta$ for all ω. If ε is less than δ we can apply the previous considerations with G being any of the G_n^ε from paragraph (d). Using the definition of P_2 and the strong Markov property of the original process we obtain

$$P_2(g \cdot f \circ \theta_T) = \lim_{\varepsilon \to 0} \sum_n E^b(e^{-G_n^\varepsilon} g \circ \theta_{G_n^\varepsilon} \cdot f \circ \theta_{G_n^\varepsilon + T\circ\theta_{G_n^\varepsilon}})$$
$$= \lim_{\varepsilon \to 0} \sum_n E^b \left(e^{-G_n^\varepsilon} (gP^{X_T}(f)) \circ \theta_{G_n^\varepsilon} \right)$$
$$= P_2(g \cdot P^{X_T}(f)).$$

This proves the first assertion of the theorem for T satisfying the additional condition, $T \geq \delta$. For the general case define a stopping time T_δ by

$$T_\delta = \max(T, \delta).$$

What we have done already, along with the fact that $\{T > \delta\}$ is in $\mathcal{F}_{T+}^0$ and $T = T_\delta$ there implies that

$$P_2(gI_{T>\delta}f \circ \theta_{T_\delta}) = P_2(gI_{T>\delta}P^{X_{T_\delta}}(f))$$

but by the definition we may in this equality replace T_δ with T and then the result follows by monotone convergence. The second assertion in (3.28) follows immediately by applying the first assertion with g replaced by $g I_{T<\sigma}$, f replaced by $f \circ \tau$ and then observing that if g is $\mathcal{F}^0_{T+}$ measurable then the function $g I_{T<\sigma}$ is unchanged by composing it with τ.

Let $P^0(t, x, A)$ denote the transition function for the process X killed at time σ; that is

$$P^0(t, x, A) = P^x(X_t \in A, t < \sigma).$$

Equivalently, $P^0(t, x, A)$ is the transition function for the stopped process, but restricted to the $\mathcal{E}$ sets which are contained in $E - \{b\}$. For t strictly positive let η_t be the measure

$$\eta_t(A) = \hat{P}(X_t \in A)$$

restricted to $\mathcal{E}$ sets contained in $E - \{b\}$. From the Markov property of the measure $\hat{P}$ it follows that

$$\eta_{t+s}(A) = \int \eta_s(dx) P^0(t, x, A)$$

or $\eta_{t+s} = \eta_s P^0_t$. A family $\{\eta_t; t > 0\}$ of σ-finite measures satisfying this relationship is said to be an *entrance law* for the semi group $\{P^0_t; t \geq 0\}$. For the example we are considering, the total mass of η_t is $\hat{P}(\sigma > t)$ and we have

$$1 \geq \hat{P}(1 - e^{-\sigma}) \geq (1 - e^{-t})\|\eta_t\|$$

so that each η_t is a finite measure. If μ is a measure on $\mathcal{E}$ then the family $\{\eta_t; t > 0\}$ defined by

$$\eta_t(A) = \mu P^0_t = P^\mu(X_t \in A, t < \sigma)$$

is an example of an entrance law for $\{P^0_t; t \geq 0\}$ provided the measures are σ-finite; but there are others. The finite dimensional distributions of the coordinate process relative to $\hat{P}$ are determined by the entrance law and the transition function P^0_t in the usual way but we cannot in general specify an initial distribution η_0 such that $\eta_0 P^0_t = \eta_t$.

(g) **The non-recurrent case.** We will keep the hypotheses of the previous sections except for allowing the possibility that for some x in E, $P^x(\sigma < \infty) < 1$. A good example is Brownian motion with a drift: $X_t = B_t + t$ where B_t is ordinary Brownian motion. Every point is regular for itself, but if we take the point b to be the origin then

$$
\begin{aligned}
P^x(\sigma < \infty) &= 1 && x \le 0 \\
&= e^{-2x} && x > 0.
\end{aligned}
$$

Let $\psi(x) = P^x(\sigma < \infty)$. If $\psi(x) < 1$ for some x then except for the trivial situation in which $\{\psi < 1\}$ is not reached by the process starting at b, the expression $P^b(\sigma \circ \theta_t = \infty)$ will increase to 1 as t increases to ∞ and so $P^b(L_\infty = L_t)$ will increase to 1. Thus L_∞ will be finite. The point process of excursions away from b will have no domain points with values exceeding L_∞ and hence can not possibly be a Poisson point process. On the other hand the strong Markov property of X and the fact that $X_{\beta_t} = b$ if $\beta_t < \infty$ yields for any A in $\mathcal{F}^0$ and Γ in $\mathcal{F}_{\beta_t}$

$$
\begin{aligned}
P^b\big(\theta_{\beta_t}^{-1}(A); \Gamma \cap \{\beta_t < \infty\}\big) \\
= P^b(A)P^b\big(\Gamma \cap \{\beta_t < \infty\}\big).
\end{aligned}
$$

This yields the fact that relative to P', the conditional probability given $\beta_t < \infty$, the distribution of the shifted point process $\theta_t Z$ is the same as that of Z under P^b and that relative to P' $\theta_t Z$ and $\mathcal{F}_{\beta_t}$ are independent. For most purposes this is an effective substitute.

At any rate since the events $\{\beta_t < \infty\}$ and $\{L_\infty > t\}$ are the same and since $\beta_{t+s} = \beta_t + \beta_s \circ \theta_{\beta_t}$ if β_t is finite we have

$$
\begin{aligned}
P^b(L_\infty > t + s) &= P^b\big(P^{X_{\beta_t}}(L_\infty > s); \beta_t < \infty\big) \\
&= P^b(L_\infty > s)P^b(L_\infty > t).
\end{aligned}
$$

Hence L_∞ has an exponential distribution, with rate λ. We will assume λ is strictly positive since otherwise $L_\infty = \infty$, the case discussed in the previous section.

So we will assume that X is a standard process, that b is a point regular for itself and that L_∞ is exponentially distributed with rate $\lambda > 0$. The argument from the previous sections for the existence of the measures P_2 and $\hat{P}$, as well as the relationships (3.26) and (3.27), in no way depend on the additional recurrence hypothesis. Thus they remain valid, as do the Markov properties from section (f). In this case the decomposition of the excursion measure $\hat{P}$ into its restriction to the set of paths with $\sigma = \infty$ and to those with a finite σ is of some use. Before setting this out we need a preliminary observation.

Recall the definition $\psi(x) = P^x(\sigma < \infty)$; and let us set $\theta(x) = 1 - \psi(x) = P^x(\sigma = \infty)$. These functions are $\mathcal{E}$ measurable because we may replace σ by $\bar{\sigma}$ (paragraph (e)) in the definitions. We have

$$\int P^0(t, x, dy)\theta(y) = P^x(\sigma \circ \theta_t = \infty, t < \sigma) = \theta(x)$$

and

$$\int P^0(t, x, dy)\psi(y) = P^x(\sigma \circ \theta_t < \infty, t < \sigma)$$

$$= P^x(t < \sigma < \infty) \le \psi(x).$$

Thus if $h(x)$ denotes either θ or ψ we have $P^0(t, x, h) \le h(x)$ and so

$$Q_t(x, f) = \frac{1}{h(x)} P^0(t, x, hf) \qquad\qquad h(x) > 0$$
$$= 0 \qquad\qquad\qquad\qquad\quad h(x) = 0$$

yields a semi-group of sub-Markov operators on the set of positive $\mathcal{E}$ measurable functions which vanish on $\{h = 0\}$. Let Q^1 and Q^2 denote the semi groups obtained by taking h to be θ and ψ respectively. In both cases the point b should be removed from the state space. In the example of Brownian motion with a positive drift, Q_t^1 would be regarded as operating on functions defined only on $(0, \infty)$, whereas Q_t^2 would operate on functions defined on $R - \{0\}$. Let $\hat{P}_i$ be the measures on $\mathcal{F}^0$ defined by

$$\hat{P}_1(A) = \hat{P}(A \cap \{\sigma = \infty\})$$
$$\hat{P}_2(A) = \hat{P}(A \cap \{\sigma < \infty\}).$$

Let $\{\eta_t; t > 0\}$ denote the entrance law for the excursion measure $\hat{P}$.

(3.29) Theorem. *The measure $\hat{P}_1$ has total mass λ. Relative to $\hat{P}_1$ the coordinate process $\{X_t; t > 0\}$ is a time homogeneous Markov process with transition function $Q^1(t, x, A)$ and entrance law $\eta_t(dx)\theta(x)$. Relative to the measure $\hat{P}_2$ the coordinate process is a time homogeneous Markov process on $t > 0$ with transition function $Q^2(t, x, A)$ and entrance law $\eta_t(dx)\psi(x)$.*

Proof. For the first assertion, in (3.27) take f to be the indicator of $\{\sigma = \infty\}$ and take Z_s identically equal to 1. There is exactly one s in G with $\sigma \circ \theta_s = \infty$ so the left side of (3.27) is 1. We then have

$$1 = \hat{P}(\sigma = \infty)E^b L_\infty$$

and since $\hat{P}(\sigma = \infty)$ is the total mass of $\hat{P}_1$ and L_∞ is exponential of rate λ the assertion follows. For the second assertion note first that if t is strictly positive and f is a positive $\mathcal{F}_t^0$ measurable function vanishing on $\{\sigma \leq t\}$ then

$$\hat{P}_1(f) = \hat{P}(f, t < \sigma, \sigma \circ \theta_t = \infty)$$
$$= \hat{P}(f \cdot \theta(X_t); t < \sigma).$$

Let g be a positive $\mathcal{E}$ measurable function vanishing at b. Applying the computation just made and the Markov properties of $\hat{P}$ we have

$$\hat{P}_1(f \cdot g(X_{t+r})) = \hat{P}(f \, P_0(r, X_t, \theta g); t < \sigma)$$

where f is any positive $\mathcal{F}_t^0$ measurable function. We can multiply and divide the integrand by $\theta(X_t)$, there being no danger of ambiguity because if $\theta(x)$ is zero then so is $P_0(r, x, \theta g)$. Then the right side of this last equation will read

$$\hat{P}(f \cdot \theta(X_t)Q^1(r, X_t, g); t < \sigma)$$

and this is equal to

$$\hat{P}_1(f \cdot Q^1(r, X_t, g)).$$

This establishes the Markov property and identifies the transition function under $\hat{P}_1$. The identification of the entrance law comes from the identity

$$\hat{P}_1(h(X_t)) = \hat{P}(h(X_t)\theta(X_t); t < \sigma)$$
$$= \eta_t(h\theta).$$

The argument for the structure of the coordinate process under $\hat{P}_2$ is exactly the same.

The next chapter is devoted to computations involving excursion laws. There we will work out an illustration of Theorem (3.29).

(3.30) Exercise. Let Y be the Poisson point process of excursions and let $\hat{P}$ be its excursion measure. Let

$$T(t) = \Sigma\sigma(Y_s) = \Sigma\beta_s - \beta_{s-}$$

where the first sum is over all s in D_Y with $s \le t$ and the second is over all $s \le t$. Use the exponential formula for Poisson measures to show that

$$E^b e^{-\lambda T(t)} = \exp\Big\{ -t \int_{(0,\infty)} (1 - e^{-\lambda x}) \hat{P}(\sigma \in dx) \Big\}.$$

(3.31) Exercise. Let G^r be the subset of G defined by

$$G^r = \{ s \in G | X_s = b \}.$$

(a) Show that the following variant of (3.27) holds:

$$(3.27)' \qquad E^x \sum_{s \in G^r} Z_s f \circ \theta_s = \hat{P}(f I_{X_0 = b}) E^x \int Z_s dL_s$$

where R is an $\{\mathcal{F}_t\}$ stopping time and

$$Z_s = I_{(0,R]}(s).$$

(b) Show that in $(3.27)'$ we may replace Z_s with

$$Z'_s = I_{(0,R)}(s).$$

(Hint: $P^x(X_R = b, R \in G) = 0$.)

(c) Observe that if the original process is a diffusion then $G = G^r$ and $\hat{P}(X_0 \ne b) = 0$, so that (3.27) is valid when Z is replaced by Z'.

(d) Show by example (any holding and jumping process will suffice) that in general one may not use Z' in (3.27).

Note: Processes such as Z are called *previsible*, those such as Z' are called *optional*. The substance of (3.31) is that (3.27) is valid for any positive previsible process, but that a restatement is needed to get a version valid for optional processes.

(3.32) Exercise. Argue that, given the probabilities $\bar{P}^x$ for the stopped process, the measure $\hat{P}$ is determined by a knowledge of the expressions $\hat{P}(\int_0^\sigma e^{-r} h(X_r) dr)$, h ranging over the bounded $\mathcal{E}$ measurable function vanishing at b. Hint: for t strictly positive, f bounded and $\mathcal{E}$-measurable and vanishing at b take $h(x) = \bar{P}^x f(X_t)$. Then

$$e^{-t} \hat{P}\Big(\int_0^\sigma e^{-r} h(X_r) dr \Big)$$

$$= \int_t^\infty e^{-u}\hat{P}(f(X_u); u < \sigma)du.$$

Argue that the expressions on the right determine the entrance measures $\eta_t(f) = \hat{P}(f(X_t); t < \sigma)$ and these in conjunction with $\{\overline{P}^x, x \in E\}$ determine $\hat{P}$. (This result should be compared with the uniqueness theorem for the operator U^α from I-9.)

(3.33) Exercise. (a) Let $K \in \mathcal{E}$ and $b \notin \bar{K}$. Let $J = \inf\{t | t \in G, X_t \in K\}$. Show that J is an $\{\mathcal{F}_t\}$ stopping time. (Hint: consider events of the form $\{0 < \sigma_K \circ \theta_p - \sigma_b \circ \theta_p < 1/n\}$ with p rational and $p + \sigma_b \circ \theta_p \le a$.)

(b) Fix $\epsilon > 0$ and let J^ϵ be J from part (a) where $K = \{x | \rho(x, b) \ge \epsilon\}$, ρ a metric for E. Let $J^\epsilon_{n+1} = J^\epsilon_n + J^\epsilon \circ \theta_{J^\epsilon_n}$, $(J^\epsilon_0 = 0)$. Use the fact that $t \in G$ and $X_t \ne b$ if and only if $t = J^\epsilon_n$ for all small ϵ and some n (depending on ϵ) to show that if f is positive and $\mathcal{F}^0$ measurable and g is positive and $\mathcal{F}^0_{0+}$ measurable then

$$P_2(fg; X_0 \ne b) = P_2(gP^{X_0}(f); X_0 \ne b).$$

(Hint: the left side is equal to

$$\lim_{\epsilon \to 0} E^b\left(\sum_n e^{-J^\epsilon_n} fg \circ \theta_{J^\epsilon_n}\right);$$

apply the strong Markov property of the original process.)

(c) Conclude that in (3.28) the condition $\{T > 0\}$ can be dropped provided $\{T = 0\}$ is contained in $\{X_0 \ne b\}$.

IV Brownian Excursion

In this chapter we will give various descriptions of Brownian excursion measure; that is the measure $\hat{P}$ one obtains when the basic process X is Brownian motion in R^1 and the distinguished point b is the origin. Also we will describe the excursion structure of some processes closely related to Brownian motion.

1. Brownian excursion.

We will in fact focus attention mainly on the case where X is reflecting Brownian motion, since this is simpler and yields the results for Brownian motion by symmetry considerations. In studying an excursion measure when the underlying process is a diffusion, that is it has continuous paths, we will without special mention take Ω to be the space of continuous functions rather than those which are only right continuous. This is legitimate since all the excursion paths of the original process are continuous; and it has the advantage that hitting times like σ_x, the time of hitting a point x, are $\mathcal{F}^0$ measurable so some measurability discussions can be avoided. Recall the notation $p^-(t, x, y)$ for the transition density of Brownian motion killed at time $\sigma(=\sigma_0)$. Recall also calculations from earlier chapters such as $E^x e^{-\lambda\sigma} = e^{-x\sqrt{2\lambda}}$. The relevant facts are summed up in the following theorem.

(1.1) Theorem. *If X is reflecting Brownian motion and $b = 0$ then*

a) $\hat{P}(\sigma_x < \sigma) = \dfrac{1}{x\sqrt{2}}, x > 0;$

b) $\eta_t(dx) = (1/t^{3/2}\pi^{1/2})xe^{-x^2/2t}dx, x > 0.$

Remark: Note that the information (b) characterizes $\hat{P}$ since we know already that the coordinate process is Markovian with $p^-(t, x, y)$ as transition density.

Proof. For assertion (a) note that for $0 < x < y$ we have, by virtue of the strong Markov property established in III-3.28

$$\hat{P}(\sigma_y < \sigma) = \hat{P}(\sigma_y \circ \theta_{\sigma_x} < \sigma \circ \theta_{\sigma_x}; \sigma_x < \sigma)$$
$$= P^x(\sigma_y < \sigma_0)\hat{P}(\sigma_x < \sigma_0).$$

For Brownian motion we know that

$$P^x(\sigma_y < \sigma_0) = x/y$$

and it follows that $x\hat{P}(\sigma_x < \sigma_0)$ is constant in $x > 0$. To evaluate the constant first take $f = 1 - e^{-\sigma}$ and $Z_s = e^{-s}$ in III-3.27. Obviously the left side is the potential, $E^0 \int_0^\infty e^{-t} I(X_t) dt$, where I is the indicator of $(0, \infty)$, and the right side is $\hat{P}(1 - e^{-\sigma})$. It follows that $\hat{P}(1 - e^{-\sigma}) = 1$. Next we have

$$\hat{P}(1 - e^{-\sigma}; \sigma_x < \sigma) = \hat{P}(1 - e^{-\sigma_x}; \sigma_x < \sigma)$$
$$+ \hat{P}\big(e^{-\sigma_x}(1 - e^{-\sigma \theta_{\sigma_x}}); \sigma_x < \sigma\big).$$

The left side increases to $\hat{P}(1 - e^{-\sigma})$ as x decreases to 0. The first term on the right approaches 0 with x since the integrand does so and the integrand is bounded by $1 - e^{-\sigma}$, whose integral relative to $\hat{P}$ is finite. The second term on the right is

$$\hat{P}(e^{-\sigma_x} P^x (1 - e^{-\sigma}); \sigma_x < \sigma)$$
$$= (1 - e^{-x\sqrt{2}})\hat{P}(e^{-\sigma_x}; \sigma_x < \sigma).$$

This last expression differs from

$$(1 - e^{-x\sqrt{2}})\hat{P}(\sigma_x < \sigma_0)$$

by the expression

$$(1 - e^{-x\sqrt{2}})\hat{P}(e^{-\sigma_x} - 1; \sigma_x < \sigma),$$

which is in absolute value less than $(1 - e^{x\sqrt{2}})$ and hence approaches 0 with x. Thus as x approaches 0, $(1 - e^{-x\sqrt{2}})\hat{P}(\sigma_x < \sigma)$ approaches 1 and so the constant in question is identified and (a) is established.

Coming to assertion (b) let g be a continuous function with compact support in $(0, \infty)$. Then

$$\eta_t(g) = \lim_{x \to 0} \hat{P}\big(g(X_{t+\sigma_x}); t + \sigma_x < \sigma\big)$$
$$= \lim_{x \to 0} \hat{P}\bigg(P^x\big(g(X_t); t < \sigma\big); \sigma_x < \sigma\bigg)$$
$$= \frac{1}{\sqrt{2}} \lim_{x \to 0} \int_0^\infty \frac{p^-(t, x, y)}{x} g(y) dy.$$

Now using the explicit form of the Gauss kernel and the fact that $p^-(t, x, y) = p(t, x, y) - p(t, -x, y)$ the result follows immediately. The interchange of limit and integral is justified by dominated convergence and a simple estimate.

For future reference we note that

$$\hat{P}(\sigma > t) = \|\eta_t\| = \frac{1}{\sqrt{\pi t}}.$$

It is obvious what alterations are needed to obtain a description of $\hat{P}$ when X is Brownian motion: in (a) of (1.1) we should have $\hat{P}(\sigma_x < \sigma_0) = \frac{1}{2\sqrt{2}|x|}$ and in (b) the entrance law, call it η'_t, is given by $\eta'_t(A) = \frac{1}{2}\eta_t(A)$ if the Borel set A is contained in $(0, \infty)$ and $\eta'_t(A) = \frac{1}{2}\eta_t(-A)$ if A is contained in $(-\infty, 0)$. We will leave the argument to the reader.

The term "Brownian excursion" is applied not just to the coordinate process under $\hat{P}$ for reflecting Brownian motion but also to the non-time-homogeneous process over the interval $[0, 1]$ described in example (d) of II-1. There we specified a transition density and absolute probabilities, and then established the existence of a corresponding process with continuous paths by making a transformation of the Bessel process. We will use the term Brownian excursion to describe either object. This possible ambiguity is supported by tradition; the context should make it clear which meaning is intended. The relationship between these two uses of the term is as follows. Let $\{e(t); 0 \le t \le 1\}$ denote the process called Brownian excursion in II-1 example (d). Extend the definition by setting $e(t) = 0$ if $t > 1$; this still yields a process with continuous paths since $e(1) = 0$. Let μ_1 denote the measure this process induces on the σ-algebra $\mathcal{F}^0$ in the space of continuous functions on $[0, \infty)$. For $a > 0$ let $\{e_a(t); t \ge 0\}$ be the process defined by

$$e_a(t) = \sqrt{a}\, e(t/a)$$

and let μ_a be the measure it induces on $\mathcal{F}^0$.

(1.2) Theorem. $\hat{P} = \int_0^\infty \mu_a \hat{P}(\sigma \in da)$.

Proof: For y and b strictly positive let $\psi(y, b) = P^y(\sigma > b)$, where the probabilities P^y refer to Brownian motion. By exercise (4.7) of Chapter I we have

$$\psi(y, b) = 1 - 2 \int_{y/\sqrt{b}}^\infty \frac{1}{\sqrt{2\pi}} e^{-x^2/2} dx.$$

Take a strictly positive number t and let F be a positive bounded $\mathcal{F}^0_t$-measurable function which vanishes on $\{\sigma \le t\}$. If a is any number strictly exceeding t then

$$\hat{P}(F; \sigma > a) = \hat{P}\big(\psi(X_t, a - t); F\big).$$

We will differentiate this with respect to $\hat{P}(\sigma > a)$. The formal manipulations are easy to justify, and using the explicit form of ψ we obtain for any c exceeding t

$$(1.3) \qquad \hat{P}(F; c < \sigma) = \int_c^\infty \gamma_a(F)\hat{P}(\sigma \in da)$$

with

$$\gamma_a(F) = \hat{P}\left(F \cdot \sqrt{2}X_t e^{-X_t^2/2(a-t)}\left(\frac{a}{a-t}\right)^{3/2}\right).$$

Take $a = 1$ and F of the form $\Pi f_i(X_{t_i})$ with $0 < t_1 < \cdots < t_n < 1$ and the f_i bounded and vanishing at 0. From the general description of $\hat{P}$ we obtain

$$(1.4) \qquad \begin{aligned}\gamma_1(F) = \int \eta_{t_1}(dx_1)\prod_1^n f_i(x_i)\prod_2^n p^-(t_i - t_{i-1}, x_{i-1}, x_i)\cdot \\ \cdot \sqrt{2}x_n e^{-x_n^2/2(1-t_n)}\left(\frac{1}{1-t_n}\right)^{3/2}dx_1 \cdots dx_n.\end{aligned}$$

Upon putting in the explicit formula for η_t from (1.1) a tedious but straightforward calculation and comparison with the expression in Chapter II describing Brownian excursion shows that

$$\gamma_1(F) = \mu_1(F).$$

Use of the scaling properties of the components of (1.4) shows that γ_a is obtained from γ_1 by use of the change of variables $X_t \to \sqrt{a}X_{t/a}$, and it follows that $\gamma_a(F) = \mu_a(F)$ at least when applied to function F of this special sort. From this a monotone class argument yields the validity of Theorem (1.2).

One could state Theorem (1.2) by saying that if $e(\cdot)$ denotes Brownian excursion then $\hat{P}$ is the distribution of $\sqrt{\sigma}e(\cdot/\sigma)$ where σ is a random quantity independent of e and with (infinite measure) distribution $d\sigma/2\sqrt{\pi}\sigma^{3/2}$.

Itô and McKean [IM, 2] give the following excursion decomposition of Brownian motion. Let $\{X_t; t \geq 0\}$ be Brownian motion starting at 0 and consider as usual $Z = \{t|X_t = 0\}$ and the excursion intervals making up the complement of Z. Number these $\mathcal{Z}_1, \mathcal{Z}_2$, etc. in some measurable way that depends only on Z. (For example, $\mathcal{Z}_1$ could be the excursion interval containing the time point 1 or the left-most interval of length exceeding 1, but not the left-most interval in which the height of the excursion exceeds 1.) Define paths $e_1, e_2, \cdots$ on $0 \leq t \leq 1$ and random variables $e_1, e_2, \cdots$ by

$$e_n(t) = |X_{t|\mathcal{Z}_n|+\inf \mathcal{Z}_n}|/\sqrt{|\mathcal{Z}_n|} \qquad\qquad 0 \leq t \leq 1$$

$$e_n = \operatorname{sgn} X_t \qquad\qquad t \in \mathcal{Z}_n.$$

Then the processes $\{e_n(t); 0 \leq t \leq 1\}$ are equal in law to Brownian excursion and are mutually independent; the random variables $\{e_n; n \geq 1\}$ are mutually independent and satisfy $P(e_n = 1) = P(e_n = -1) = 1/2$; and $\{e_n; n \geq 1\}$, $\{e_n; n \geq 1\}$ and Z are independent.

We will give a proof of this. The idea is to note first that the validity of the assertion is independent of the way the Brownian motion is constructed, and then to construct it in such a way that the truth of the assertion is clear. Specifically we will construct Brownian excursion measure $\hat{P}$ in an appropriate way and then will hook together the paths of the corresponding Poisson point process to make up Brownian motion. The justification for the last step is Itô's synthesis theorem; so our argument will not be complete until this result is established in the next chapter.

First we need a more precise statement of the theorem so we can identify the way in which the excursions are to be labelled. Let $\{X_t; t \geq 0\}$ be Brownian motion, $Z = \{t|X_t = 0\}$, $\sigma = \inf\{t > 0|X_t = 0\}$. Let $\mathcal{I}$ denote the collection of maximal open intervals making up $[0, \infty) - Z$, and G denote the collection of left-hand end points of these intervals. Let L_{nk} denote the kth smallest among the left hand end points of those I in $\mathcal{I}$ such that $|I| > 1/n$ $(L_{nk} = G_k^{1/n}$ in the notation of Chapter III) and let $R_{nk} = L_{nk} + \sigma \circ \theta_{L_{nk}}$. Thus $\mathcal{I}$ consists of the intervals (L_{nk}, R_{nk}) and G consists of the points L_{nk} as n and k range over $1, 2, \cdots$. We will denote by $\mathcal{Z}$ the σ-algebra, $\sigma\{L_{nk}, R_{nk}; n \geq 1, k \geq 1\}$. Obviously any random variable which is defined in terms of Z only is $\mathcal{Z}$ measurable, and such random variables generate $\mathcal{Z}$. For a point $\ell(= \ell(\omega))$ in $G(= G(\omega))$ set $r = \ell + \sigma \circ \theta_\ell$ and let e^ℓ denote the excursion

$$e^\ell(t) = X_{\ell+t(r-\ell)}/\sqrt{r-\ell} \qquad\qquad 0 \leq t \leq 1,$$

and let e^ℓ denote the sign

$$e^\ell = 1 \qquad\qquad \text{if } e^\ell(t) > 0 \qquad\qquad 0 < t < 1$$

$$ = -1 \qquad\qquad\qquad < 0 \qquad\qquad 0 < t < 1.$$

(1.5) Theorem. *Let $L_1, L_2, \cdots, L_n$ be $\mathcal{Z}$ measurable random variables such that for each k $L_k(\omega) \in G(\omega)$ for all ω and such that $L_i(\omega) \neq L_j(\omega)$ for all ω and all $i \neq j$. Then for each i the law of e^{L_i} is Brownian excursion; $P(e^{L_i} = 1) = P(e^{L_i} = -1) = 1/2$; and the collection*

$$e^{L_1}, \cdots e^{L_n}, e^{L_1}, \cdots e^{L_n}, \mathcal{Z}$$

is independent.

Proof. Let $(\Omega, \mathcal{F}, P)$ be a complete probability space supporting the following objects:

(a) a process $\{R(t); t \geq 0\}$ which is a stable subordinator of index $1/2$ (that is the process has stationary independent non-negative increments, the paths are right continuous, with $R(0) = 0$ and $P(e^{-\lambda R(t)}) = e^{-tg(\lambda)}$ where

$$g(\lambda) = \int_0^\infty (1 - e^{-\lambda x}) dx / 2\sqrt{\pi} x^{3/2};$$

(b) an independent family $\{e_{nk}(t); 0 \leq t \leq 1\}_{n,k=1}^\infty$ of Brownian excursion processes;

(c) an independent family $\{e_{nk}\}_{n,k=1}^\infty$ of random variables such that $P(e_{nk} = 1) = P(e_{nk} = -1) = 1/2$ for all n, k; and

(d) these quantities are to be mutually independent.

Now let

$$D = \{t | R(t) - R(t-) > 0\}$$

and for t in D set

$$Y_t = R(t) - R(t-).$$

Of course all of these quantities depend on ω. The process $\{R'(s); s \geq 0\}$ defined by $R'(s) = R(s+t) - R(t)$ is equal in law to the original subordinator and is independent of $\sigma\{R(s); s \leq t\}$ and so obviously the process Y with domain points D and values Y_t for t in D is a Poisson point process with values in $[0, \infty)$. Let

$$D_n = \{t \in D | Y_t \in [1/n, 1/n - 1)\}$$

for $n = 1, 2, \cdots$ $(1/0 = \infty)$. The domain restriction Y^n of Y to D_n also is a Poisson point process and the processes $Y^1, Y^2, \cdots$ are mutually independent because their ranges are disjoint. The process Y^n is discrete:

let d_{nk} be the kth smallest point in D_n. Now make up for each n a point process U^n with values in $\mathcal{C}[0, \infty)$ by taking as domain points the set D_n and associating with d_{nk} the continuous function

$$\sqrt{Y_{d_{nk}}}\, e_{nk} e_{nk}(\cdot/Y_{d_{nk}}).$$

(We assume that e_{nk} has been extended from $[0, 1]$ to all of $[0, \infty)$ by setting $e_{nk}(t) = 0$ if $t \geq 1$.) Once again it is quite obvious that each U^n is a Poisson point process and the family $\{U^n; n \geq 1\}$ is independent. We will make up a Poisson point process U as the direct sum of the processes U^n; that is the set of domain points for U is $\underset{n}{\cup}D_n$, these being disjoint, and U_t is equal to U_t^n if t is in D_n. If η is the characteristic measure for U then $\eta = \overset{\infty}{\underset{1}{\sum}}\eta_n$ where η_n is the characteristic measure for U^n. Let A be a Borel subset of $\mathcal{C}[0, \infty)$. Then

$$(1.6) \qquad \eta_n(A) = P\big(\sqrt{Y_{d_{n1}}}\, e_{n1} e_{n1}(\cdot/Y_{d_{n1}}) \in A\big) \cdot E_n$$

where E_n is the expected number of points in D_n which lie in $[0, 1]$. We have

$$E_n = \int_{n^{-1}}^{(n-1)^{-1}} d\sigma/2\sqrt{\pi}\sigma^{3/2}.$$

The distribution of $Y_{d_{n1}}$ is given by

$$E_n^{-1}(1/2\sqrt{\pi}\sigma^{3/2})d\sigma$$

and so we have

$$\eta_n(A) = (1/2) \overset{2}{\underset{k=1}{\sum}} \int_0^\infty P\big((-1)^k \sqrt{\sigma}\, e_{n1}(\cdot/\sigma) \in A\big)d\sigma/2\sqrt{\pi}\sigma^{3/2}.$$

It follows from Theorem (1.2) that

$$\eta = \frac{1}{2}(\hat{P} + \hat{P}_-)$$

where $\hat{P}$ is the measure for excursions away from 0 of reflecting Brownian motion and $\hat{P}_-$ is the image of $\hat{P}$ under the mapping $f \to -f (f \in \mathcal{C}[0, \infty))$ which sends positive excursions to negative ones. We have noted already that this is the excursion measure for excursions away from 0 of Brownian motion.

We will piece together the paths U_t in the point process as follows: set

$$\tau(t) = \sum_{r \le t} \sigma(U_r)$$

where the sum is over those domain points r of U with $r \le t$. Of course $\tau(t)$ is the same as $R(t)$. Set

$$\varphi(t) = \inf\{r|\tau(r) > t\} = \inf\{r|\tau(r) \ge t\}.$$

To define our process X_t, suppose $\varphi(t) = s$. Then $\tau(s-) \le t \le \tau(s)$. If $\tau(s-)$ is strictly less than $\tau(s)$ then s is a domain point of Y and hence also of U. In this case we set

$$X_t = U_s\big(t - \tau(s-)\big).$$

If $\tau(s-) = \tau(s)$ or if $s = 0$ we set $X_t = 0$. We will take as already established the fact that $\{X_t; t \ge 0\}$ is indeed Brownian motion. (Of course X_t differs from 0 if and only if t is for some s strictly between $\tau(s-)$ and $\tau(s)$. These are the points of non-increase of φ which will be local time at 0 for X and so all this simply amounts to reversing the steps in the construction of a point process from the excursions.)

Now we can complete the proof of (1.5). First consider the case $n = 1$. Since L_1 is in G we have $\varphi(L_1)$ in D and hence $\varphi(L_1) = d_{nk}$ for some n and k (random, of course). Let A be a Borel set in $C[0, \infty)$, δ denote ± 1 and Γ be a set in $\mathcal{Z}$. Clearly the σ-algebra $\mathcal{Z}$ is contained in $\sigma\{R(t); t \ge 0\}$ and hence Γ is independent of anything involving the excursion or sign variables. We have, if $R_1 = L_1 + \sigma \circ \theta_{L_1}$

$$P(|X_{L_1 + (R_1 - L_1)}/\sqrt{R_1 - L_1}| \in A, e^{L_1} = \delta, \Gamma)$$

$$= \sum_{n,k} P\big(e_{nk} \in A, e_{nk} = \delta, \Gamma, L_1 = d_{nk}\big)$$

$$= P\big(e_{11} \in A\big) \frac{1}{2} \sum_{n,k} P(\Gamma, L_1 = d_{nk})$$

$$= P(e = \delta)P(e \in A)P(\Gamma)$$

as required. In considering n greater than 1 we will consider only $n = 2$ as this illustrates all the difficulties, except the typographical ones, of the general situation. We have

$$P(e^{L_i} \in A_i, e^{L_i} = \delta_i, \Gamma; i = 1, 2)$$

$$= \sum_{(n,k)\,(m,j)} P(\cdots; L_1 = d_{nk}, L_2 = d_{mj}).$$

The hypothesis that $L_1 \neq L_2$ implies that the only non-zero terms in this sum are those with (n, k) unequal to (m, j). From this, the mutual independence of the collection $\{e_{nk}, e_{mj}; n, k, m, j \geq 1\}$, and the fact that Γ, L_1, L_2 are $\sigma\{R(t); t \geq 0\}$ measurable it follows that each summand is equal to

$$\frac{1}{4} P(e_{11} \in A_1) P(e_{11} \in A_2) P(\Gamma, L_1 = d_{nk}, L_2 = d_{mj})$$

so summing on (n, k) and (m, j) the proof is complete.

2. Path decomposition.

Several other descriptions of excursion measure $\hat{P}$ for reflecting Brownian motion away from 0 have been obtained, notably by Itô and McKean [IM, 2] and by David Williams [W, 1], using schemes that involve piecing together of Bessel processes. The description given by Williams is quite interesting: take two processes $\{B_i(t); t \geq 0\}$ each with continuous paths and equal in law to a 3-dimensional Bessel process starting at 0. Suppose the two processes are independent. Pick a point x strictly positive and let W_t^x be the process

$$W_t^x = B_1(t) \qquad\qquad t \leq \sigma_x^1$$
$$= x - B_2(t - \sigma_x^1), \qquad t > \sigma_x^1$$

where σ_x^1 is the time the process $B_1(t)$ reaches the level x, and where we set $W_t^x = 0$ for all t greater than the first time W_t^x returns to 0. Let $\mathcal{P}^x$ be the measure induced on $\mathcal{C}[0, \infty)$ by this process. Then

$$\hat{P} = \int_0^\infty \mathcal{P}^x \, dx / x^2 \sqrt{2}.$$

In words, independently of the Bessel processes we pick a positive number x according to the "distribution" $dx / x^2 \sqrt{2}$, run $B_1(t)$ up until it reaches the level x and then run $B_2(t)$ down from that level. A picture of this is on the dust jacket of Williams' book.

Following Rogers [R, 1] with slight alterations we will give a proof of this description. It makes use of Williams' fascinating path decomposition of a diffusion process, (see Williams [W, 2]), which finds application in other parts of excursion theory as well.

First we will describe briefly what path decomposition involves: consider a diffusion process X_t with state space (A, B), $-\infty \leq A < B \leq \infty$. We assume the process is such that starting at a point x in (A, B) its infimum over all time, call it γ, is strictly larger than A and that with P^x probability 1 either $X_t = B$ for some finite t (which we regard as the lifetime of the process) or at least X_t approaches B as t tends to ∞. Examples are a 3-dimensional Bessel process with $A = 0$, $B = \infty$, Brownian motion with a positive drift (as in paragraph (g) of III-3) with $A = -\infty$, $B = +\infty$, or Brownian motion killed when it reaches 0, with $A = -\infty$, $B = 0$. In the first example if the Bessel process starts at x then γ has the uniform distribution on $(0, x)$; in the second if the Brownian motion with drift starts at 0 then γ has density $2e^{2x}$ on $(-\infty, 0)$; in the third if the Brownian motion starts at x less than 0 then γ has the density $|x|dy/y^2$ on $(-\infty, x)$. The path decomposition theorem says that if the diffusion process starts at a point x in (A, B) then there are two other diffusion processes X_t^1 and X_t^2 independent of each other and of γ such that the process described by

$$Y_t = X_t^1 \qquad\qquad t \le \rho (= \inf\{t > 0 | X_t^1 = \gamma\})$$
$$= X_{(t-\rho)}^2 + \gamma \qquad t > \rho$$

modified by being killed upon reaching B, is equal in law to the original diffusion started at x. The theorem includes a calculus for determining in terms of the analytic data for the original process, the analytic data that characterizes the component diffusions X^1 and X^2. In the first example X^1 is Brownian motion started at x and X^2 is the 3-dimensional Bessel process started at 0. In the third example X^1 is the negative of a 3-dimensional Bessel process started at the starting point x of the Brownian motion and X^2 is a 3-dimensional Bessel process started at 0. This is the result we will use to make our calculations. We refer the reader to Williams [W, 2] for more information on path decomposition.

Coming to the details of Williams's description of $\hat{P}$, let Ω_1 denote the set of continuous functions ω_1 from $[0, \infty)$ to $[0, \infty)$ such that $\omega_1(0) = 0$ and $0 = \inf\{t | \omega_1(t) > 0\}$. Let $\mathcal{F}^{01}$ be the σ-algebra generated by the coordinate functions and let P_1 denote the measure on $\mathcal{F}^{01}$ relative to which the coordinate functions form a 3-dimensional Bessel process starting at 0. Let $(\Omega_2, \mathcal{F}^{02}, P_2)$ be a copy of this measure space, and form the measurable space $\Omega = \Omega_1 \times \Omega_2 \times (0, \infty)$, $\mathcal{F}^0 = \mathcal{F}^{01} \times \mathcal{F}^{02} \times \mathcal{B}(0, \infty)$ and on $\mathcal{F}^0$ put the measure $\mathcal{P} = P_1 \times P_2 \times dx/\sqrt{2}x^2$. If $\omega = (\omega_1, \omega_2, r)$ set

$$x_t^1(\omega) = \omega_1(t)$$
$$x_t^2(\omega) = \omega_2(t)$$
$$\gamma(\omega) = r$$

and set

$$\sigma_\varepsilon(\omega) = \inf\{t | x_t^1(\omega) = \varepsilon\}$$
$$\rho(\omega) = \inf\{t | x_t^1(\omega) = \gamma(\omega)\}$$
$$\overline{\sigma}(\omega) = \inf\{t | x_t^2(\omega) = \gamma(\omega)\}.$$

Now define a process $\{X_t; t \geq 0\}$ over this measure space by

$$
\begin{aligned}
X_t(\omega) &= x_t^1(\omega) & t &\leq \rho(\omega) \\
&= \gamma(\omega) - x_{t-\rho(\omega)}^2(\omega) & \rho(\omega) &< t \leq \rho(\omega) + \overline{\sigma}(\omega) \\
&= 0 & t &> \rho(\omega) + \overline{\sigma}(\omega).
\end{aligned}
$$

The measure that the process X induces on $C[0,\infty)$ is obviously just exactly Williams' measure. A formal statement of Williams' description is:

(2.1) Theorem. *The distribution in $C[0,\infty)$ of X under $\mathcal{P}$ is equal to $\hat{P}$.*

Proof: A number of bookkeeping points need to be noted: first of all the various quantities we have defined are shown easily to be appropriately measurable; we will not give details. Secondly the paths of a 3-dimensional Bessel process leave the origin immediately, never to return, and tend to infinity with t. So our choice of Ω_1 as basic space is legitimate; and also σ_ε decreases to 0 with ε, and the quantities ρ and $\overline{\sigma}$ are almost surely finite. Now fix ε strictly positive, let $\{X_t^\varepsilon; t \geq 0\}$ be the process

$$
X_t^\varepsilon(\omega) = X_{t+\sigma_\varepsilon(\omega)}(\omega).
$$

Note that for each t and ω, $X_t^\varepsilon(\omega)$ approaches $X_t(\omega)$ as ε approaches 0. Let ω^ε be the point of Ω defined by

$$
\omega^\varepsilon = \big((\omega^\varepsilon)_1, \omega_2, r\big)
$$

where $\omega = (\omega_1, \omega_2, r)$ and

$$
(\omega^\varepsilon)_1(t) = \omega_1\big(t + \sigma_\varepsilon(\omega)\big).
$$

Let $\mathcal{P}_\varepsilon$ denote the restriction of $\mathcal{P}$ to the set $\{\gamma > \varepsilon\}$ so that $\varepsilon\sqrt{2}\mathcal{P}_\varepsilon$ is a probability measure, which we will denote P_ε. It is quite clear that relative to P_ε the quantities

$$
x_t^1(\omega^\varepsilon), x_t^2(\omega^\varepsilon), \gamma(\omega^\varepsilon)
$$

define respectively a Bessel process starting at ε, a Bessel process starting at 0 and a random quantity γ distributed like the maximum of a Brownian motion started at ε and killed upon reaching the origin; and the three are mutually independent. In other words these are just the ingredients in the path decomposition of our third example, except turned upside down to fit the present situation. Thus if we consider over Ω the process $\{Y_t^\varepsilon; t \geq 0\}$ defined by

$$Y_t^\varepsilon(\omega) = X_t(\omega^\varepsilon)$$

then under P_ε the Y^ε process is just Brownian motion started at ε and killed when it reaches 0. Let δ be strictly positive, ε less than δ and G^ε a function of the form

$$G^\varepsilon = g_1(Y_{t_1}^\varepsilon) \cdots g_n(Y_{t_n}^\varepsilon)$$

where the g's are bounded and continuous and vanish in $[0, \delta]$, and $0 < t_1 < \cdots < t_n$. Now the integrand vanishes unless $\gamma(\omega^\varepsilon)$ $(= \gamma(\omega))$ exceeds ε. But if $\gamma(\omega)$ exceeds ε then quite obviously $\rho(\omega^\varepsilon) = \rho(\omega) - \sigma_\varepsilon(\omega), \overline{\sigma}(\omega) = \overline{\sigma}(\omega^\varepsilon)$, and a glance at the definitions shows that for such an ω

$$X_t^\varepsilon(\omega) = X_t(\omega^\varepsilon)$$

for all t. Thus we have

$$\begin{aligned}
\mathcal{P}\big(\Pi g_i(X_{t_i}^\varepsilon)\big) &= \mathcal{P}_\varepsilon\big(\Pi g_i(X_{t_i}^\varepsilon)\big) \\
&= (1/\varepsilon\sqrt{2})P_\varepsilon\big(\Pi g_i(Y_{t_i}^\varepsilon)\big) \\
&= (1/\varepsilon\sqrt{2})P_0^\varepsilon\big(\Pi g_i(x_{t_i})\big)
\end{aligned}$$

where P_0^ε denotes probabilities on $\mathcal{C}[0, \infty)$ for Brownian motion started at ε and killed at the origin, and the x_t denote the coordinate functions. Now let ε approach 0. The last term in the display was shown in section 1 to approach $\hat{P}$ applied to the integrand. On the other side of the display the integrand approaches $\Pi g_i(X_{t_i})$. The integrands all are dominated by J, the indicator function of the set where X_{t_1} is strictly positive. We have

$$\begin{aligned}
\mathcal{P}(J) &= \mathcal{P}\big(\underline{\lim} I_{(\varepsilon, \infty)}(X_{t_1}^\varepsilon)\big) \\
&\leq \underline{\lim} \mathcal{P}\big(I_{(\varepsilon, \infty)}(X_{t_1}^\varepsilon)\big) \\
&\leq \underline{\lim}(1/\varepsilon\sqrt{2})P_0^\varepsilon(t_1 < \sigma),
\end{aligned}$$

and

$$\begin{aligned}
P_0^\varepsilon(t_1 < \sigma) &\leq 1/(1 - e^{-t_1})P_0^\varepsilon(1 - e^{-\sigma}) \\
&= (1 - e^{-\varepsilon\sqrt{2}})/(1 - e^{-t_1}).
\end{aligned}$$

So $\mathcal{P}(J)$ is finite and limit and integral may be interchanged. Thus we have

$$\mathcal{P}\big(\Pi g_i(X_{t_i})\big) = \hat{P}\big(\Pi g_i(x_{t_i})\big)$$

and now the validity of (2.1) is obvious.

3. The non-recurrent case.

We have discussed already the situation in which $P^x(\sigma < \infty) < 1$ for some values of x. Then the greatest domain point for the point process of excursions away from b is the one whose value is L_∞, and the corresponding path is the one excursion path for which $\sigma = \infty$. The point process can not be a Poisson point process; it is what Meyer [M, 2] calls an *absorbed* Poisson *point process*. As already noted, formulas such as III-3.26 remain valid but the interpretation of $\hat{P}$ as a "characteristic measure" is lost. For example if Δ is a set in $\mathcal{B}(0, \infty) \times \mathcal{U}$ and $N(\Delta)$ denotes as usual the number of domain points t for the point process Y of excursions such that (t, Y_t) is in Δ then when $\Delta = (0, t] \times B$, III-3.27 with Z_s the indicator of $\{s \le \beta_t\}$ and f the indicator of B yields

$$(3.1) \qquad E^b\left(N\big((0, t] \times B\big) \right) = \hat{P}(f) E^b L_{\beta_t}.$$

Now L_{β_t} equals t for $t < L_\infty$ and equals L_∞ otherwise and since L_∞ has the exponential distribution with rate λ the right side of (3.1) is

$$\hat{P}(f)(1 - e^{-\lambda t})/\lambda.$$

If we take $B = I_{\{\sigma = \infty\}}$ and let t increase to ∞ then the left side of (3.1) increases to 1 and the right side approaches $\lambda^{-1}\hat{P}(\sigma = \infty)$. Thus the restriction to $\{\sigma = \infty\}$ of $\hat{P}$, which we called $\hat{P}_1$ in Chapter III, has total mass λ^{-1}.

This situation is related to a second part of Williams' path decomposition of a diffusion process started at a point b in (A, B) in which the path started up from its minimum is decomposed into the part after the last time it visits the initial point b and the segment before then. We will explain the relationship.

First of all the left side of (3.1) uniquely characterizes the measure, $E\big(N(\Delta)\big)$, on $\mathcal{B}\big((0, \infty)\big) \times \mathcal{U}$ (which is, of course, the product of lebesgue measure and the characteristic measure when one is dealing with a genuine Poisson point process.) The fact of the matter is that even in the present situation this measure uniquely determines the probabilistic structure of the family $\{N(\Delta); \Delta \in \mathcal{B}(0, \infty) \times \mathcal{U}\}$. The proof of this is given by Meyer in [M, 2]. His technique is to make up a Poisson point process by hooking together independent copies of the point process at hand, to argue that the characteristic measure of this new process is determined by the quantities in (3.1), to use the fact, developed in III-2, that for a Poisson point process the characteristic measure determines the probabilistic structure, and then to note that the absorbed process is an initial segment of this extended process. We want to use this fact. Recall Chapter III Theorem (3.29) and the surrounding discussion: it says that relative to $\hat{P}_2$, the restriction of $\hat{P}$ to $\{\sigma < \infty\}$, the coordinate process is a time homogeneous Markov process on $t > 0$ and (3.29) describes the transition function $Q^2(t, x, A)$ in terms of analytic data from the original process. For example if the original process is the 3-dimensional Bessel process and b is strictly positive then the transition data for the coordinate process under $\hat{P}_2$ is on (b, ∞) that of a Brownian motion killed at σ_b and on $(0, b)$ that of a 3-dimensional Bessel process killed at σ_b. This follows from the fact that the function $\psi(x)$ from III-3.29 is identically 1 for $x < b$ and is equal to b/x for $x > b$ followed by some infinitesimal generator calculations as described in Williams [W, 2]. This is the set-up to which Itô's synthesis theorem applies; according to that theorem if we make up a Poisson point process W with characteristic measure $\hat{P}_2$ and then use the domain points and corresponding paths to construct a process $\{w_t; t \geq 0\}$ exactly as was done in section 1 of this chapter to construct Brownian motion we will obtain a time homogeneous Markov process recurrent at b and whose behavior away from b is described by the transition function $Q^2(t, x, A)$ from III-3.29. Admittedly our proof in Chapter V of the synthesis theorem will require smoothness hypotheses on Q^2. These are more than satisfied in the specific example at hand; and we are quite content to restrict our attention to this special case only. So let us suppose we have established that the process $\{w_t; t \geq 0\}$ is Markovian, so that with a suitable normalization for local time, W is its Poisson point process of excursions. Make up a new point process Z as follows: let S^λ denote a random variable independent of W and having an exponential distribution of rate λ and let $z = \{z_t; t \geq 0\}$ denote a Markov process independent of W and of S^λ whose law is given by $\lambda^{-1}\hat{P}_1$. (In fact we have established the Markovian properties only for $t > 0$ so we must restrict attention to special cases such as the Bessel process example, where inspection shows that $t = 0$ can be included.) Now set

$$t \in D_Z \qquad \text{and} \qquad Z_t = W_t \quad \text{if } t \in D_W \cap [0, S^\lambda)$$

$$S^\lambda \in D_Z \qquad \text{and} \qquad Z_{S^\lambda} = z$$

$$D_Z \cap (S^\lambda, \infty) = \emptyset.$$

Let $M(\Delta)$ denote the number of domain points t for the Z process such that (t, Z_t) is in Δ.

(3.2) Theorem. *For all Δ in $\mathcal{B}(0, \infty) \times \mathcal{U}$*

$$E^b\big(M(\Delta)\big) = E^b\big(N(\Delta)\big).$$

Proof. We need consider only $\Delta = (0, t] \times B$ with B in $\mathcal{U}$. We have

$$E^b\big(M(\Delta)\big) = E^b\big(M(\Delta); t < S^\lambda\big) + E^b\big(M(\Delta); t \geq S^\lambda\big).$$

For Δ as above the first integrand on the right involves only the W point process and hence that term is

$$t\hat{P}_2(B)e^{-\lambda t}.$$

The second term on the right is

$$E^b\big(M\big((0, S^\lambda) \times B\big); t \geq S^\lambda\big) + (1 - e^{-\lambda t})\lambda^{-1}\hat{P}_1(B).$$

Of these the first term again involves only W and it is

$$\int_0^t y\hat{P}_2(B)\lambda e^{-\lambda y}\, dy.$$

Upon calculating and putting these together we obtain

$$[(1 - e^{-\lambda t})/\lambda]\{\hat{P}_2(B) + \hat{P}_1(B)\}$$

which is what we obtained from (3.1).

If we start with the process X, take the corresponding point process of excursions and then reconnect them as in section 1 we obtain X back, of course. If we follow the process $\{w_t; t \geq 0\}$ but, when its local time reaches a level S^λ adjoin a single path from the independent process $\{z_t; t \geq 0\}$ then we obtain another process. The conclusion of (3.2) and Meyer's theorem is that these two processes are equal in law. Thus when Itô's synthesis theorem is applicable we obtain the following result:

(3.3) Theorem. *Take three mutually independent quantities, (a) a process $\{w_t; t \geq 0\}$ whose excursion law is $\hat{P}_2$, (b) a process $\{z_t; t \geq 0\}$ whose probability law is $\lambda^{-1}\hat{P}_1$, (c) a random variable S^λ, exponential of rate λ. Let $\tau = \inf\{t | L_t = S^\lambda\}$ where $\{L_t; t \geq 0\}$ is local time at b for the process w. Then the process $\{x_t; t \geq 0\}$ defined by*

$$
\begin{aligned}
x_t &= w_t & t &< \tau \\
&= z_{t-\tau} & t &\geq \tau
\end{aligned}
$$

is equal in law to $\{X_t; t \geq 0\}$.

Remark: This theorem is of course to be applied only in particular cases when the applicability of synthesis can be checked and where specific formulas are obtainable. In the Bessel process case, for example, the process $\{z_t; t \geq 0\}$ is $b + U_t$ where $\{U_t; t \geq 0\}$ is a 3-dimensional Bessel process started at 0.

Upon comparing this with Theorem 3.5 of Williams [W, 2] we see that this decomposition gives the third piece of his path decomposition.

We will leave to the interested reader the task of making a more thorough and rigorous development of this material.

4. Feller Brownian motions.

Describing the excursion measures for the Feller Brownian motions is only a simple exercise; but the decomposition into a part attributable to paths which leave the origin continuously and a part coming from paths which leave by a jump gives a good illustration of what happens also in the most general case.

Recall the material, including notation, from section 3 of Chapter II. We will let X be a Feller Brownian motion, $\hat{P}$ its excursion measure for excursions away from 0, $\hat{R}$ the excursion measure for reflecting Brownian motion, and P_0^x the measures for Brownian motion started at x and killed upon reaching 0, so that $P_0^\mu = \int P_0^x \mu(dx)$.

(4.1) Theorem.

$$\hat{P} = q\hat{R} + P_0^\eta$$

where q and η are the parameter and measure from II-3.9. The measure η is the restriction of $\hat{P}(X_0 \in \cdot)$ to $(0, \infty)$.

Proof. First note that from the normalization $p + q + \int (1 - e^{-x\sqrt{2}})\eta(dx) = 1$, the fact that the third summand is $P_0^\eta(1 - e^{-\sigma})$ and that $\hat{R}(1 - e^{-\sigma}) = 1$, it will follow that

$$\hat{P}(1 - e^{-\sigma}) = 1 - p.$$

For ε strictly positive let

$$\Gamma_\varepsilon = \{X_{\tau_\varepsilon} = \varepsilon, \tau_\varepsilon < \sigma\}$$
$$\Delta_\varepsilon = \{X_{\tau_\varepsilon} > \varepsilon, \tau_\varepsilon < \sigma\}$$

where τ_ε denotes the first time the excursion path reaches $[\varepsilon, \infty)$. All the excursion paths of a Feller Brownian motion are continuous even though the paths of the corresponding process may have jumps away from 0. This fact implies, quite clearly, that

$$\Delta_\varepsilon = \{X_0 > \varepsilon, \tau_\varepsilon < \sigma\}$$

so that as ε decreases to 0, Δ_ε increases to $\{X_0 > 0\}$. Let g be a positive bounded continuous function on $[0, \infty)$ and

$$f = \int_0^\sigma e^{-t} g(X_t)\, dt.$$

Then

$$(4.2) \qquad \hat{P}(f) = \lim_{\varepsilon \to 0} \left(\hat{P}\left(e^{-\tau_\varepsilon} \int_{\tau_\varepsilon}^{\sigma} e^{-t} g(X_t) dt; \Gamma_\varepsilon\right) + \hat{P}(\cdots; \Delta_\varepsilon) \right)$$

where the integrand in the second piece is the same as in the first. Applying the strong Markov property we can evaluate the first $\hat{P}$ integral on the right as

$$\left[(1/\varepsilon\sqrt{2}) P_0^\varepsilon \left(\int_0^\sigma e^{-t} g(X_t) dt \right) \right] \left(\varepsilon\sqrt{2} \hat{P}(\Gamma_\varepsilon) \right).$$

As ε approaches 0 the product in brackets converges to $\hat{R}(f)$ as we saw in section 1. Since the expressions are uniformly bounded we may assume that as ε approaches 0 through a subsequence $\varepsilon\sqrt{2}\hat{P}(\Gamma_\varepsilon)$ approaches a positive number β. The second term on the right of (4.2) can be written as

$$\hat{P}\left(P_0^{X_{\tau_\varepsilon}}(f); \Delta_\varepsilon \right)$$

once we supply a simple argument to deal with the fact that on Δ_ε, τ_ε is in fact equal to 0. This can be written also as

$$\hat{P}\left(V^1 g(X_0); X_0 > \varepsilon, \tau_\varepsilon < \sigma \right).$$

Letting ε approach 0 we have

$$\hat{P}(f) = \beta\hat{R}(f) + P_0^\mu(f)$$

where μ is the restriction to $(0, \infty)$ of the distribution under $\hat{P}$ of X_0. Except for the identification of β and μ this establishes (4.1) at least for integrands f of the special form at hand. We will leave to the reader the argument that this is all that is needed to establish the equality for all positive $\mathcal{F}^0$ measurable f. As for the parameters, if the function g in (4.2) vanishes at the origin then

$$E^0 \Sigma_{s \in G} e^{-s} f \circ \theta_s = U^1 g(0) = \hat{P}(f)$$

where U^1 is the potential operator for our Feller Brownian motion. By II-3.9, (replace f by g), and the uniqueness assertion in the theorem it follows that $\beta = q$ and $\mu = \eta$ as asserted.

5. Reflecting Brownian motion.

We have given two descriptions of reflecting Brownian motion, one as the absolute value of Brownian motion and another as Brownian motion with the current minimum subtracted off. Here, as an application of excursion theory, is a third description, which we will use in Chapter VI. Let $\{x_t; t \geq 0\}$ denote any Brownian motion process and assume $x_0 = 0$. Set

$$\theta(t) = \text{leb meas}\{s \leq t | x_s \geq 0\}$$
$$\rho(t) = \inf\{s | \theta(s) > t\}$$

and define a process $\{y_t; t \geq 0\}$ by

$$y_t = x_{\rho(t)}.$$

(5.1) Theorem. *The process $\{y_t; t \geq 0\}$ is reflecting Brownian motion starting at 0.*

Proof. The validity of the assertion does not depend on how the original Brownian motion was obtained. We will take it to be one constructed by hooking together excursions as in section 1. Of course this means that the complete justification requires Itô's synthesis theorem from Chapter V. Coming to the details, let $\hat{P}$ be the characteristic measure for excursions away from 0 of Brownian motion and let X, over a probability space $(\Omega, \mathcal{H}, P)$, be a Poisson point process with $\hat{P}$ as characteristic measure. Let D (depending on ω) denote the domain points of X, let $D^+(D^-)$ denote those points s of D such that the corresponding path $X_s(\cdot)$ is strictly positive (negative) in $(0, \sigma(X_s))$, so that D is a sum of D^+ and D^-. Set

$$\tau(r) = \Sigma\sigma(X_s),$$

the sum being over all s in D with s not exceeding r, and define τ^+ and τ^- similarly using D^+ and D^-. Let $\varphi(t) = \inf\{r | \tau(r) \geq t\} = \inf\{r | \tau(r) > t\}$, with φ^+ and τ^+ related similarly. We set

$$x_t = X_r\big(t - \tau(r-)\big)$$

if $r = \varphi(t)$ is a point of D and otherwise we set $x_t = 0$. We will assume Itô's synthesis theorem is at our disposal and so we may assume known that $\{x_t; t \geq 0\}$ is Brownian motion starting at 0. If we construct a process

$\{x_t^+; t \geq 0\}$ in exactly the same way but using τ^+, φ^+ and D^+ instead, then quite clearly we obtain reflecting Brownian motion because it is based on the restriction of $\hat{P}$ to the set of positive paths which, except for an unimportant normalization constant (see V-2.12), is the characteristic measure for reflecting Brownian motion. We will show that $y_t = x_t^+$ and this will provide the proof of (5.1). Doing this is little more than an exercise in tracking the definitions. Specifically, suppose $x_{\rho(t)}$ is strictly positive. Set $\rho(t) = u$. If $\varphi(u) = r$ then r is in D and in fact it must be in D^+ and

$$(5.2) \qquad\qquad \tau(r-) < u < \tau(r).$$

It is clear that

$$\theta(u) = u - \sum_{\substack{s \in D^- \\ s < r}} \sigma(X_s) = u - \tau^-(r-)$$

$$= u - \tau^-(r).$$

Since always $\theta\big(\rho(t)\big) = t$ we can substitute $\rho(t)$ for u and obtain

$$(5.3) \qquad\qquad t = \rho(t) - \tau^-(r).$$

Using (5.2) and $\tau - \tau^- = \tau^+$ we have

$$\tau^+(r-) < \rho(t) - \tau^-(r) < \tau^+(r)$$

and so

$$r = \varphi^+\big(\rho(t) - \tau^-(r)\big)$$

or

$$(5.4) \qquad\qquad \varphi^+(t) = \varphi\big(\rho(t)\big).$$

The set of t values for which we have established (5.4) is dense in $[0, \infty)$ and, since all the expressions are right continuous, (5.4) holds for all t. We have established also as an identity in t that

$$(5.5) \qquad\qquad t = \rho(t) - \tau^-\big(\varphi^+(t)\big).$$

Since $\rho(t)$ always is a point of right increase of θ, if $x_{\rho(t)}$ is not zero then $\varphi\big(\rho(t)\big) = r$ with r in D^+ and

$$x_{\rho(t)} = X_r\big(\rho(t) - \tau(r-)\big)$$

$$= X_r\big(t - \tau^+(r-)\big) = x_t^+.$$

In any other case $x_{\rho(t)}$ and x_t^+ are both zero, so the proof is complete.

There is a consequence of the argument we have just completed that is worth noting: let us change the notation surrounding (5.1) to θ^+, ρ^+ and y^+, and define

$$\theta^-(t) = \mathrm{leb}\{s \le t | x_s \le 0\}$$

with ρ^- and then y^- defined in terms of θ^-. The argument from (5.1) applies equally well to these quantities, and we conclude that the process $\{y_t^-; t \ge 0\}$ is the negative of a reflecting Brownian motion. It is customary to call the σ-algebras $\sigma\{y^+\}$ and $\sigma\{y^-\}$ the σ-algebras of *events depending only on the positive excursions* and *negative excursions* respectively. If X is the Poisson point process appearing in the proof of (5.1), the processes y^+ and y^- are defined in terms of "points" X_t with t in D^+ and in D^- respectively. The corresponding paths lie in disjoint sets (positive paths and negative paths) and so the basic independence property III-2.1 implies that $\sigma\{y^+\}$ and $\sigma\{y^-\}$ are independent. Sometimes this is phrased, rather loosely, in the following way:

(5.6) Theorem. *Excursions above and below a fixed level are independent.*

Note that the definition of $\sigma\{y^+\}$ and $\sigma\{y^-\}$ just involve the Brownian motion x so that the Poisson point process is only an auxiliary device for establishing the independence assertion.

Theorem (5.6) is not of much use. But a more refined statement along the same lines will be quite important to our applications of excursion theory. See Theorems (2.9) and (3.4) of Chapter VI.

V Itô's Synthesis Theorem

In this chapter we will prove Itô's synthesis theorem to the effect that under certain hypotheses the paths of a Poisson point process of excursions can be linked together to make up a Markov process.

1. Introduction.

We will assume that we are given a standard process $(\bar{X}, \bar{P}^x)_{x \in E}$ and a fixed point b in E, and that $\bar{X}$ is recurrent at b in the sense that

$$\bar{P}^x(\sigma < \infty) = 1, \qquad x \in E$$

where as usual $\sigma = \sigma_b = \inf\{t > 0 | \bar{X}_t = b\}$. Let us assume also that b is a *trap* relative to $\bar{X}$ so that

$$\bar{P}^x(\bar{X}_t = b \text{ for all } t \geq \sigma) = 1$$

for all x.

Let $\{P_t^0; t \geq 0\}$ denote the transition semigroup for $\bar{X}$ killed at time σ; that is

$$P^0(t, x, A) = \bar{P}^x(X_t \in A; t < \sigma).$$

The semigroup $\{P_t^0\}$ is our basic piece of data; the primary purpose of $\bar{X}$ is to provide a convenient way of saying that this semigroup is associated with a well-behaved process. Thus the behavior of $\bar{X}_t$ for $t > \sigma$ is not important.

The basic problem to which Itô's construction is relevant is this: find all the standard processes $(X, P^x)_{x \in E}$ which are *recurrent extensions* of $\bar{X}$ in the sense that

$$(1.1) \qquad P^x(X_t \in A, t < \sigma) = P^0(t, x, A)$$

for all $t \geq 0$, x in E and A in $\mathcal{E}$. We will include also the requirement that relative to X the point b is regular for $\{b\}$. One speaks of the original $\bar{X}$ killed at σ as being the *minimal process*. And so we are looking for all recurrent extensions of a given minimal process. Properly the state space for the P_t^0 semigroup is E with the point b removed, but that issue should not cause any confusion. The term "recurrent" means that $P^x(\sigma < \infty) = 1$ for all x; but that will follow automatically from the corresponding property for $\bar{X}$. A basic example is the problem of finding all the Feller Brownian motions; there the minimal process is Brownian motion on $[0,\infty)$ killed when it reaches the origin.

Half of the extension problem has been solved in Chapter III. There we associated with each such process an "excursion measure" which turned out to be the characteristic measure of the Poisson point process of excursions of X away from b. We obtained formulas like

$$U^1 g(b) = \hat{P}\Big(\int_0^{\sigma} e^{-t} g(X_t)\, dt\Big)$$

where U^1 denotes the 1-potential operator for the process X and g is a bounded $\mathcal{E}$ measurable function vanishing at b. Since $U^1 1(b) = 1$ this determines $U^1 g(b)$ for all bounded g. By the strong Markov property for X we have

$$(1.2) \qquad\qquad U^1 g(x) = V^1 g(x) + E^x e^{-\sigma} U^1 g(b)$$

for any x in E, where

$$V^1 g(x) = \int_0^{\infty} e^{-t} P_t^0 g(x)\, dx$$

is the potential operator for the minimal process. The first term on the right of (1.2) and the factor $E^x e^{-\sigma} = \overline{E}^x e^{-\sigma}$ are determined by the minimal process. Hence $\hat{P}$ determines $U^1 g(x)$ for all x and bounded g; and according to I-9 this determines the law of the X process.

Let X denote a recurrent extension and let $\hat{P}$ denote the excursion (characteristic) measure it induces on the σ-algebra $\mathcal{F}^0$ in the space U of right continuous left limit functions. We showed, or it follows easily from what we did show, that $\hat{P}$ has the following properties:

(i) σ is $\hat{P}$ measurable and $\hat{P}$ is carried on the set $\{0 < \sigma(u) < \infty, u(t) = b$ for all $t \geq \sigma(u)\}$,

(ii) $\hat{P}\big(u(0) \notin V\big) < \infty$ for every neighborhood V of b,

(iii) $\hat{P}(1 - e^{-\sigma}) \leq 1$,

(iv)
$$\hat{P}\big(g(u(t+s)); \Lambda \cap \{\sigma > s\}\big)$$
$$= \hat{P}\big(P_t^0 g(u(s)); \Lambda \cap \{\sigma > s\}\big)$$

for all $t \geq 0$, $s > 0$, Λ in $\mathcal{F}_s^0$ and bounded $\mathcal{E}$ measurable g vanishing at b,

(v)
$$\hat{P}\big(g(u(t)); u(0) \in B\big)$$
$$= \hat{P}\big(P_t^0 g(u(0)); u(0) \in B\big)$$

for g and t as in (iv) and $B \in \mathcal{E}$ such that $b \notin B$.

Of course by the usual iteration procedure, relationship (iv) implies the apparently more general

(1.3)
$$\hat{P}\big(F \circ \theta_s; \Lambda \cap \{\sigma > s\}\big)$$
$$\hat{P}\big(\bar{P}^{u(s)}(F); \Lambda \cap \{\sigma > s\}\big)$$

for F bounded and $\mathcal{F}^0$ measurable. Relationship (v) is the version of (iv) appropriate for the case $s = 0$ and can be extended in the same way. Together these conditions will be phrased as "$\hat{P}$ is compatible with the minimal semigroup $\{P_t^0; t \geq 0\}$." The existence of $\hat{P}$ and the fact that these conditions necessarily hold provides a solution to the first half of the extension problem. The second, and in some ways more interesting, half of the problem is to start with some necessary analytic data such as a characteristic measure compatible with the minimal semigroup, and then to show that there is indeed a recurrent extension whose law is determined uniquely by the minimal semigroup and given analytic data. And if the construction is made according to some probabilistic prescription such as hooking together the paths in a Poisson point process of excursions then so much the better. In the next section we will show that under some additional assumptions on the minimal semigroup this is indeed possible.

A few comments on hypotheses and other work on the subject are in order. The idea of using local time to parametrize the excursions and thus to give them an analytically interesting and usable structure came from Itô [I, 1]. In that paper he pointed out also the possibility of reversing the construction, so that a Markov process could be constructed from the paths of a suitable Poisson point process. Itô did not give a formal statement and proof indicating exactly when this could be done. In [B, 1] Blumenthal gave a proof (essentially what appears in the next section) that in certain cases Itô's program can indeed be carried out. Rogers [R, 2], inspired by Itô's program, gave a theorem on the existence of recurrent extensions which is more general than Blumenthal's; but his paper did not treat the actual probabilistic construction. Then in two demanding papers Salisbury [S, 1-2] gave the definitive treatment of the subject. He started with an excursion measure compatible with a minimal process, linked together the excursion paths according to Itô's prescription, and then analyzed the properties— measurability, Markovian character, etc. of the resulting object. One can view his papers as a study of the Markovian character of a compound process made up of components which themselves have Markov properties; so the methods are applicable in the many situations where such issues must be dealt with. He showed that the basic necessary conditions (i)-(v) need to be strengthened, primarily by the inclusion of a sort of (necessary) minimality condition on the given excursion measure; and that under these hypotheses Itô's program can indeed be carried out.

By contrast, the theorem we will give is quite simple. It uses smoothness hypotheses on the minimal semigroup to suppress the measure theoretic issues. Our hypotheses cover many cases of interest, so this approach has the good feature of developing something usable without requiring a reader new to the subject to deal with a formidable theorem. But even with a view towards applications Salisbury's work has the big advantage that his hypotheses are insensitive to transformations, while our smoothness hypotheses are not. The situation described in section 3 of Chapter IV is an illustration.

2. Construction.

First we will develop some preliminary facts.

(a) *Approximating processes.* We always can obtain recurrent extensions by the linking process described in paragraph (e) of II-1. Recall that given a probability measure γ carried on the Borel sets of $E - \{b\}$ and a strictly positive number β the prescription is that the process starting at b remains there for a length of time J having the exponential distribution with rate β, then jumps to a position X_J distributed according to γ and independent of J, then proceeds according to the rules for the minimal process until it reaches b from which position it starts anew.

Obviously the local time at $\{b\}$ for this process is the occupation time

$$c\ell\{r \le t | X_r = b\}$$

where ℓ denotes lebesgue measure and c is chosen so as to satisfy the normalizing requirement

$$E^b \int_0^\infty e^{-t} dL_t = 1.$$

The potential operators $\{U^\lambda\}$ satisfy

$$(2.1) \qquad U^\lambda g(x) = V^\lambda g(x) + \overline{E}^x e^{-\lambda \sigma} U^\lambda g(b).$$

To determine $U^\lambda g(b)$ when g is a positive or bounded $\mathcal{E}$ measurable function we write

$$U^\lambda g(b) = g(b) E^b \int_0^J e^{-\lambda t} dt + E^b e^{-\lambda J} E^{X_J} \int_0^\sigma e^{-\lambda t} g(X_t) dt$$
$$+ E^b e^{-\lambda(J + \sigma \circ \theta_J)} U^\lambda g(b).$$

The first summand is $g(b)$ times

$$\lambda^{-1} E^b (1 - e^{-\lambda J}) = (\lambda + \beta)^{-1}.$$

The second summand is

$$\beta/(\lambda + \beta)\langle \gamma, V^\lambda g \rangle$$

and the third is $U^\lambda g(b)$ times

$$\beta/(\lambda + \beta)\overline{E}^\gamma(e^{-\lambda \sigma}).$$

Collecting the coefficients of $U^\lambda g(b)$ on the left and dividing we obtain

$$(2.2) \qquad U^\lambda g(b) = \frac{g(b) + \beta\langle \gamma, V^\lambda g\rangle}{\lambda(1 + \beta\langle \gamma, V^\lambda 1\rangle)}.$$

When $\lambda = 1$ and g is the indicator of $\{b\}$ we obtain

$$E^b \int_0^\infty e^{-t} I_{\{b\}}(X_t)dt = (1 + \beta\langle \gamma, V^1 1\rangle)^{-1}$$

and so the proper multiplier for the occupation time is

$$c = 1 + \beta\langle \gamma, V^1 1\rangle.$$

If g vanishes at b then III-(3.27) with $f = \int_0^\sigma e^{-t} g(X_t)dt$ and $Z_s = e^{-s}$ says that

$$U^1 g(b) = \hat{P}(\int_0^\sigma e^{-t} g(X_t)dt).$$

By (2.2) this is the same as

$$k\overline{P}^\gamma(\int_0^\sigma e^{-t} g(X_t)dt)$$

where k is the constant

$$k = \beta/(1 + \beta\langle \gamma, V^1 1\rangle).$$

And so $\hat{P} = k\overline{P}^\gamma$. In particular when g is the indicator of $E - \{b\}$ we obtain

$$(2.3) \qquad \hat{P}(1 - e^{-\sigma}) = \beta\overline{E}^\gamma(1 - e^{-\sigma})/(1 + \beta\overline{E}^\gamma(1 - e^{-\sigma}))$$

which is less than 1.

The primary interest in such processes is due to their use in approximation schemes. It is worth noting once again that we know of no uncomplicated argument that the linking process actually yields a Markov process. We gave in Chapter II a reference to Meyer [M, 1], and of course the result also is a special case of Salisbury's work. Our acceptance of this fact as obvious accounts for some of the simplification in our approach.

(b) *Entrance laws.* A family $\{\eta_s; s > 0\}$ of measures on the Borel sets of $E - \{b\}$ is called an *entrance law* for the semigroup $\{P_t^0\}$ if

$$\eta_s P_t^0 = \eta_{t+s} \qquad\qquad s > 0, t \geq 0,$$

and

$$\langle \eta_s, V^1 1\rangle = \overline{E}^{\eta_s}(1 - e^{-\sigma}) \leq 1 \qquad\qquad s > 0.$$

Since $\overline{E}^x(1 - e^{-\sigma})$ is strictly positive for x not equal to b it follows that each η_s is σ-finite. For us the most important example of an entrance law is

$$\eta_s(A) = \hat{P}(X_s \in A; s < \sigma)$$

for an excursion measure $\hat{P}$ compatible with the minimal semigroup. For any strictly positive λ the ratio $(1-e^{-\sigma})/(1-e^{-\lambda\sigma})$ is bounded and bounded away from 0 and so $\overline{E}^{\eta_s}(1 - e^{-\lambda\sigma})$ is bounded over all s also.

Let r, s and t be strictly positive with $r < s$. Then

$$
\begin{aligned}
(2.4) \qquad \overline{P}^{\eta_r}(\sigma > t) &\geq \overline{P}^{\eta_r}(\sigma > t + s - r; \sigma > s - r)\\
&= \overline{P}^{\eta_r}(\sigma \circ \theta_{s-r} > t; \sigma > s - r)\\
&= \overline{P}^{\eta_r}(\overline{P}^{X_{s-r}}(\sigma > t); \sigma > s - r)\\
&= \overline{P}^{\eta_s}(\sigma > t).
\end{aligned}
$$

Thus $\overline{P}^{\eta_s}(\sigma > t)$ increases as s decreases. Also we have

$$
\begin{aligned}
(2.5) \qquad \overline{E}^{\eta_s}(1 - e^{-\lambda\sigma}) &= \overline{E}^{\eta_s}(\lambda \int_0^\infty e^{-\lambda t} I_{t<\sigma}\, dt)\\
&= \lambda \int_0^\infty e^{-\lambda t}\overline{P}^{\eta_s}(\sigma > t)\, dt,
\end{aligned}
$$

and so this increases also as s decreases. Since $\overline{E}^x(1 - e^{-\lambda\sigma})$ is equal to $\lambda V^\lambda 1(x)$ we have also

$$
\begin{aligned}
e^{-\lambda s}\overline{E}^{\eta_s}(1 - e^{-\lambda\sigma}) &= \lambda e^{-\lambda s} \int_0^\infty e^{-\lambda t}\eta_s P_t^0 1\, dt\\
&= \lambda \int_s^\infty e^{-\lambda r}\eta_r(1)\, dr.
\end{aligned}
$$

This shows that $\|\eta_r\|$ is finite for almost all r and hence for all r as it is monotonic.

The usefulness to us of holding and jumping processes as approximations comes from the following result.

(2.6) Theorem. *Let $\{\eta_s; s > 0\}$ be an entrance law and let $\{\alpha_s; s > 0\}$ be a set of strictly positive numbers such that $\lim\limits_{s \to 0}\alpha_s$ exists in $(0, \infty]$. Let*

U_ϵ^λ denote the resolvent operator for the recurrent extension with holding parameter $\alpha_\epsilon \|\eta_\epsilon\|$ and jumping in measure $\eta_\epsilon/\|\eta_\epsilon\|$. Then for each $\lambda > 0$

$$\lim_{\epsilon \to 0} U_\epsilon^\lambda g(b)$$

exists uniformly in all Borel measurable g on E with $0 \le g \le 1$.

Proof. According to (2.2) and some cancellation and division,

$$U_\epsilon^\lambda g(b) = \left(\frac{g(b)}{\alpha_\epsilon} + \langle \eta_\epsilon, V^\lambda g \rangle\right)\bigg/\left(\frac{\lambda}{\alpha_\epsilon} + \overline{E}^{\eta_\epsilon}(1 - e^{-\lambda\sigma})\right)$$

and so from what we have developed already we need consider only $\langle \eta_\epsilon, V^\lambda g \rangle$. We have, for any positive s and t

$$\langle \eta_s, V^\lambda g \rangle = \overline{E}^{\eta_s}\left(\int_0^\sigma e^{-\lambda r} g(X_r)dr; \sigma \le t\right)$$
$$+ \overline{E}^{\eta_s}\left(\int_0^t e^{-\lambda r} g(X_r)dr; t < \sigma\right) + \overline{E}^{\eta_s}\left(\int_t^\sigma e^{-\lambda r} g(X_r)dr; t < \sigma\right)$$
$$= I + II + III.$$

Obviously what we have labelled III is simply

$$e^{-\lambda t} < \eta_{s+t}, V^\lambda g \rangle$$

and so subtracting we get

$$e^{-\lambda t}|\langle \eta_s, V^\lambda g \rangle - \langle \eta_{s+t}, V^\lambda g \rangle|$$
$$\le (1 - e^{-\lambda t})\langle \eta_s, V^\lambda g \rangle + I + II.$$

In the first summand $\langle \eta_s, V^\lambda g \rangle \le \lambda^{-1}\overline{E}^{\eta_s}(1 - e^{-\lambda\sigma})$ which is bounded in s and hence its supremum over all s and all g with $0 \le g \le 1$ approaches 0 with t. For the other summands we have

$$I \le \lambda^{-1}\overline{E}^{\eta_s}(1 - e^{-\lambda\sigma}; \sigma \le t),$$

$$II \le \lambda^{-1}\overline{E}^{\eta_s}(1 - e^{-\lambda t}; t < \sigma).$$

To estimate $I + II$ write $\overline{E}^{\eta_s}(1 - e^{-\lambda\sigma})$ as the sum of an integral over $\sigma \le t$ and one over $\sigma > t$. The second of these is

$$\overline{E}^{\eta_s}(1 - e^{-\lambda t}; \sigma > t) + e^{-\lambda t}\overline{E}^{\eta_{s+t}}(1 - e^{-\lambda\sigma}),$$

and so $\lambda(I + II)$ does not exceed

$$\overline{E}^{\eta_s}(1 - e^{-\lambda\sigma}) - e^{-\lambda t}\overline{E}^{\eta_{s+t}}(1 - e^{-\lambda\sigma})$$

which approaches 0 as s and t approach 0. So the proof is complete.

Of course because of the validity of (2.1) with U_ε^λ in place of U^λ it follows that as $\varepsilon \to 0$ $U_\varepsilon^\lambda g(x)$ approaches a limit and the approach is uniform over all x in E and all $\mathcal{E}$ measurable g with $0 \le g \le 1$.

(c) *Resolvents*. We need to establish properties of the transformations $U^\lambda = \lim_{\varepsilon \to 0} U_\varepsilon^\lambda$ obtained in (2.6); but first we must introduce and derive consequences of hypotheses on the minimal process and semigroup which will be in force throughout the rest of this section. Let C_0^b denote the set of continuous functions which vanish at ∞ in the metric space $E - \{b\}$. Thus a real valued function g on $E - \{b\}$ is in C_0^b if it is continuous and if for each strictly positive ε there is a compact subset K of $E - \{b\}$ such that $|g(x)| < \varepsilon$ for every x not in K. We will assume the following:

(a) if $g \in C_0^b$ then $P_t^0 g \in C_0^b$ and $P_t^0 g \to g$ uniformly as $t \to 0$.

(b) $\overline{E}^x e^{-\lambda\sigma}$ is continuous in x for each λ. $\overline{E}^x e^{-\sigma} \to 1$ as $x \to b$. For each $\varepsilon > 0$ there is a compact subset K of E such that $\overline{E}^x e^{-\sigma} < \varepsilon$ for all x not in K.

These hypotheses are satisfied in most of the familiar situations, which, for our purposes, is satisfactory. Later on we will consider examples in which they are violated.

Conditions (a) and (b) have a few simple consequences which will be useful later on. They are:

(c) the function $x \to \overline{E}^x(1 - e^{-\sigma})$ is bounded away from 0 on the complement of any neighborhood of b.

(d) if K is a compact subset of E then for any positive number r, $\overline{P}^x(\sigma_K \le r) \to 0$ as $x \to \infty$ in E.

(e) if $\{\eta_s; s > 0\}$ is an entrance law for $\{P_t^0\}$ there is a σ-finite measure η on the Borel sets of $E - \{b\}$ such that

$$\eta_s(\cdot) = \theta_s(\cdot) + \overline{P}^\eta(X_s \in \cdot, s < \sigma)$$

where $\{\theta_s; s > 0\}$ is an entrance law for $\{P_t^0\}$ with the additional property that

$$(2.7) \qquad\qquad \theta_s(V^c) \to 0 \qquad\qquad s \to 0$$

for every neighborhood V of b. The validity of (c) follows from the fact that $\overline{E}^x(1 - e^{-\sigma})$ is continuous in x, strictly positive unless $x = b$ and is near 1 outside of large compact subsets of E. For (d), given the compact set K pick ε strictly positive and such that $\overline{E}^x e^{-\sigma} \geq \varepsilon$ for all x in K. Then

$$\overline{E}^x e^{-\sigma} \geq \overline{E}^x(e^{-\sigma_K + \sigma \cdot \theta_{\sigma_K}}; \sigma_K \leq r)$$
$$\geq \varepsilon e^{-r}\overline{P}^x(\sigma_K \leq r),$$

and by condition (b) the left side approaches 0 as x leaves large compact subsets of E. As to (e), property (c) and the fact that $\overline{E}^{\eta_s}(1 - e^{-\sigma})$ remains bounded implies that $\eta_s(V^c)$ is bounded in s for any neighborhood V of b. Thus by letting s approach 0 through a subsequence $\{s_k\}$ we obtain a measure η as a limit on $E - \{b\}$ of the measures η_{s_k} in the sense that

$$\lim_{k \to \infty} \langle \eta_{s_k}, g \rangle = \langle \eta, g \rangle$$

for all continuous g vanishing off a compact subset of $E - \{b\}$. Let g be a positive function in C_0^b and h be positive continuous with compact support in $E - \{b\}$ and $0 \leq h \leq 1$. Then $(P_t^0 g)h$ has compact support in $E - \{b\}$ and

$$\langle \eta, (P_t^0 g)h \rangle \leq \liminf_k \langle \eta_{s_k}, P_t^0 g \rangle = \liminf_k \langle \eta_{s_k + t}, g \rangle$$
$$= \liminf_k \langle \eta_t, P_{s_k}^0 g \rangle = \langle \eta_t, g \rangle,$$

and replacing the fixed h with a sequence of such function increasing to 1 on $E - \{b\}$ we have $\langle \eta P_t^0, g \rangle \leq \langle \eta_t, g \rangle$. Thus $\eta P_t^0 \leq \eta_t$. The family defined by

$$\theta_t = \eta_t - \eta P_t^0$$

obviously is an entrance law. The reader should have no trouble showing that the property in (2.7) holds, and that η is determined uniquely by the requirements of (e). Finally we note that if the entrance law is such that the total mass of η_s remains bounded as s approaches 0 then the measures θ_s are all zero. To see this take a positive function g in C_0^b and a sequence h_n of continuous functions with compact support increasing to 1 on $E - \{b\}$ so that $h_n P_t^0 g$ approaches $P_t^0 g$ uniformly. Then

$$\langle \eta, P_t^0 g \rangle = \lim_n \lim_k \langle \eta_{s_k}, h_n P_t^0 g \rangle$$
$$= \lim_k \lim_n \langle \eta_{s_k}, h_n P_t^0 g \rangle$$
$$= \lim_k \langle \eta_t, P_{s_k}^0 g \rangle = \langle \eta_t, g \rangle$$

where the interchange of limits is justified by the fact that the masses of the η_s are uniformly bounded. Thus $\eta P_t^0 = \eta_t$.

Now let us return to (2.6) and the family of operators $\{U^\lambda; \lambda > 0\}$ on bounded $\mathcal{E}$ measurable functions defined by the limit

$$U^\lambda g(x) = \lim_{\varepsilon \to 0} U_\varepsilon^\lambda g(x)$$

whose existence was established in Theorem (2.6) and the sentence following its proof. Because of the uniformity over all g with $0 \le g \le 1$ of the limit the family satisfies the resolvent equation as each of the approximating families $\{U_\varepsilon^\lambda; \lambda > 0\}$ does. For our next theorem we will make, as we must, the assumption that if $\sup\{\|\eta_s\|; s > 0\}$ is finite then $\lim_{s \to 0} \alpha_s$ is finite also. The notation $\mathcal{C}_0$ will stand for the space of continuous real valued functions on E vanishing at ∞ and given the supremum norm.

(2.8) Theorem. $\{U^\lambda; \lambda > 0\}$ *is the resolvent of a strongly continuous probability semigroup on* $\mathcal{C}_0$.

Proof. According to the Hille-Yosida theorem it is enough to prove that each U^λ maps $\mathcal{C}_0$ into itself and that for each g in $\mathcal{C}_0$, $\lambda U^\lambda g$ approaches g uniformly as $\lambda \to \infty$. By standard Banach space reasoning, for this last requirement it is enough to verify that $\lambda U^\lambda g$ approaches g pointwise. Now hypothesis (a) at the beginning of this section implies that V^λ maps $\mathcal{C}_0$ to $\mathcal{C}_0$. And then (2.1) and the hypothesis that $\overline{E}^x e^{-\lambda\sigma}$ is in $\mathcal{C}_0$ implies that U^λ maps $\mathcal{C}_0$ to itself. Now suppose g is a bounded continuous function on E and $x \ne b$. Then

$$\lambda V^\lambda g(x) = \overline{E}^x \int_0^{\lambda\sigma} e^{-t} g(X_{t/\lambda}) dt$$

which clearly approaches $g(x)$ as $\lambda \to \infty$. For the second term in the expression defining $\lambda U^\lambda g(x)$ we note that $|\lambda U^\lambda g(x)|$ is less than $\|g\|$ since obviously this holds with U^λ replaced by any U_ε^λ. The factor $\overline{E}^x e^{-\lambda\sigma}$ approaches 0 as $\lambda \to \infty$ since $\overline{P}^x(\sigma = 0) = 0$. This checks the pointwise approach on $E - \{b\}$. If g has the constant value k then $\lambda U^\lambda g(b) = k$ for all λ since clearly each U_ε^λ has this property. Thus we can replace g with $g - g(b)$ and restrict our attention to a bounded continuous function which vanishes at b. Assuming g has this additional property and given $\varepsilon > 0$ write $g = u + v$ where $\|v\| \le \varepsilon$ and u vanishes on a neighborhood V of b.

Then $|\lambda U^\lambda v(b)| \leq \varepsilon$ so clearly all we must show is that $\lambda U^\lambda u(b) \to 0$ as $\lambda \to \infty$. Assume first that $\|\eta_\varepsilon\|$ remains bounded as $\varepsilon \to 0$ so that α, the limit of the α_ε, is required to be finite. Then

$$(2.9) \qquad \lambda U_\varepsilon^\lambda u(b) = \frac{\langle \eta_\varepsilon, \lambda V^\lambda u \rangle}{\lambda/\alpha_\varepsilon + \overline{E}^{\eta_\varepsilon}(1 - e^{-\lambda\sigma})}.$$

Assuming $0 \leq u \leq 1$ as we may, the right side is less than

$$k\|u\|/(\lambda/\alpha_\varepsilon + \overline{E}^{\eta_\varepsilon}(1 - e^{-\sigma}))$$

if $\lambda \geq 1$, with k denoting an upper bound for $\{\|\eta_\varepsilon\|; \varepsilon > 0\}$. The second term in the denominator increases as ε decreases and clearly we get

$$\lambda U^\lambda u(b) \leq k\|u\|/(\alpha^{-1}\lambda + \overline{E}^\eta(1 - e^{-\sigma}))$$

so $\lambda U^\lambda u(b)$ approaches 0 as $\lambda \to \infty$. So let us restrict attention to the case where $\|\eta_\varepsilon\|$ increases to ∞ as ε approaches 0. Focusing our attention on the numerator in (2.9) we have

$$\begin{aligned}
\langle \eta_\varepsilon, \lambda V^\lambda u \rangle &= \lambda \overline{E}^{\eta_\varepsilon} \int_0^\sigma e^{-\lambda t} u(X_t)\,dt \\
&= \lambda \overline{E}^{\eta_\varepsilon} \left(\int_T^\sigma e^{-\lambda t} u(X_t)\,dt; T < \sigma \right) \\
&= \overline{E}^{\eta_\varepsilon} \left(e^{-\lambda T} \lambda V^\lambda u(X_T); T < \sigma \right)
\end{aligned}$$

where T is the time of hitting $\overline{V}^c$, u vanishing on $\overline{V}$. This last expression is no greater than $\|u\|\overline{P}^{\eta_\varepsilon}(T < \sigma)$ and we assert that this is bounded over all ε. Indeed

$$\begin{aligned}
\overline{E}^{\eta_\varepsilon}(1 - e^{-\sigma}) &\geq \overline{E}^{\eta_\varepsilon}(1 - e^{-\sigma}; T < \sigma) \\
&\geq \overline{E}^{\eta_\varepsilon}(1 - e^{-\sigma\circ\theta_T}; T < \sigma) \\
&\geq \inf_{x \notin V} \overline{E}^x(1 - e^{-\sigma}) \cdot \overline{P}^{\eta_\varepsilon}(T < \sigma)
\end{aligned}$$

and the left side is bounded in ε while the first factor on the right is strictly positive. Thus it will be enough to focus on the denominator in (2.9) and in fact to show that for every $L < \infty$ there are numbers $\varepsilon_0 > 0, \lambda_0 < \infty$ such that

$$\overline{E}^{\eta_\varepsilon}(1 - e^{-\lambda\sigma}) > L$$

for all $\varepsilon \leq \varepsilon_0$ and $\lambda \geq \lambda_0$. To do this simply pick ε_0 so that $\|\eta_{\varepsilon_0}\| > L$. Since $\overline{P}^{\eta_{\varepsilon_0}}(\sigma = 0)$ is zero and, as $\lambda \to 0$, $1 - e^{-\lambda\sigma}$ increases to 1 on $\{\sigma > 0\}$ we can pick λ_0 so large that $\overline{P}^{\eta_{\varepsilon_0}}(1 - e^{-\lambda_0\sigma}) > L$. Obviously this expression increases as we increase λ, and we observed earlier that also it increases as we decrease ε. Thus the proof is complete.

The Hille-Yosida theorem guarantees then the existence of a transition function

$$P(t, x, A)$$

such that the operators P_t map C_0 to itself and such that for each g in C_0

$$P_t g \to g$$

uniformly as $t \to 0$, and with

$$U^\lambda g(x) = \int_0^\infty e^{-\lambda t} P_t g(x) dt.$$

The conditions on the semigroup guarantee that it can be realized as the semigroup of a standard process. In the next paragraph we will construct that process by Itô's method.

(d) *Markov processes.* Let $\hat{P}$ be a measure on the σ-algebra $\mathcal{F}^0$ in the space U such that conditions (i), (iii) and (1.3) of section 1 are satisfied. Also we will assume throughout this section that the minimal semigroup $\{P_t^0; t \geq 0\}$ satisfies conditions (a) and (b) in the previous paragraph (c). Define m, the *delay coefficient*, by

$$m = 1 - \hat{P}(1 - e^{-\sigma})$$

and if $\|\hat{P}\|$ is finite assume that the additional condition, $\hat{P}(1 - e^{-\sigma}) < 1$, is satisfied, that is $m > 0$. Let $\{\eta_s; s > 0\}$ denote the entrance law associated with $\hat{P}$; that is $\eta_s(A) = \hat{P}(X_s \in A, s < \sigma)$.

Consider the Poisson point process having $\hat{P}$ as characteristic measure. We will denote this process by Y, and will use Y_s to denote the path associated with the domain point s. Thus Y_s is a path, $Y_s(t)$. Let

$$\sigma(Y_s) = \inf\{t > 0 | Y_s(t) = b\}.$$

We will let P and E stand for probability and expectation over the probability space Ω on which Y is defined and will, insofar as possible, suppress any notation for points in Ω. Set

$$\tau(t) = mt + \sum_{r \leq t} \sigma(Y_r)$$

where the notation means that the sum is over those domain points r of Y with $r \leq t$. From our earlier discussion of Poisson point processes it follows that the stochastic process $\{\tau(t); t \geq 0\}$ has stationary independent non-negative increments and that its paths are right continuous. Thus it is a subordinator. If γ is its Lévy measure then $\|\eta_s\| = \hat{P}(\sigma > s) = \gamma(s, \infty)$. In fact the paths of $\tau(t)$ are strictly increasing. This is certainly true if m is strictly positive. Otherwise the Lévy measure, whose total mass is that of $\hat{P}$, has infinite mass and so the corresponding process has jumps in every time interval. Let φ denote the right continuous inverse of τ,

$$\varphi(t) = \inf\{r | \tau(r) \geq t\} = \inf\{r | \tau(r) > t\}.$$

We will define a process $\{x_t; t \geq 0\}$ by linking together the paths $Y_r, r \in D_Y$, as follows: Let $\varphi(t) = s$. Then $\tau(s-) \leq t \leq \tau(s)$. If $\tau(s-) < \tau(s)$ then s is a domain point for Y. In that case we set

$$x_t = Y_s\big(t - \tau(s-)\big).$$

If $\tau(s-) = \tau(s)$ or if $s = 0$ set

$$x_t = b.$$

(2.10) Theorem. $\{x_t; t \geq 0\}$ *is a recurrent extension of* $\overline{X}$*. Its excursion measure is equal to* $\hat{P}$.

Proof. Let us first consider the case where $\hat{P}$ is a finite measure. Then in the decomposition, $\eta_s = \eta P_s^0 + \theta_s$, from paragraph (c) the second part is absent and quite clearly

$$\hat{P} = \overline{P}^\eta.$$

In the associated Poisson point process the first domain point has an exponential distribution with rate $\|\hat{P}\|$; and the corresponding path is distributed as that of the minimal process with initial distribution $\eta/\|\eta\|$.

Obviously then the x_t process is just one of holding and jumping type with rate $\|\eta\|/m$ and jumping in measure $\eta/\|\eta\|$. We worked out the characteristic measure for such a process in paragraph (a) of this section. Substituting in $\|\eta\|/m$ for β and $\eta/\|\eta\|$ for γ we obtain $\overline{P}^\eta$ as required.

In the rest of the proof we will assume that $\hat{P}$ is an infinite measure.

For $\varepsilon > 0$ we will define a process $\{x_t^\varepsilon; t > 0\}$ exactly as x_t was defined except using only those domain points r of the Y process such that $\sigma(Y_r) > \varepsilon$, using at such an r the path $Y_r^\varepsilon(t) = Y_r(t + \varepsilon)$, and replacing the delay coefficient m by $m + \delta_\varepsilon$ with $\delta_\varepsilon > 0$ and $\delta_\varepsilon = o(\varepsilon)$. Thus

$$\tau_\varepsilon(t) = (m + \delta_\varepsilon)t + \sum_{r \leq t, \sigma(Y_r) > \varepsilon} \big(\sigma(Y_r) - \varepsilon\big).$$

Let φ_ε be the function inverse to τ_ε and if $\varphi_\varepsilon(t) = s$ and $\tau_\varepsilon(s-) < \tau_\varepsilon(s)$ then

$$x_t^\varepsilon = Y_s\big(t + \varepsilon - \tau_\varepsilon(s-)\big)$$

and $x_t^\varepsilon = b$ in the remaining cases. It is quite clear that $\{x_t^\varepsilon; t \geq 0\}$ is a holding and jumping process with holding parameter and jumping in measure given by

$$\|\eta_\varepsilon\|/(m + \delta_\varepsilon), \qquad \eta_\varepsilon/\|\eta_\varepsilon\|$$

respectively. The considerations of Theorem 2.6 apply; the resolvent of the x_t^ε process converges to a limiting resolvent U^λ associated with a standard process. We will show next that there is a sequence $\varepsilon_n \to 0$ such that for each fixed t, $x_t^{\varepsilon_n} \to x_t$ with probability one. It will follow immediately that x_t is a Markov process with U^λ as resolvent. Then we will argue that its paths are right continuous and so it will be a standard process.

Coming to this, let $J(\varepsilon, t)$ denote the sum $\Sigma \sigma(Y_r)$ where the sum is over all $r \leq t$ such that $\sigma(Y_r) \leq \varepsilon$; let $N(\varepsilon, t)$ denote the number of domain points $r \leq t$ such that $\sigma(Y_r) > \varepsilon$ and let $E(\varepsilon, t)$ be equal to $J(\varepsilon, t) + \varepsilon N(\varepsilon, t)$. Then

$$\tau_\varepsilon(t) = \delta_\varepsilon t + \tau(t) - E(\varepsilon, t).$$

The term $E(\varepsilon, t)$ increases in t for fixed ε, but for t fixed it converges in probability to 0 as ε approaches 0. Let $\{\varepsilon_n\}$ denote a fixed sequence decreasing to 0 and having the property that with probability one $E(\varepsilon_n, t) \to 0$ for each t. Note that we have the uniform estimate

$$\sup_{s \leq t} |\tau_\varepsilon(s) - \tau(s)| \leq \max\big(E(\varepsilon, t), \delta_\varepsilon t\big).$$

It follows that τ_{ε_n} approaches τ uniformly on compacts. Also $\tau_{\varepsilon_n}(r) - \delta_{\varepsilon_n} r \leq \tau(r)$ and so

$$t + \varepsilon_n - \tau_{\varepsilon_n}(s-) \geq t - \tau(s-)$$

as soon as ε_n exceeds $\delta_{\varepsilon_n} s$. And finally

$$\tau_\varepsilon(t) - \tau(t) \leq \varepsilon\big((\delta_\varepsilon/\varepsilon)t - N(\varepsilon, t)\big),$$

with the term in parentheses approaching $-\infty$ as $\varepsilon \to 0$; and so $\tau_{\varepsilon_n}(t) < \tau(t)$ for all large n. Now let t be fixed and suppose $s = \varphi(t)$, so that s is random. There are four possibilities: (1) $\tau(s-) < t < \tau(s)$, (2) $\tau(s-) = t < \tau(s)$, (3) $\tau(s-) = t = \tau(s)$, (4) $\tau(s-) < t = \tau(s)$. We will suppose in the following that $\varepsilon \to 0$ through the sequence ε_n. Case (1): $x_t = Y_s(t - \tau(s-))$ and for small ε, $\tau_\varepsilon(s-) < t < \tau_\varepsilon(s)$ so $x_t^\varepsilon = Y_s(t + \varepsilon - \tau_\varepsilon(s-))$. Since $t + \varepsilon - \tau_\varepsilon(s-)$ approaches $t - \tau(s-)$ ultimately from above, the right continuity of the path $Y_s(\cdot)$ will imply that $x_t^\varepsilon \to x_t$. Case (2): we noted that for small ε $\tau_\varepsilon(q) - \tau(q)$ is strictly negative for $q > 0$, and a glance at the argument shows that this holds uniformly in q restricted to compact subsets of $(0, \infty)$. Thus $\tau_\varepsilon(s-) < t < \tau_\varepsilon(s)$ for small ε. For such an ε we have $x_t^\varepsilon = Y_s(t - \tau_\varepsilon(s-) + \varepsilon)$ and of course $x_t = Y_s(0)$. The fact that $t - \tau_\varepsilon(s-) + \varepsilon$ approaches 0 from above together with the right continuity of the path again gives the approach of x_t^ε to x_t. Cases (3) and (4): in case (3) $x_t = b$ by definition and in case (4) $t - \tau(s-) = \sigma(Y_s)$ and so $x_t = b$ in either case. Now $\varphi(t)$ is a stopping time for the filtration $\{\sigma(\alpha_t Y)\}_t$ and in either of the two cases we have $\tau(\varphi(t)) = t$. If $s = \varphi(t)$ and $\tau_\varepsilon(s) < \tau(s) = t$ then $\varphi_\varepsilon(t) > s$ and so either $x_t^\varepsilon = b$ or the path Y_r used in defining x_t^ε belongs to a domain point r with $r > \varphi(t)$. More specifically for any $\delta > 0$ we have $\tau(s + \delta) > t$ and so $\tau_\varepsilon(s + \delta) > t$ for small ε, that is $\varphi_\varepsilon(t) \leq s + \delta$. If the inequalities $\tau_\varepsilon(s) < t$ and $\varphi_\varepsilon(t) \leq s + \delta$ both hold then either $x_t^\varepsilon = b$ or x_t^ε is defined using a path Y_r corresponding to a domain point r with $\varphi(t) < r < \varphi(t) + \delta$. In cases (3) and (4) both of these inequalities hold for small ε. Let V be a neighborhood of b and let $\Lambda(\delta, V)$ denote the event that $Y_r(q)$ is not in V for some domain point r in $(\varphi(t), \varphi(t) + \delta)$ and some $q > 0$. To show that $x_t^\varepsilon \to x_t$ in cases (3) and (4) it will be enough to show that for each V, $P(\Lambda(\delta, V)) \to 0$ as $\delta \to 0$. The point process $\theta_{\varphi(t)} Y$ is a probabilistic replica of Y itself, so in proving this we may replace $\varphi(t)$ by 0. We will separate out the estimate as a lemma.

(2.11) Lemma. *If V is a neighborhood of b then $\lim\limits_{\delta \to 0} P\big(\wedge(\delta, V)\big) = 0$.*

Proof. Pick a number $\alpha > 0$ such that $\overline{P}^x(\sigma > \alpha) > \alpha$ for all x not in V. Since $\overline{E}^x(1 - e^{-\sigma})$ is bounded away from 0 on V^c this is possible. Let $\wedge(\delta, V, \varepsilon)$ denote the event that there is a domain point $r < \delta$ with $\sigma(Y_r) > \varepsilon$ and $Y_r(q) \notin V$ for some $q > \varepsilon$. This is of course just the event $\wedge(\delta, V)$ but defined using the point process Y^ε. As ε decreases to 0 $\wedge(\delta, V, \varepsilon)$ increases to $\wedge(\delta, V)$. If there is a strictly positive θ such that $P\big(\wedge(\delta, V)\big) > \theta$ for all $\delta > 0$ then for each such δ there is a strictly positive ε with $P\big(\wedge(\delta, V, \varepsilon)\big) > \theta$. With such a δ and ε fixed let $T_1, T_2, \cdots$ be the successive domain points r of Y with $\sigma(Y_r) > \varepsilon$ and let $y_t^i = Y_{T_i}(t + \varepsilon)$. The law of y^i is simply that of a Markov process with probabilities $\overline{P}^x$ and initial distribution η_ε. Set

$$\wedge_1 = \{T_1 < \delta, y_t^1 \notin V \text{ some } t\}$$
$$\wedge_{n+1} = \{T_{n+1} < \delta, y_t^{n+1} \notin V, \text{ some } t; y_t^i \in V \text{ all } t, \text{ all } i \leq n\}$$

The $\wedge_n$ are disjoint and their union is $\wedge(\delta, V, \varepsilon)$. Let Γ stand for the event that $\sigma(Y_r) > \alpha$ for some $r < \delta$. Then $\wedge_n \cap \{\sigma(y^n) > \alpha\} \subset \wedge_n \cap \Gamma$ where $\sigma(y^n) = \inf\{t > 0 | y_t^n = b\}$. Thus

$$P(\Gamma) \geq \sum_n P\big(\wedge_n \cap \{\sigma(y^n) > \alpha\}\big).$$

Fix n and let $R = \inf\{t | y_t^n \notin V\}$. Then $y_R^n \notin V$ on the set $\wedge_n$ and so

$$\alpha P(\wedge_n) \leq E\big(\overline{P}^{y_R^n}(\sigma > \alpha); T_n < \delta; y_t^i \in V, t \geq 0, i < n; R < \infty\big).$$

The expectation on the right of this inequality is $P\big(\sigma(y^n) > R + \alpha; \wedge_n\big)$ which is less than $P\big(\wedge_n \cap \{\sigma(y^n) > \alpha\}\big)$. And so

$$P(\Gamma) \geq \sum_m \alpha P(\wedge_n) \geq \alpha\theta.$$

But clearly $\tau(\delta)$ exceeds α on the set Γ, and so this inequality violates the fact that with probability one $\tau(\delta) \to 0$ as $\delta \to 0$. This completes the proof of the lemma.

The argument that $\{x_t; t \geq 0\}$ is a Markov process whose transition function is the one associated with the resolvent $\{U^\lambda; \lambda > 0\}$ is straightforward: specifically suppose we are given points $0 \leq t_1 < \cdots < t_n \leq r$ and functions $g_1, \cdots, g_n$ and h all in C_0. Write Π for $\overset{n}{\underset{1}{\Pi}} g_i(x_{t_i})$ and Π_ε for the same expression with x replaced by x^ε. The convergence of x_t^ε to x_t implies that

$$\lim_{\varepsilon \to 0} E\Big(\int_r^\infty e^{-\lambda t} h(x_t^\varepsilon) dt \cdot \Pi_\varepsilon\Big)$$
$$= E\Big(\int_r^\infty e^{-\lambda t} h(x_t) dt \cdot \Pi\Big).$$

The term on the left side of this relation is $e^{-\lambda r} E\big(U_\varepsilon^\lambda h(x_r^\varepsilon) \cdot \Pi_\varepsilon\big)$ since $\{x_t^\varepsilon; t \geq 0\}$ is a Markov process and U_ε^λ is its resolvent. Obviously we have $|U_\varepsilon^\lambda h(x) - U^\lambda h(x)| \leq |U_\varepsilon^\lambda h(b) - U^\lambda h(b)|$ regardless of x and so

$$E\Big(\int_r^\infty e^{-\lambda t} h(x_t) dt \cdot \Pi\Big)$$
$$= e^{-\lambda r} E\big(U^\lambda h(x_r) \cdot \Pi\big)$$
$$= E \int_r^\infty e^{-\lambda t} P_{t-r} h(x_r) dt \cdot \Pi.$$

We will argue in a bit that the x_t process has right continuous sample functions. However a much simpler fact is that for each fixed t $x_{t+\varepsilon} \to x_t$ with probability one as $\varepsilon \to 0$. This is shown by essentially the argument for the convergence of x_t^ε to x_t. This implies that $Eh(x_t)$ is right continuous in t and then the uniqueness theorem for Laplace transforms implies that

$$E\big(P_{t-r} h(x_r) \cdot \Pi\big) = E\big(h(x_t) \cdot \Pi\big)$$

for all $t \geq r$. Thus $\{x_t; t \geq 0\}$ is Markovian and $\{P_t\}$ is its semigroup.

There remain two tasks: One is to establish that the paths of the x_t process are right continuous. Then the fact that its semigroup maps C_0 to itself and is strongly continuous will yield the other properties of a standard process (see the arguments in Chapter I). The other task is to show that the original $\hat{P}$ is indeed the characteristic measure of the process we have constructed.

For the right continuity, let V be a neighborhood of b and let δ denote the supremum of $E^x e^{-\sigma}$ as x ranges over the complement of V. According to property (c) in the discussion of resolvents δ is strictly less than 1. Let $\{\eta_\varepsilon\}$ denote the entrance law associated with $\hat{P}$. The obvious inequality

$$E^{\eta_\varepsilon}(1 - e^{-\sigma}) \geq (1 - \delta) \overline{P}^{\eta_\varepsilon}(\sigma_{V^c} < \sigma)$$

implies that $\overline{P}^{\eta_\varepsilon}(\sigma_{V^c} < \sigma) \leq (1-\delta)^{-1}$ and letting ε approach 0 we conclude that

$$\hat{P}(\sigma_{V^c} < \sigma) < \infty.$$

Now consider a point t where the path x_s is not continuous from the right. Obviously neither cases (1) nor (2) in our earlier argument can apply to t and so $x_t = b$ and there is a neighborhood V of b and two sequences $\{s_n\}$ and $\{r_n\}$ decreasing to t from strictly above and such that $x_{s_n} = b$ and $x_{r_n} \notin V$. Clearly this means that in the point process Y there is with positive probability a finite value of t such that for infinitely many domain points $r \leq t$ we have $\sigma_{V^c}(Y_r) < \sigma(Y_r)$. But the expected number of such points is $t\hat{P}(\sigma_{V^c} < \sigma)$ which is finite. This contradiction implies that the paths of $\{x_t; t \geq 0\}$ are with probability one everywhere right continuous.

Obviously $\{x_t; t \geq 0\}$ is a recurrent extension of $\overline{X}$. Let $\hat{\mathcal{P}}$ denote its excursion measure. Suppose that g is a positive $\mathcal{E}$ measurable function vanishing at b. Then

$$\hat{\mathcal{P}}\left(\int_0^\sigma e^{-t} g(x_t) dt\right) = U^1 g(b)$$

since this relationship between the excursion measure and potential always holds. But clearly we have

$$\hat{P}\left(\int_0^\sigma e^{-t} g(X_t) dt\right)$$

$$= \lim_{\varepsilon \to 0} \hat{P}\left(\int_\varepsilon^\sigma e^{-t} g(X_t) dt\right)$$

$$= \lim_{\varepsilon \to 0} \int_\varepsilon^\infty \langle \eta_t, g \rangle e^{-t} dt$$

$$= \lim_{\varepsilon \to 0} e^{-\varepsilon} \int_0^\infty \langle \eta_\varepsilon, P_t^0 g \rangle e^{-t} dt$$

$$= \lim_{\varepsilon \to 0} \langle \eta_\varepsilon, V^1 g \rangle.$$

This last limit is simply $U^1 g(b)$ as one sees upon referring back to paragraph (c). Thus $\hat{\mathcal{P}}$ and $\hat{P}$ agree on functions of the form $\int_0^\sigma e^{-t} g(x_t) dt$. To show that $\hat{\mathcal{P}}$ and $\hat{P}$ are the same it is enough to show that they have the same entrance law, as the rest of the structure is determined by Markov properties using the same semigroup $\{P_t^0\}$. Let $\{\mu_s; s > 0\}$ denote the entrance law associated with the measure $\hat{\mathcal{P}}$. We know already that

$$\hat{\mathcal{P}}\left(\int_0^\sigma e^{-t} g(x_t) dt\right) = \int_0^\infty e^{-t} \langle \eta_t, g \rangle dt$$

and that the same equality holds when $\hat{P}$ is replaced by $\check{P}$ and η_t by μ_t. That is we know that

$$\int_0^\infty e^{-t}\langle \eta_t, g\rangle dt = \int_0^\infty e^{-t}\langle \mu_t, g\rangle dt.$$

Replace g with $P_s^0 h$ for some $s > 0$ and h bounded and $\mathcal{E}$ measurable. Then $\langle \eta_t, P_s^0 g\rangle = \langle \eta_{t+s}, g\rangle$ and we conclude that

$$\int_s^\infty e^{-t}\langle \eta_t, h\rangle dt = \int_s^\infty e^{-t}\langle \mu_t, h\rangle dt.$$

If also h is in C_0^b the integrands are continuous in t and we conclude that

$$\langle \eta_t, h\rangle = \langle \mu_t, h\rangle$$

for all t. Thus $\hat{P} = \check{P}$. This completes the proof of Itô's synthesis theorem.

(2.12) Exercise. Assume $\|\hat{P}\|$ is infinite and set $c = \hat{P}(1 - e^{-\sigma})$, but drop the requirement $c \leq 1$. If Y is the Poisson point process with characteristic measure $\hat{P}$ make Itô's construction but starting with

$$\tau(t) = \sum_{r \leq t} \sigma(Y_r);$$

in other words take $m = 0$. Show that the resulting recurrent extension has excursion measure $c^{-1}\hat{P}$. (This fact is needed in situations such as the one in IV–5 where $\hat{P}$ is obtained by restricting an excursion measure to a subset of the set of all paths.)

3. Examples and complements.

(a) **Feller Brownian motions.** Recall the representation

$$U^1 f(0) = pf(0) + qR^1 f(0) + \langle \eta, V^1 f \rangle$$

from Chapter II. Let $\hat{P}$ denote the excursion measure for reflecting Brownian motion and $\overline{P}^x$ the probabilities for Brownian motion in $[0, \infty)$ stopped upon reaching the origin. Then $\hat{P} = q\hat{P} + \overline{P}^\eta$ is an excursion measure compatible with the minimal semigroup $P^0(t, x, dy) = p^-(t, x, y)dy$, $x > 0$, $y > 0$, and $\hat{P}(1 - e^{-\sigma})$ is equal to $1 - p$. The hypotheses of Theorem 2.10 are satisfied by the minimal semigroup; and so Itô's construction using p for delay coefficient and $\hat{P}$ for excursion measure yields a Feller Brownian motion (X, P^x) such that $E^0 \int_0^\infty e^{-t} f(X_t)dt$ is equal to $U^1 f(0)$ as given above. Thus it is the desired process. This construction surely appears preferable to the tricky construction of Chapter II. The appeal of the earlier construction is its imaginative use of interesting probability theory and the fact that similar prescriptions can be used in cases where the single boundary point $\{b\}$ is replaced by a curve.

(b) **Skew Brownian motions.** This term is applied to a number of situations. The most common one is where the minimal process is Brownian motion on the real line killed upon hitting 0 and where the term "skew Brownian motion" means a recurrent extension which is a diffusion process (i.e. continuous paths) and has no "sojourn" at $\{0\}$ in the sense that $E^x \int_0^\infty I_{\{0\}}(X_t)dt = 0$. Let $\hat{P}_+$ denote excursion measure for reflecting Brownian motion on $[0, \infty)$ and $\hat{P}_-$ denote excursion measure for reflecting Brownian motion on $(-\infty, 0]$. Given a skew Brownian motion the hypothesis of continuous paths rules out jumps away from the origin and it follows (as we will show shortly) that its excursion measure $\hat{P}$ satisfies

$$\hat{P} = a\hat{P}_+ + b\hat{P}_-$$

with $a + b \leq 1$. The hypothesis of no sojourn at $\{0\}$ implies that $a + b = 1$ so we write $a = p$ and $b = 1 - p$ with $0 \leq p \leq 1$. The measure $\hat{P}$ is compatible with the minimal semigroup and so there does exist such a process. The case $p = \frac{1}{2}$ yields ordinary Brownian motion on the real line. A glance at our proof in Chapter IV of the Lévy-Itô-McKean excursion description of Brownian motion (Theorem IV-1.5) shows that the excursion structure in the general case is obtained by taking the "choice of direction" variables e_{nk} to satisfy $P(e_{nk} = 1) = p = 1 - P(e_{nk} = -1)$, but leaving everything else unchanged. This is the best one can do by way of justifying the description that "upon reaching 0 the process chooses from the two possible exit directions by tossing a possibly biased coin."

Obviously this example can be generalized to the situation in which $\hat{P}$ is obtained as a rather general mixture, $\int \hat{\mathcal{P}}_\theta \mu(d\theta)$, of excursion laws all compatible with a given minimal process, provided appropriate measurability hypotheses are made. One example of this is the so-called "Brownian hedgehog", where the state space is R^2 with points labelled (x, θ) with $x \geq 0$ and θ a point on the unit circle. The minimal process is described by saying that the process $\overline{X}$ starting at the point (x, θ) with $x > 0$ performs Brownian motion on that ray until it reaches the origin, which is our point b. If, in an obvious notation, $\hat{\mathcal{P}}_\theta$ denotes excursion measure for reflecting Brownian on $\{(x, \theta) | x \geq 0\}$ and μ is uniform distribution on the circle then $\int \hat{\mathcal{P}}_\theta \mu(d\theta)$ is an appropriate excursion measure. The minimal semigroup satisfies the hypotheses of Theorem 2.10. The process obtained by Itô's construction is called the Brownian hedgehog. In the excursion description analogous to that of Chapter IV the e_{nk} now take values in the unit circle according to the uniform distribution (or some other distribution if one wants to consider a "generalized hedgehog").

The minimal processes in these examples are rather trivial compoundings of Brownian motion. Other minimal processes which arise naturally give rise to challenging analytic problems as we will see in a bit.

(c) **Skew product diffusions.** K. B. Erickson [E, 1] considers the following general class of processes with state space R^d. Let $\{R_t; t \geq 0\}$ denote a diffusion process whose state space is $[0, \infty)$. We will let σ be the time R_t reaches 0 and will assume that the minimal process $\{R_t; t < \sigma\}$ satisfies the conditions of section 2. Let $\{\theta_t; t \geq 0\}$ be a non-degenerate diffusion process independent of R and taking values in S^{d-1}, the $d - 1$ dimensional unit sphere ($d \geq 2$). Finally let $\{K_t; t \geq 0\}$ be an additive functional of the R process killed at time σ which is continuous and strictly increasing and finite in $[0, \sigma)$ almost surely. The situation in which $d = 2$, R is reflecting Brownian motion, $\theta_t = e^{iB_t}$ where B_t is Brownian motion independent of R, and

$$(3.1) \qquad K_t = \int_0^{t \wedge \sigma} k(R_s)ds$$

with $k : (0, \infty) \to (0, \infty)$ continuous is adequate to illustrate all the possibilities in the general set-up. One can show without much difficulty that the process (in polar coordinates)

$$X_t^0 = \{(R_t, \theta_{K_t}); t < \sigma\}$$

is a diffusion process defined up until the time it reaches the origin in R^d. This is of course the time, σ, when R reaches 0. Erickson takes X_t^0 as minimal process, so that $E = R^d, b = 0$ in R^d, and then asks under what conditions is there a recurrent extension of X, and when is there only one. Since the minimal process is a diffusion it is natural to require that the extension be a diffusion also, and thus he rules out holding and jumping extensions. One always can obtain a diffusion extension by making the origin a trap, that is setting $X_t \equiv 0$ for $t \geq \sigma$. To rule this out and to eliminate non-uniqueness caused by making the origin a sticky point he requires in addition that the extension have no sojourn at 0, that is almost surely $\int_0^\infty I_{\{0\}}(X_t)dt = 0$, where X is the extension. With this restriction on the meaning of recurrent extension Erickson obtains the following interesting fact:

(3.2) Theorem. (a) *If $P^x(K(\sigma) = \infty) = 1$ for all x there is a unique extension of X^0.* (b) *If $P^x(K(\sigma) < \infty) = 1$ for each x then there is a one-to-one correspondence between extensions of X^0 and probability measures on the Borel sets of S^{d-1}.*

Intuitively condition (a) says that the wandering path of X^0 winds around the origin infinitely often as t increases to σ while in case (b) there is a limiting direction. Also a statement like the conclusion of (b) obviously describes the general extension as a mixture of minimal ones, the latter being labelled by points on the unit sphere. In the next section we will give some general theorems on the existence and uniqueness of extensions. Then, assuming as known some basic theory of diffusion processes in R, we will see what is involved in verifying the hypotheses of these theorems in Erickson's set-up.

In case R and θ are reflecting and circular Brownian motion and $K = t \wedge \sigma$, then (b) holds, so $\{R_t e^{iB_t}; t < \sigma\}$ has many extensions. But the process $\{R_t e^{iB_t}; t \geq 0\}$ is not one of them as it is not strong Markov—the path remembers how it approached 0 at time σ.

4. Existence and uniqueness.

Given a minimal process it is natural to ask if there is any non-trivial recurrent extension, and if so is there only one. In many of the interesting special cases one wishes to rule out processes whose paths jump from b into E. Thus we introduce the notion of continuous entrance (into E). An extension of the minimal process has *continuous entrance* if almost surely $X_s = b$ for every $s \in G$; that is every excursion path has b as its initial position. Clearly this is equivalent to the statement that

$$\hat{P}(X_0 \neq b) = 0.$$

(Some confusion is possible as some authors call this continuous exit (from b).) Clearly if the minimal process is a diffusion then an extension will be a diffusion only when it has continuous entrance.

Let $(\bar{X}, \bar{P}^x)_{x \in E}$ be a standard process as in section 1, let $\{P_t^0; t \geq 0\}$ be the minimal semi-group and let $\hat{P}$ be a characteristic measure compatible with the minimal semi-group. Assume that the function $\bar{E}^x(1 - e^{-\sigma}) = V^1 1(x)$ is bounded away from 0 outside any neighborhood of b. Then the expressions

$$\eta_s^c(A) = \hat{P}(X_s \in A, X_0 = b; s < \sigma)$$
$$\eta_s^j(A) = \hat{P}(X_s \in A, X_0 \neq b; s < \sigma)$$

for $s > 0$ and A a Borel subset of $E - \{b\}$ define $\{P_t^0\}$ entrance laws which add up to the entrance law, $\eta_s(A) = \hat{P}(X_s \in A; s < \sigma)$ associated with $\hat{P}$ and clearly $\eta_s^j = \eta P_s^0$ with $\eta(A) = \hat{P}(X_0 \in A), b \notin A$. The function $1 - e^{-\sigma}$ is $\hat{P}$ integrable and $\hat{P}$ is carried by the set of right continuous functions, and so if V is any neighborhood of b we have

$$0 = \lim_{s \to 0} \hat{P}(1 - e^{-\sigma}, X_0 = b, X_s \notin V)$$
$$\geq \lim_{s \to 0} e^{-s} \int_{V^c} \eta_s^c(dx) \bar{E}^x(1 - e^{-\sigma}).$$

Since $\bar{E}^x(1 - e^{-\sigma})$ is bounded away from 0 on V^c we have $\eta_s^c(V^c) \to 0$ as $s \to 0$. Suppose on the other hand that $\{\mu_s; s > 0\}$ is an entrance law such that $\eta_s^j \geq \mu_s$ for all s and also $\mu_s(V^c) \to 0$ as $s \to 0$ whenever V is a neighborhood of b. Then μ_s is identically 0. Indeed if t is less than s and U is a neighborhood of b we have

$$\|\mu_s\| \leq \int_U \eta_t^j(dx) P_{s-t}^0(x, E) + \mu_t(U^c).$$

The first term on the right is equal to $\hat{P}(X_0 \neq b, X_t \in U, s < \sigma)$. The fixed event $\{s < \sigma\}$ has finite $\hat{P}$ measure and so this expression can be made small for all small t by taking U small. By hypothesis the second term approaches 0 with t and the assertion $\mu_s = 0$ follows. Thus we have proved the following simple fact.

(4.1) Theorem. *Assume that* $\bar{E}^x(1-e^{-\sigma})$ *is bounded away from* 0 *outside neighborhoods of* b. *Let* X *be a recurrent extension with* $\{\eta_s; s > 0\}$ *as entrance law. Then* X *has continuous entrance if and only if for every neighborhood* V *of* b, $\eta_s(V^c) \to 0$ *as* $s \to 0$.

For discussing existence and uniqueness a basic piece of data is the ratio $\hat{H}^\lambda g(x)$ defined by

$$\hat{H}^\lambda g(x) = \frac{V^\lambda g(x)}{V^1 1(x)} \qquad x \in E - \{b\}$$

defined for $\lambda > 0$, $x \in E - \{b\}$ and g a bounded Borel function on E. Here V^λ is the λ potential operator for the minimal process. Of course only the restriction of g to $E - \{b\}$ enters into the definition. If for a given g, $\hat{H}^\lambda g(x)$ approaches a limit as x approaches b we will denote the limit by $\hat{H}^\lambda g(b)$ and say "$\hat{H}^\lambda g(b)$ exists." For example if the minimal process is Brownian motion on $(0, \infty)$ killed at 0 then for every bounded continuous g on $(0, \infty)$, $\hat{H}^1 g(0)$ exists and equals $R^1 g(0)$ where R^1 is the one-potential for reflecting Brownian motion. If the minimal process is Brownian motion on all of $R - \{0\}$ then there are two limits for $\hat{H}^1 g(x)$ according to whether x approaches 0 from the right or the left. The basic fact in establishing uniqueness assertions is the following.

(4.2) Theorem. *Suppose* X *is a recurrent extension with continuous entrance into* $E - \{b\}$ *and no sojourn at* b. *If* g *is positive and* $\hat{H}^1 g(b)$ *exists then*

$$E^b \int_0^\infty e^{-t} g(X_t) dt = \hat{H}^1 g(b).$$

Proof. Let ρ be a metric on E giving rise to the topology. Fix $\varepsilon > 0$; let $O_\varepsilon = \{x \mid \rho(x, b) > \varepsilon\}$ and let σ_ε denote σ_{O_ε}. There are only finitely many excursion intervals in a finite time interval with $\sigma_\varepsilon < \sigma$. Let $G_1^\varepsilon, G_2^\varepsilon, \cdots$ denote the left end points of the excursion intervals with $\sigma_\varepsilon < \sigma$ arranged in increasing order and let $T_n^\varepsilon = G_n^\varepsilon + \sigma_\varepsilon \circ \theta_{G_n^\varepsilon}$. The T_n^ε are stopping times for X and every s in G is of the form G_n^ε for some n if ε is small enough. Let h denote a positive $\mathcal{E}$ measurable function bounded by 1 and consider the two expressions

$$E^b \sum_{s\in G} e^{-s}\big(h(X_{\sigma_\varepsilon})(1-e^{-\sigma})\big)\circ\theta_s$$

(4.3)

$$E^b \sum_n e^{-T_n^\varepsilon}\big(h(X_{\sigma_\varepsilon})(1-e^{-\sigma})\big)\circ\theta_{T_n^\varepsilon}.$$

For $s\in G$ with $\sigma_\varepsilon\circ\theta_s=\infty$ the summand in the first expression is 0. All others are of the form G_n^ε for some n, and the corresponding terms in the two sums are

$$e^{-s}(1-e^{-\sigma\circ\theta_s})h(X_{T_n^\varepsilon})$$

and

$$e^{-T_n^\varepsilon}(1-e^{-\sigma\circ\theta_{T_n^\varepsilon}})h(X_{T_n^\varepsilon}).$$

Of these the first always is larger so the difference between the two sums is no more than

$$1-E^b\sum_n e^{-T_n^\varepsilon}(1-e^{-\sigma})\circ\theta_{T_n^\varepsilon},$$

as the sum $E^b\sum_{s\in G}e^{-s}(1-e^{-\sigma})\circ\theta_s$ is equal to 1. Now suppose g is positive and bounded by 1. Then since X spends no time at b we have

$$E^b\int_0^\infty e^{-t}g(X_t)dt$$

(4.4)

$$= E^b\sum_{s\in G}e^{-s}\Big(\int_0^\sigma e^{-t}g(X_t)dt\Big)\circ\theta_s$$

$$= \lim_{\varepsilon\to 0}E^b\sum_n e^{-T_n^\varepsilon}\Big(\int_0^\sigma e^{-t}g(X_t)dt\Big)\circ\theta_{T_n^\varepsilon}.$$

Applying the strong Markov property twice to the individual summands in the last line we see it can be rewritten as

$$E^b\sum_n e^{-T_n^\varepsilon}\hat{H}^1g(X_{T_n^\varepsilon})(1-e^{-\sigma})\circ\theta_{T_n^\varepsilon}.$$

Let us bring in the hypothesis that $\hat{H}^1g(b)$ exists and rewrite this as

$$E^b\sum_n e^{-T_n^\varepsilon}\big(\hat{H}^1g(X_{T_n^\varepsilon})-\hat{H}^1g(b)\big)(1-e^{-\sigma})\circ\theta_{T_n^\varepsilon}$$

(4.5)

$$+\,\hat{H}^1g(b)E^b\sum_n e^{-T_n^\varepsilon}(1-e^{-\sigma})\circ\theta_{T_n^\varepsilon}.$$

The expectation factor in the second of these summands approaches 1 as ε appproaches 0 since there is no sojourn at $\{b\}$, and so to complete the proof it suffices to show that the first term approaches 0 with ε. Given $\delta > 0$ let V be a neighborhood of b such that $\hat{H}^1 g(x)$ is within δ of $\hat{H}^1 g(b)$ if x is in the closure of V. Then the first sum in (4.5) is less than

$$\delta E^b \sum_n e^{-T_n^\varepsilon}(1 - e^{-\sigma}) \circ \theta_{T_n^\varepsilon}$$

$$+ 2E^b \sum_n e^{-T_n^\varepsilon} I_{\bar{V}^c}(X_{T_n^\varepsilon})(1 - e^{-\sigma}) \circ \theta_{T_n^\varepsilon}$$

so we need show only that the second sum approaches 0 with ε. Bring in the considerations from the beginning of the proof with h denoting the indicator of $\bar{V}^c$. By what we established there the second summand above differs only by a term approaching 0 from

$$E^b \sum_{s \in G} e^{-s} \left(I_{\bar{V}^c}(X_{\sigma_\varepsilon})(1 - e^{-\sigma}) \right) \circ \theta_s,$$

and this is simply

$$\hat{P}\left((1 - e^{-\sigma}); X_{\sigma_\varepsilon} \notin \bar{V} \right).$$

Since, almost surely relative to $\hat{P}$, X_{σ_ε} is in V for all small ε and $1 - e^{-\sigma}$ is $\hat{P}$ integrable this last expression approaches 0 with ε, so the proof is complete.

(4.6) Remark. Because the law of a recurrent extension is determined by $E^b \int_0^\infty e^{-t} g(X_t) dt$ as g ranges over the continuous functions with compact support vanishing near b, a consequence of the theorem is that if $\hat{H}^1 g(b)$ exists for all such functions then there is at most one recurrent extension with continuous entrance and no sojourn at b. But it can be that there is no such extension: for example if the minimal process is uniform motion to the left on $(0, \infty)$ then $\hat{H}^1 g(0) = 0$ for any bounded function vanishing near 0. Thus there is no sawtooth process with these additional properties, as is obvious enough to start with.

We will turn next to the question of existence. One way to establish the existence of a recurrent extension with continuous entrance and no sojourn is to find a characteristic measure with $\hat{P}(1 - e^{-\sigma}) = 1$, $\hat{P}(X_0 \neq b) = 0$, and then apply Itô's synthesis theorem. Note that we have proved this theorem only under additional hypotheses on the minimal semi-group so we will have to verify these as well. The most obvious way to construct a characteristic measure is by constructing first an entrance law, so we need a theorem saying that under proper hypotheses a characteristic measure can be obtained from an entrance law.

(4.7) Theorem. *Suppose the minimal semigroup $\{P_t^0\}$ satisfies the additional conditions (a) and (b) of V–2 paragraph (c); and let $\{\eta_s; s > 0\}$ be an entrance law for $\{P_t^0\}$. Then there is a unique characteristic measure whose entrance law is the given one.*

Proof. We know already that a characteristic measure is determined by its entrance law so no further argument is needed on that point. We will carry out the proof under the additional assumption that in the decomposition

$$\eta_s = \theta_s + \int P^0(s, x, \cdot)\eta(dx)$$

with

$$\theta_s(V^c) \to 0 \qquad\qquad s \to 0$$

for every neighborhood V of b, only the θ_s part is present. This is the only case we need to consider because the sum of two characteristic measures compatible with $\{P_t^0\}$ is again one, and the entrance law $\int P^0(s, x, \cdot)\eta(dx)$ is the entrance law corresponding to $\bar{P}^\eta$, which is a characteristic measure as it stands.

Given $\varepsilon > 0$ let $\{x_t^\varepsilon; t \geq 0\}$ denote a Markov process whose paths are right continuous and have left limits and whose law is $\bar{P}^{\eta_\varepsilon}$. Such a process exists since $\{\bar{P}^x; x \in E\}$ are the laws of a standard process. Let $\{y_t^\varepsilon; \varepsilon \leq t < \infty\}$ be the process defined by $y_t^\varepsilon = x_{t-\varepsilon}^\varepsilon$. Note η_ε is the law of $y_\varepsilon^\varepsilon$, and since η_ε is carried on $E - \{b\}$ the event $y_\varepsilon^\varepsilon = b$ has measure 0. If $\delta < \varepsilon$ then a simple consequence of the entrance law property, $\eta_\delta P_{\varepsilon-\delta}^0 = \eta_\varepsilon$, is that the law of the process $\{y_t^\delta; \varepsilon \leq t < \infty\}$ restricted to the set $y_\varepsilon^\delta \neq b$ is equal to that of the process $\{y_t^\varepsilon; \varepsilon \leq t < \infty\}$. Now pick a strictly positive number t_0; it will be fixed for awhile and will not be displayed in the notation. For $\varepsilon < t_0$ we will let $\{Z_t^\varepsilon; \varepsilon \leq t\}$ denote the process $\{y_t^\varepsilon; t \geq \varepsilon\}$ restricted to the set $\{y_{t_0}^\varepsilon \neq b\}$. The various processes might be defined on measure spaces that change with ε but that is of no importance. If $0 < t_1 < \cdots < t_k$ is a finite set of time points and $0 < \varepsilon < t_1$ then the joint distribution of $(Z_{t_1}^\varepsilon, \ldots, Z_{t_k}^\varepsilon)$ is independent of ε and has total mass

$\|\eta_{t_0}\|$. Obviously this allows us to prescribe finite dimensional distributions $\mu_{t_1,\ldots,t_k}$, in a consistent way (with the total mass being perhaps different than one, but finite at any rate.) By Kolmogorov's consistency theorem there is then a stochastic process $\{x_t; t > 0\}$ whose state space is E and whose finite dimensional distributions $\mu_{t_1,\ldots,t_k}$ are those of $\{Z_{t_1}^\varepsilon, \cdots Z_{t_k}^\varepsilon\}$ for all $\varepsilon < t_1$.

Let $(\Omega, \mathcal{G}, P)$ be the underlying finite measure space over which the process $\{x_t; t > 0\}$ is defined and let Δ denote the set of ω such that for some $t > 0$ either

$$\lim_{q\uparrow t, q\in Q} x_q(\omega) \quad \text{or} \quad \lim_{q\downarrow t, q\in Q} x_q(\omega)$$

fails to exist, Q denoting the rationals in $(0, \infty)$. We assert that Δ is in $\mathcal{G}$, that $P(\Delta) = 0$, and that upon deleting Δ from Ω and setting

$$\bar{x}_t(\omega) = \lim_{q\downarrow t, q\in Q} x_q(\omega)$$

we obtain a process $\{\bar{x}_t; t > 0\}$ equal in law to the x process and having paths which are right continuous and have left limits on $(0, \infty)$. As to the proof, whether or not ω is in Δ is determined by looking at expressions of the form $\rho\big(x_q(\omega), x_r(\omega)\big) > \frac{1}{k}$ with k an integer and q and r elements of Q, ρ being a metric on E compatible with the topology and such that a closed subset of E is compact if and only if it is ρ bounded. We will leave to the reader the task of arguing from this that indeed Δ is in $\mathcal{G}$. For the evaluation of $P(\Delta)$ note that Δ is a countable union, $\Delta = \bigcup_\varepsilon \Delta_\varepsilon$, where Δ_ε denotes those paths for which the one-sided limits fail at some time exceeding ε. The set Δ_ε is measurable on the process $\{x_t; t > \varepsilon\}$ and hence its P measure will be the same as what one obtains when the process x is replaced with $\{Z_t^\varepsilon; t > \varepsilon\}$, as these two have the same finite dimensional distributions. But the Z^ε process has paths which are everywhere right continuous and with left limits and so the assertion $P(\Delta_\varepsilon) = 0$ is clear. The same sort of reasoning yields the equivalence in law of the x and $\bar{x}$ processes. The continuity properties of the $\bar{x}$ process are obvious. We will assume that x has been replaced with $\bar{x}$, that is that the x process paths are all right continuous at every point of $(0, \infty)$ and have left hand limits there.

Next we will argue that for almost all ω

$$\lim_{t \to 0} x_t(\omega) = b$$

so that setting $x_0 \equiv b$ extends the domain of x_t to $[0, \infty)$ and maintains the right continuity. The argument for this requires a Kolmogorov-inequality sort of argument, which we will separate out.

(4.8) Lemma. *If the entrance law $\{\eta_s\}$ has the property that for each neighborhood V of b, $\eta_s(V^c) \to 0$ as $s \to 0$ then for every neighborhood V of b and every $\varepsilon > 0$ there is an $r > 0$ such that*

$$\bar{P}^{\eta_s}(\sigma_{V^c} \leq r) \leq \varepsilon$$

for all $s \leq r$.

Proof. Let V and ε be given and let U be a neighborhood of b whose closure is compact and is contained in V. Recall we are assuming the hypotheses (a) and (b) from V–2–(c). These assure the existence of a compact subset, J, of E such that $\bar{P}^x(\sigma_{\bar{U}} \leq 1) < \varepsilon$ if x is not in J. Let g be a positive continuous function vanishing at ∞ and inside the closure of U and equal to 1 on $J - V$. Then as t approaches 0, $P_t^0 g$ approaches 1 uniformly on $J - V$. If we take q less than 1 and small enough that $P_t^0 g(x)$ exceeds $1 - \varepsilon$ for all $t \leq q$ and x in $J - V$ then clearly we have

$$\bar{P}^x(X_t \notin U) \geq 1 - \varepsilon, \qquad t \leq q, x \in V^c.$$

Now
$$\eta_{s+t}(U^c) = \bar{P}^{\eta_s}(X_t \notin U)$$
$$\geq \bar{P}^{\eta_s}\left(\bar{P}^{X_{\sigma_{V^c}}}(X_{t-\sigma_{V^c}} \notin U); \sigma_{V^c} \leq t\right)$$

and if $t \leq q$ the right side of this display exceeds

$$(1 - \varepsilon)\bar{P}^{\eta_s}(\sigma_{V^c} \leq t).$$

According to our hypothesis on the entrance law we can choose a number $r \leq q$ and such that $\eta_t(U^c)$ is less than $\varepsilon(1 - \varepsilon)$ if t is less than $2r$. Then for s less than r we have

$$(1 - \varepsilon)\varepsilon \geq \eta_{s+r}(U^c) \geq (1 - \varepsilon)\bar{P}^{\eta_s}(\sigma_{V^c} \leq r)$$

which yields the assertion of the lemma.

With the conclusion of (4.8) in hand the assertion that except for a null set of paths we have $x_t \to b$ as $t \to 0$ is obvious. To be specific if V is a neighborhood of b and we have numbers s and r with $0 < s < r$ then

$$P\{x_t \notin V \text{ some } t \in [s,r]\}$$
$$\leq \bar{P}^{\eta_s}(\sigma_{V^c} \leq r)$$

and by (4.8) the right side will be small if r is close to zero.

Now we may regard the measure P as a measure on the space $(U, \mathcal{F}^0)$. Let us reintroduce the fixed parameter t_0 and call this measure P_{t_0}. From the construction it is clear that if $0 < t_1 < t_0$ then P_{t_0} is the restriction of P_{t_1} to the set $\{X_{t_0} \neq b\}$. It follows that as t decreases to 0, P_t increases to a measure $\hat{P}$ on $(U, \mathcal{F}^0)$. The fact that $\hat{P}$ is compatible with the minimal semigroup and has $\{\eta_s\}$ as entrance law is easy to establish. We will leave the details to the reader.

Now we need to consider what conditions on the ratios $\hat{H}^\lambda g$ guarantee the existence of an entrance law having the additional property required for the corresponding characteristic measure to put all its mass on paths that start at b. Since we then will want to apply (4.7) we will assume throughout the rest of this discussion that the minimal semigroup satisfies the conditions (a) and (b) of V-2. We know that if $\hat{H}^1 g(b)$ is 0 for every continuous g with compact support in $E - \{b\}$ then no non-trivial recurrent extension exist. So let us assume that there is a continuous g with compact support in $E - \{b\}$ and a sequence $\{x_n\}$ approaching b such that

$$\lim_n \hat{H}^1 g(x_n) = \delta > 0.$$

Let η^n denote the measure $\varepsilon_{x_n}/V^1 1(x_n)$ and let $\{\eta_s^n; s > 0\}$ denote the entrance law

$$\eta_s^n(A) = \bar{P}^{\eta_n}(X_s \in A, s < \sigma).$$

The inequalities

$$1 = \bar{E}^\eta(1 - e^{-\sigma}) \geq e^{-s} \int \eta_s(dx) \bar{E}^x(1 - e^{-\sigma})$$
$$\geq e^{-s} \eta_s(K) \inf_{x \in K} \bar{E}^x(1 - e^{-\sigma})$$

and

$$1 = \int_0^\infty \eta_s(E) e^{-s} ds \geq (1 - e^{-s}) \|\eta_s\|$$

$(\eta = \eta^n, \eta_s = \eta_s^n)$ show that $\eta_s^n(K)$ is bounded uniformly over all s and n if K is a compact subset of $E - \{b\}$ and that for each $s > 0$, $\|\eta_s^n\|$ is bounded over all n. Thus by passing to a subsequence of $\{x_n\}$ we may assume that for each rational q, η_q^n converges as a sequence of functionals on the continuous function with compact support in $E - \{b\}$ to a measure η_q. Given a number $r > 0$ and $q < r$ let $t = r - q$. If f is continuous with compact support in $E - \{b\}$ then $P_t^0 f$ vanishes at b and is continuous, and

$$\eta_r^n(f) = \eta_q^n P_t^0 f \to \eta_q P_t^0 f,$$

where we are using the fact that $\|\eta_q^n\|$ is bounded. This shows that for all $r > 0$ the measures η_r^n converge as $n \to \infty$ to a measure η_r. One checks immediately that $\{\eta_r; r > 0\}$ is an entrance law for $\{P_t^0\}$. We need now to rule out the possibility that $\eta_r = 0$.

Take g continuous with compact support K contained in $E - \{b\}$ and with $0 \leq g \leq 1$ and such that $\hat{H}^1 g(x_n) \geq \alpha > 0$ for all large n. Our initial hypothesis guarantees that such a function exists. If θ denotes the infimum of $\bar{E}^x(1 - e^{-\sigma})$ as x ranges over K then $\theta > 0$ and $\eta_s^n(K)$ is less than θ^{-1} for all n and s. Thus $\eta_s^n(g)e^{-s}$ is dominated by the integrable function $\theta^{-1}e^{-s}$ and so

$$\alpha \leq \overline{\lim}_n \hat{H}^1 g(x_n) = \overline{\lim}_n \int_0^\infty \eta_s^n(g)e^{-s}ds$$
$$\leq \int_0^\infty \overline{\lim}_n \eta_s^n(g)e^{-s}ds = \int_0^\infty \eta_s(g)e^{-s}ds.$$

It follows that $\eta_s(g)$ is strictly positive for some positive s and so the entrance law $\{\eta_s; s > 0\}$ is not the trivial one, $\eta_s \equiv 0$. We would like to conclude that the entrance law has the additional property that $\eta_s(V^c)$ approaches 0 with s so that it provides the basis for a recurrent extension with continuous entrance. This requires additional hypotheses which, however, cover many cases of interest. In the next theorem $\{\eta_s; s > 0\}$ refers to the entrance law we have just constructed. We will let $\varphi(x)$ denote the function $\rho(x, b)$ where ρ is a metric on E giving rise to its topology and such that a closed subset of E is compact if and only if it is ρ-bounded.

(4.9) Theorem. *If the minimal process $\{\bar{X}_t; t < \sigma\}$ has the property that for some choice of metric ρ almost surely the paths $t \to \varphi(\bar{X}_t)$ are continuous then $\eta_s(V^c) \to 0$ as $s \to 0$ for every neighborhood V of b.*

Proof. For c strictly positive set

$$J_c = \{\varphi \geq c\}, \quad K_c = \{\varphi = c\}$$

and set $\sigma_c = \sigma_{K_c}$. Now $K_{c/2}$ is a compact set disjoint from J_c and so clearly from conditions (a) and (b) of V–2 given any strictly positive ε we can find a strictly positive r such that

$$P_t^0(y, J_c) < \varepsilon \cdot e^{-1} \cdot \inf_{x \in K_{c/2}} \bar{E}^x(1 - e^{-\sigma}) = \delta$$

for all y in $K_{c/2}$ and t less than r. If $\varphi(x)$ is strictly less than $c/2$ then a path of the minimal process proceeding from x meets $K_{c/2}$ before reaching J_c and so the usual strong Markov property argument yields

$$P_t^0(x, J_c) \leq \bar{P}^x(\sigma_{c/2} \leq t) \cdot \delta$$

if $\varphi(x) < c/2$ and $t \leq r$. Now

$$\bar{E}^x(1 - e^{-\sigma}) \geq \bar{E}^x(e^{-\sigma_{c/2}} - e^{-(\sigma_{c/2} + \sigma \circ \theta_{\sigma_{c/2}})}; \sigma_{c/2} \leq r)$$
$$\geq e^{-r} \bar{P}^x(\sigma_{c/2} \leq r) \inf_{x \in K_{c/2}} \bar{E}^x(1 - e^{-\sigma})$$

and so if $\varphi(x) < c/2$ and $t < \min(r, 1)$ we have

$$P_t^0(x, J_c) \leq \bar{E}^x(1 - e^{-\sigma})\big(e / \inf_{x \in K_{c/2}} \bar{E}^x(1 - e^{-\sigma})\big) \cdot \delta$$
$$= \bar{E}^x(1 - e^{-\sigma}) \cdot \varepsilon.$$

This implies that $\eta_t^n(J_c)$ is less than ε if $\varphi(x_n) < c/2$ and $t < \min(r, 1)$. From this the conclusion of (4.9) follows immediately.

(4.10) Remarks. A consequence of (4.9) and the rest of the discussion is that under the conditions (a) and (b) of V–2 if the minimal process is a diffusion (or satisfies the weaker condition in (4.9)) and the obviously necessary condition

$$\overline{\lim}_{x \to b} \hat{H}^1 g(x) > 0$$

for some continuous g with compact support in $E - \{b\}$, is satisfied then there is a recurrent extension having continuous entrance and no sojourn at $\{b\}$. To obtain it we take a non-trivial entrance law, whose existence is guaranteed by the discussion preceeding (4.9). Then we refer to (4.7) for the existence of a characteristic measure $\hat{P}$ having the given entrance law, multiplying by a constant so that $\hat{P}(1 - e^{-\sigma}) = 1$. Then we invoke Itô's synthesis theorem to obtain a recurrent extension. Then (4.9) together with (4.1) implies that the process has continuous entrance, and of course the fact that $\hat{P}(1 - e^{-\sigma}) = 1$ implies that there is no sojourn at b.

Note, using the notation of (4.9), that if we replace the original sequence by a subsequence which we will denote $\{x_n\}$ also so that for each s the sequence η_s^n converges to η_s as functionals on $C_K(E - \{b\})$ then for a continuous function g with compact support in $E - \{b\}$ we will have

$$\lim_n \hat{H}^1 g(x_n) = \lim_n \int_0^\infty \eta_s^n(g) e^{-s} ds$$
$$= \int_0^\infty \eta_s(g) e^{-s} ds = \hat{P}\left(\int_0^\sigma e^{-s} g(X_s) ds\right)$$

where $\hat{P}$ is the characteristic measure (Theorem 4.7) arising from $\{\eta_s\}$. Thus if sequences $\{x_n\}$ and $\{y_n\}$ approach b, and the limits of $\hat{H}^1 g(x_n)$ and $\hat{H}^1 g(y_n)$ exist for all g continuous with compact support in $E - \{b\}$ then the procedure leading to (4.9) will yield characteristic measures which will be the same if and only if

$$\lim_n \hat{H}^1 g(x_n) = \lim_n \hat{H}^1 g(y_n) \qquad g \in C_K(E - \{b\}).$$

The path continuity condition of (4.9) may not be essential, but it cannot be completely eliminated from the argument even when all the other conditions are satisfied. For example let $E = R$ and $b = 0$. The minimal process starting at a point x in $(0, \infty)$ will be ordinary Brownian motion killed at 0. Starting at a point x in $(-\infty, 0)$ the moving particle waits at x for a time having an exponential distribution with rate λ_x, and then jumps to 0 or 1 with probabilities q_x and p_x respectively, and from 1 it proceeds as Brownian motion. Choose λ_x and p_x so that as $x \to 0$, $\lambda_x \to \infty$ and $\lambda_x p_x \to 1$. For g bounded continuous and vanishing near 0 easy calculation shows that as x increases to 0, $\hat{H}^1 g(x)$ approaches $V^1 g^+(1)/2 - e^{-\sqrt{2}}$ where V^1 is the resolvent for Brownian motion killed at 0 and $g^+ = g|_{[0,\infty)}$. If x decreases to 0 then $\hat{H}^1 g(x)$ of course approaches $R^1 g^+(0)$ with R^1 being the resolvent for reflecting Brownian motion. The construction leading to (4.9) will yield a nontrivial entrance law if the sequence $\{x_n\}$ either increases or decreases to 0. But only in the second case does the entrance law have the required property to yield continuous entrance.

(4.11) Skew product diffusions. We will apply the information just developed to Erickson's Theorem, (3.2). To simplify the presentation we will assume that the minimal process is a one-dimensional version of Erickson's process described as follows: the state space will be R with the origin being the point $\{b\}$. Let $R(t)$ be reflecting Brownian motion and suppose $\{\theta_t; t \geq 0\}$ is a two state Markov chain with states -1 and 1 and holding time parameters λ_{-1} and λ_1, and that θ and R are independent. The minimal process will be

$$\bar{X}_t = R(t)\theta_{K_t} \qquad t < \sigma_0$$

where $\{K_t; t \geq 0\}$ is a continuous strictly increasing additive functional of R and σ_0 is the time R or $\bar{X}$ reaches the origin. Informally $\bar{X}$ starting at x proceeds as Brownian motion until J, the time when K_t reaches a value S, where S is independent of the Brownian motion and has an exponential distribution with rate λ_- if x is negative and rate λ_+ if x is positive. If J is strictly less than σ_0 and the Brownian motion is at y at time J it jumps to $-y$ and proceeds as before. If σ_0 occurs first then the minimal process is already at the origin awaiting further instructions. The minimal process does not have continuous paths but the process $|\bar{X}_t|$ does, so the considerations of (4.9) apply. If K_{σ_0-} denotes the limit of K_t as t increases to σ_0 then as the reader will show easily either $P^x(K_{\sigma_0-} = \infty) = 1$ or $P^x(K_{\sigma_0-} < \infty) = 1$, ($P^x$ denoting Brownian motion probabilities), and which case holds is independent of x. In the first case the path of $\bar{X}_t$ jumps back and forth across the origin infinitely often as $|\bar{X}_t|$ approaches 0 and in the second case $\bar{X}_t$ approaches 0 either from the right or from the left. Erickson's alternative states that in the first case there is a unique extension of $\bar{X}$ which leaves the origin continuously and has no sojourn and in the second case there are two such (extremal) extensions, and all others are mixtures of these.

Here is his argument. From general Markov chain theory or in this special case just simple matrix theory one shows that the two state Markov chain $\{\theta_t; t \geq 0\}$ has an invariant measure μ, two eigenvalues γ_0 and γ_1, with $\gamma_0 = 0$, $\gamma_1 > 0$ and two eigenfunctions φ_0 and φ_1, with $\varphi_0 \equiv 1$, and a density $\Gamma_t(i,j)$ relative to μ for the transition function of the chain which satisfy

$$\sum_j \Gamma_t(i,j)\varphi_n(j)\mu(j) = e^{-\gamma_n t}\varphi_n(i) \qquad i = \pm 1.$$

(4.12)

$$\sum_j \varphi_n(j)\varphi_m(j)\mu(j) = \delta_{nm}.$$

Suppose a bounded Borel function f on $R - \{0\}$ is of the form $f(x) = f_+(|x|)f_-(\text{sgn } x)$ where f_+ is a bounded Borel function on $(0,\infty)$. Then for the potential operator $W^\lambda f$ of the minimal process we have

$$W^1 f(x) = E^{|x|}\int_0^{\sigma_0} e^{-t}f_+(X_t)\sum_j \Gamma_{K_t}(\text{sgn } x, j)f_-(j)dt$$

where E^x refers to expectation for Brownian motion. Putting in the two cases $f_- = \varphi_i, i = 0,1$ we obtain (with $F(x) = f_+(|x|)$ and $G(x) = f_+(|x|)\varphi_1(\text{sgn } x)$)

$$W^1 F(x) = E^{|x|}\int_0^{\sigma_0} e^{-t}f_+(X_t)dt = V^1 f_+(|x|),$$

(4.13)

$$W^1 G(x) = \varphi_1(\text{sgn } x)E^{|x|}\int_0^{\sigma_0} e^{-t-\gamma_1 K_t}f_+(X_t)dt,$$

where V^1 is the one-potential for Brownian motion on $(0,\infty)$. Note that we have

$$\bar{E}^x(1 - e^{-\sigma}) = E^x(1 - e^{-\sigma}).$$

Of course the common value is $(1 - e^{-x\sqrt{2}})$, which we will denote by $s(x)$. We need a simple preliminary fact. First recall the notation E^x denotes expectations for Brownian motion ($x > 0$), V^1 the one-potential operator for Brownian motion killed at σ_0 and R^1 the one-potential operator for reflecting Brownian motion. Let $\{K_t; t \geq 0\}$ denote a continuous strictly increasing additive functional for Brownian motion killed at time σ_0 which is finite for $t < \sigma_0$. Given a bounded Borel function φ on $(0,\infty)$ and a constant $\gamma \geq 0$ set

$$V_\gamma\varphi(x) = E^x\int_0^{\sigma_0} e^{-(t+\gamma K_t)}\varphi(X_t)dt, \qquad x > 0.$$

(4.14) **Lemma.** *The limit*

$$\lim_{x\to 0} V_\gamma\varphi(x)/s(x)$$

exists.

Proof. We can restrict our attention to the case where φ vanishes in some interval $(0, c)$. Indeed if we take $\psi = I_{(0,c)}$ then

$$0 \le V_\gamma \psi \le V_0 \psi = V^1 \psi,$$

and as we know $V^1 \psi(x)/s(x) \to R^1 \psi(0)$, which can be made as small as we like by taking c close enough to 0. When φ vanishes in $(0,c)$ we have

$$V_\gamma \varphi(x) = E^x \int_{\sigma_c}^{\sigma_0} e^{-(t+\gamma K_t)} \varphi(X_t)\,dt$$
$$= E^x(e^{-J(\sigma_c)}; \sigma_c < \sigma_0) V_\gamma \varphi(c)$$

where J is the additive functional $J_t = t + \gamma K_t$. For $0 < x \le y \le c$ set $u(x,y) = E^x(e^{-J(\sigma_y)}; \sigma_y < \sigma_0)$. Then $V_\gamma \varphi(x) = u(x,c)V_\gamma \varphi(c)$ and

$$u(x,c) = E^x(e^{-J(\sigma_y)} e^{-J(\sigma_c)\circ\theta_{\sigma_y}}; \sigma_y < \sigma_0, \sigma_c \circ \theta_{\sigma_y} < \sigma_0 \circ \theta_{\sigma_y})$$
$$= u(x,y)u(y,c) \le (x/y)u(y,c)$$

where the inequality comes from the fact that

$$u(x,y) \le P^x(\sigma_y < \sigma_0) = x/y.$$

Thus $u(x,c)/x$ decreases as x decreases. This yields (4.14) since $s(x)$ is asymptotic to a constant times x as x approaches 0. The next lemma is the key to Erickson's argument. Here $\{K_t; t \ge 0\}$ will denote the continuous additive functional of Brownian motion involved in the definition of the minimal process. Recall that we have left as an exercise the verification that either $K(\sigma_0-) = \infty$ almost surely, that is $P^x(K(\sigma_0-) = \infty) = 1$ for all $x > 0$, or $K(\sigma_0-) < \infty$ almost surely. We will denote by $\hat{H}_\gamma^1 \varphi(0)$ the limit in (4.14).

(4.15) Lemma. *If $K(\sigma_0-) = \infty$ almost surely then for all $\gamma > 0$, $\hat{H}_\gamma^1 \varphi(0) = 0$. If $K(\sigma_0-) < \infty$ almost surely then for all $\gamma > 0$, $\hat{H}_\gamma^1 \varphi(0) \ne 0$ whenever φ is a positive continuous function on $(0, \infty)$ which is not identically 0.*

Proof: From the comments at the beginning at the proof of (4.14) it is clear that we need only show that for some strictly positive c the ratio $E^x(e^{-K(\sigma_c)}; \sigma_c < \sigma_0)/s(x)$ approaches 0 with x in case $K(\sigma_0-) = \infty$ but not in the case $K(\sigma_0-) < \infty$. Knowing, as we do, that these limits exist in a monotone fashion and that ratios such as $E^x(e^{-\sigma_c}; \sigma_c < \sigma_0)/s(x)$ do not approach 0 the assertions of (4.15) are quite plausible. However we have not found a way of carrying out the proofs that does not rely on considerable machinery from Feller's general theory of diffusion (in particular the classification of boundaries) as set forth in [IM, 2], and as applied by Erickson. Developing this material will take us too far afield, so we simply will proceed on the assumption that (4.15) has been established.

With (4.15) in hand we can obtain easily the conclusion of (3.2). To be specific, from (4.13) applied to an even function $f(x) = f^+(|x|)$ we have

$$W^1 f(x)/s(|x|) = V^1 f^+(|x|)/s(x) \to R^1 f^+(0)$$

which certainly is non zero for some choice of f^+ continuous and vanishing near 0 and ∞, and so (4.9) and its applications imply the existence of at least one non-trivial extension of the minimal process. Now suppose a bounded continuous function f on $R - \{0\}$ can be written as a sum

$$f(x) = f^+(|x|)\big(a\varphi_0(\text{sgn } x) + b\varphi_1(\text{sgn } x)\big).$$

Then

$$W^1 f(x) = aE^{|x|} \int_0^{\sigma_0} e^{-t} f^+(X_t)dt$$
$$+ b\varphi_1(\text{sgn } x)E^x \int_0^{\sigma_0} e^{-(t+\gamma_1 K_t)} f^+(X_t)dt.$$

Let us divide by $s(|x|)$ and let x approach 0 either from the right or from the left. We obtain

(4.16)
$$\lim_{x \to 0} W^1 f(x)/W^1 1(x) = aR^1 f^+(0)$$
$$+ b\varphi_1(\pm 1)\hat{H}^1_{\gamma_1} f^+(0)$$

where on the right we use $+1$ or -1 according to whether x approaches 0 from the right or the left.

Now let us assume $K(\sigma_0-) = \infty$ almost surely. The eigenvalue γ_1 is strictly positive and so (4.15) implies that the second term on the right of (4.16) is absent; so the two sided limit on the left exists. Any bounded continuous function on $R - \{0\}$ can be written as a linear combination of functions such as these and so the limit on the left of (4.16) exists for any such function. From the uniqueness theorem (4.2) it follows that there is only one extension of the minimal process. In fact a little algebra on the representation of a general function as a linear combination shows that

$$\lim_{x \to 0} W^1 f(x)/W^1 1(x) = R^1 g(0)$$

$$g(x) = \mu(1)f(x) + \mu(-1)f(-x) \qquad x > 0$$

with μ the invariant measure for the two state Markov chain. This is half of Erickson's result.

Now suppose $K(\sigma_0 -)$ is finite almost surely. Then the second term on the right of (4.16) does not vanish if b is non-zero and f_+ is positive continuous and not the zero function. Also $\varphi_1(+1)$ is not equal to $\varphi_1(-1)$ as φ_0 and φ_1 are linearly independent, and so following the procedure described in the discussion preceeding (4.9) we obtain two essentially different entrance laws as limits

$$\eta_t(\cdot) = \lim_{x \to 0} \bar{P}^{\epsilon_x / s(|x|)}(X_t \in \cdot, t < \sigma_0)$$

one limit obtaining as x decreases to 0 and the other as x increases to zero. Call these two entrance laws $\{\eta_t^+\}$ and $\{\eta_t^-\}$. Each of these yields, by the usual application of (4.7) and Itô's synthesis theorem an extension of the minimal process with continuous entrance and no sojourn at 0. And so does any convex combination, $p\eta_t^+ + (1-p)\eta_t^-$; and the corresponding extensions are all different as their entrance laws are. Consequently to complete the discussion we need to show only that the entrance law of any recurrent extension (satisfying the additional conditions) is a convex combination as above. To see this let $\hat{P}$ be the excursion law for an extension having continuous entrance and no sojourn at 0. Define functions ℓ and r on $(0, \infty)$ by

$$\ell(x) = \hat{P}(\theta_x < \sigma_0, X_{\theta_x} = -x)$$

$$r(x) = \hat{P}(\theta_x < \sigma_0, X_{\theta_x} = x)$$

where $\theta_x = \inf\{t \mid |X_t| = x\}$. Since the law of the absolute value of the minimal process is that of Brownian motion and paths enter continuously the discussion at the beginning of Chapter IV of the Brownian excursion law shows that

$$r(x) + \ell(x) = \frac{1}{x\sqrt{2}} \sim s^{-1}(x)$$

and that the entrance law for $\hat{P}$ is given by

$$(4.17) \qquad \begin{aligned} \eta_t(g) &= \lim_{x \to 0} \hat{P}\big(g(X_{t+\theta_x}); \theta_x < \sigma_0\big) \\ &= \lim_{x \to 0} \big(\bar{P}^x\big(g(X_t); t < \sigma_0\big)r(x) + \bar{P}^{-x}\big(g(X_t); t < \sigma_0\big)\ell(x)\big), \end{aligned}$$

with g continuous and vanishing near 0. Let x approach 0 through a sequence such that $r(x)s(x) \to p$ so that $\ell(x)s(x) \to 1 - p$. Upon replacing r and ℓ with ps^{-1} and $(1-p)s^{-1}$ in (4.17) it should now be obvious that $\eta_t = p\eta_t^+ + (1-p)\eta_t^-$.

5. A counter-example.

Salisbury [S,1] gives several examples showing that given a characteristic measure compatible with a minimal semi-group it is not always possible to construct a recurrent extension which is strong Markov with right continuous paths and has the given characteristic measure as excursion measure. The most illustrative of these examples is the following. Prior to an alteration to be made in a moment let $E = [-1, \infty)$ and $b = 0$. The minimal semigroup is described by saying that starting at x in $(0, \infty)$ the minimal process performs Brownian motion until reaching 0, and starting at x in $[-1, 0)$ the minimal process is uniform motion to the right until the time, $-x$, when it reaches 0. Clearly the hypotheses (a) and (b) of section 2 are satisfied. One obtains a recurrent extension, X, by saying that starting at 0, X performs reflecting Brownian motion in $[0, \infty)$ in the time interval $[0, \tau)$ where τ is defined by $\ell(\tau) = S$ with S exponentially distributed and independent of X and ℓ local time at 0 for the reflecting Brownian motion. Then X jumps to -1, that is $X_\tau = -1$, from where it proceeds as uniform motion until it reaches 0 where it starts over again. If $\hat{P}$ is the excursion measure for X then $\hat{P} = \alpha \hat{\mathcal{P}} + \beta \hat{Q}$ where $\hat{\mathcal{P}}$ is the excursion measure for reflecting Brownian motion, $\hat{Q}$ is unit mass on the one path x given by $x_t = -1 + t$ for $t < 1$ and $x_t = 0$ for $t \geq 1$ and α and β are strictly positive numbers chosen so that $\hat{P}(1 - e^{-\sigma}) = 1$, and otherwise dependent on the rate of S. Now let us delete -1 from E and, retopologize $(-1, 0]$ as if it were the unit circle, $x \leftrightarrow e^{2\pi i x}, -1 < x \leq 0$, change the path x to x' by setting $x'_0 = 0$ where previously $x_0 = -1$ and make exactly the same change in the process paths X_t, that is setting $X'_t = 0$ if previously $X_t = -1$. If $\hat{R}$ puts unit mass on the path x' then $\alpha \hat{\mathcal{P}} + \beta \hat{R}$ is compatible with the original minimal semigroup $\{P^0_t\}$ since $P_0(t, x, \{-1\}) = 0$ for all x and $t > 0$ so the change from $\hat{Q}$ to $\hat{R}$ is unnoticed. The process X' has continuous paths relative to the new topology on $E - \{-1\}$ and is equal in law to X. It is not strong Markov since $X'_T = 0$ at the time T when X' enters $(-1, 0)$ while for the original process starting at 0 the time T is strictly positive. It is quite clear that a recurrent extension with state space $E - \{-1\}$ and $\alpha \hat{\mathcal{P}} + \beta \hat{R}$ as excursion measure would have to be identical to X' so if the term "recurrent extension" carries the requirement of strong Markov then no such process exists. Of course retopologizing E destroys conditions (a) and (b) of section 2 on the minimal process; for example $\bar{E}^x e^{-\sigma}$ is not near 1 for all x near 0.

In this example the new characteristic measure puts all its mass on paths with $X_0 = b$ and is a sum of two characteristic measures, one having finite mass. Salisbury shows that this state of affairs always is fatal.

6. Integral Representation.

Given a characteristic measure compatible with a minimal semigroup we wish to consider the possibility of representing it as a mixture of extreme characteristic measures.

Our set-up will be that of section 1 of this chapter. Thus we have as fixed data a standard process $(\bar{X}, \bar{P}^x)_{x \in E}$ with $\sigma = \sigma_b$ denoting the hitting time of a point b in E. As usual $\{P_t^0\}$ will denote the transition function of the minimal process, $\bar{X}$ killed at time σ. The term *characteristic measure* will mean a measure η on the σ-algebra $\mathcal{F}^0$ in the space U which, with η replacing $\hat{P}$, satisfies conditions (i) through (v) and the consequent (1.3) of section 1. Various characteristic measures will arise but the minimal process and semigroup will be fixed.

To phrase the integral representation problem properly we will consider only characteristic measures with $\eta(1 - e^{-\sigma}) = 1$ (rather than less than or equal to 1.) We will say that a characteristic measure is *extreme* if whenever η is a convex combination of characteristic measures η_1 and η_2

$$\eta = p\eta_1 + (1 - p)\eta_2 \qquad 0 < p < 1$$

it follows that $\eta_1 = \eta_2 = \eta$. An integral representation of η means a representation

$$\eta = \int_A \eta_a \mu(da)$$

where (A, μ) is a probability space and $\{\eta_a; a \in A\}$ is a family of extreme characteristic measures adequately measurable in the parameter a that the integral makes sense. Itô proves in [I, 1] that any η has an integral representation, and he gives a canonical way of obtaining it. We will present his results, filling in some necessary arguments that he did not include.

First of all we will dispose of the characteristic measures of the form $\bar{P}^\mu$. Quite clearly such a measure is extreme if and only if μ puts all its mass at a single point of E different from b, and we have of course the representation $\bar{P}^\mu = \int \bar{P}^x \mu(dx)$. This aspect of the problem is uninteresting and so henceforth we will restrict our attention to characteristic measures η having the additional property that

$$\eta(\{u|u(0) \neq b\}) = 0,$$

so that the corresponding recurrent extension, if there is one, has continuous entrance (and, because of $\eta(1 - e^{-\sigma}) = 1$, no sojourn at b.)

The simple example of skew Brownian motion (section 3) illustrates the general situation: there the minimal process is Brownian motion on $R - \{0\}$ and the characteristic measures of an extension with continuous entrance and no sojourn can be written $p\hat{P}_+ + (1-p)\hat{P}_-$. The characteristic measures $\hat{P}_+$ and $\hat{P}_-$, which correspond to reflecting Brownian motion on $[0, \infty)$ and $(-\infty, 0]$ respectively, are extreme (a consequence of (6.1), which follows), and p represents the proportion of excursion paths which take their values in $(0, \infty)$.

The key to deciding whether or not a characteristic measure is extreme is the following fact.

(6.1) Theorem. *Suppose η is a characteristic measure, that f is a bounded positive $\mathcal{E}$ measurable function on E and that for some sequence $\{t_n\}$ approaching 0 and some constant a_f we have*

$$V^1 f\big(u(t_n)\big)/V^1 1\big(u(t_n)\big) \to a_f \qquad t_n \to 0$$

for almost all u relative to η. Then

$$\eta\left(\int_0^\sigma e^{-t} f\big(u(t)\big) dt\right) = a_f.$$

Proof. Using twice the simple Markov property of the coordinate process $\{u(t); t > 0\}$ relative to η we have

$$\eta\left(\int_0^\sigma e^{-t} f\big(u(t)\big) dt\right)$$

$$= \lim_n \eta\left(\int_{t_n}^{t_n + \sigma \circ \theta_{t_n}} e^{-t} f\big(u(t)\big) dt; t_n < \sigma\right)$$

$$= \lim_n e^{-t_n} \eta\left(V^1 f\big(u(t_n)\big); t_n < \sigma\right)$$

$$= \lim_n \eta\left(\frac{V^1 f\big(u(t_n)\big)}{V^1 1\big(u(t_n)\big)} (e^{-t_n} - e^{-\sigma}); t_n < \sigma\right).$$

This last limit is $a_f \eta(1 - e^{-\sigma})$ which is equal to a_f.

As an application of this result suppose η is a convex combination $\eta = p\eta_1 + (1 - p)\eta_2$ as above. If a_f exists and is constant a.e. (η) then the same thing holds relative to the η_i, and applying (6.1) we find that

$$\eta_i\left(\int_0^\sigma e^{-t}f(u(t))\,dt\right) = a_f \qquad i = 1,2.$$

In particular if for each positive bounded continuous f the limit a_f exists and is constant a.e. η then $\eta_1 = \eta_2 = \eta$ since a characteristic measure is determined by its action on such functions; and hence η must be extreme.

To apply this let $\Omega = (0,\infty) \times U$ with points $\omega = (\tau, u)$, σ-algebra $\mathcal{B}(0,\infty) \times \mathcal{F}^0$, and measure

$$Q(d\tau, du) = I_{\tau<\sigma}e^{-\tau}d\tau\eta(du),$$

where η is our given characteristic measure. As should be clear, objects like σ, which are defined on U rather than Ω are interepreted as

$$\sigma(\omega) = \sigma(u) = \inf\{t > 0|u(t) = b\} \qquad \omega = (\tau, u).$$

Define over Ω a stochastic process $\{Z_t; t > 0\}$ by

$$\begin{aligned} Z_t(\omega) &= u(t) & t &< \tau \\ &= b & t &\geq \tau \qquad \omega = (\tau, u).\end{aligned}$$

It is worth spending a moment to make some calculations with the measure Q. Suppose we have $0 < t_1 < \cdots < t_n \leq t$, bounded $\mathcal{E}$ measurable functions $f_1,\ldots,f_n$ and a bounded $\mathcal{F}^0$ measurable function g on U. If we write Π_Z for $\Pi_1^n f_i(Z_{t_i})$ and Π_U for $\Pi_1^n f_i(u(t_i))$ then we have

$$Q\big(I_{t<\tau}g(\theta_t u)\Pi_Z\big)$$
$$= \eta\left(e^{-t}((1-e^{-\sigma})g) \circ \theta_t \Pi_U; t < \sigma\right)$$

by first integrating $e^{-\tau}$ from t to σ and noting that $\Pi_Z = \Pi_U$ if $\tau > t$. The η integral is

$$e^{-t}\eta\left(\bar{P}^{u(t)}((1-e^{-\sigma})g)\Pi_U; t < \sigma\right)$$
$$= e^{-t}\eta\left(\frac{\bar{P}^{u(t)}((1-e^{-\sigma})g)}{V^11(u(t))}(1 - e^{-\sigma\circ\theta_t})\Pi_U; t < \sigma\right),$$

and reversing the first calculation this last expression becomes

$$Q\left(I_{t<\tau}\frac{\bar{P}^{u(t)}\big((1-e^{-\sigma})g\big)}{V^1 1(u(t))}\Pi_Z\right).$$

In short the integrand

$$I_{t<\tau}\bar{P}^{u(t)}\big((1-e^{-\sigma})g\big)/V^1 1\big(u(t)\big)$$

is the conditional expected value (relative to Q) of $I_{t<\tau}g(\theta_t u)$ given the σ-algebra $\sigma\{Z_s; s \leq t\}$. Note that Q is a probability measure as its total mass is computed as

$$Q(\Omega) = \int\big(\int_0^\sigma e^{-\tau}d\tau\big)d\eta = \eta(1-e^{-\sigma}) = 1.$$

The next step in establishing the integral representation is the following. Let f be a positive bounded $\mathcal{E}$ measurable function on E and set

$$\hat{H}^1 f(x) = V^1 f(x)/V^1 1(x) \qquad x \neq b$$
$$= 0 \qquad\qquad\qquad x = b.$$

(6.2) Lemma. *The process* $\{\hat{H}^1 f(Z_t); t > 0\}$ *is a supermartingale.*

Proof. We will show that

(6.3)
$$Q\big(\hat{H}^1 f(Z_{t+r}); \Pi_Z\big)$$
$$\leq Q\big(\hat{H}^1 f(Z_t); \Pi_Z\big)$$

where Π_Z and Π_U (below) have the same meaning as they did in the discussion just preceeding (6.2). The right side of (6.3) is

$$Q\left(\hat{H}^1 f\big(u(t)\big)\Pi_U I_{t<\tau}\right)$$
$$= \eta\left(e^{-t}\big(1-e^{-\sigma\circ\theta_t}\big)I_{t<\sigma}\Pi_U \hat{H}^1 f\big(u(t)\big)\right)$$
$$= \eta\left(V^1 1(u(t))\hat{H}^1 f(u(t))e^{-t}\Pi_U I_{t<\sigma}\right)$$
$$= \eta\left(V^1 f(u(t))e^{-t}\Pi_U I_{t<\sigma}\right).$$

By exactly the same calculation the left side of (6.3) is

$$\eta\left(V^1 f\big(u(t+r)\big)e^{-(t+r)}I_{t+r<\sigma}\Pi_U\right)$$

and this in turn is equal to

$$\eta\left(e^{-r}P_r^0(V^1f)\big(u(t)\big)e^{-t}\amalg_U I_{t<\sigma}\right).$$

From our original discussion of excessive functions in Chapter III we know that

$$e^{-r}P_r^0(V^1f)(x) \le f(x) \qquad x \in E$$

and the inequality in (6.3) follows.

The process $\hat{H}^1f(Z_t)$ is positive and bounded, and so a standard martingale convergence fact implies that if $\{t_n\}$ is any sequence decreasing to 0 then $\lim_n \hat{H}^1f(Z_{t_n})$ exists with Q probability one. We will use this consequence of (6.2) in a moment.

Let us regard u as being the result of a mapping $\omega = (\tau, u) \to u$ from the probability space $\big(\Omega, \mathcal{B}(0,\infty) \times \mathcal{F}^0, Q\big)$ to $(U, \mathcal{F}^0)$. The mapping is measurable, and since $(U, \mathcal{F}^0)$ has the structure of the Borel sets of a complete separable metric space (obtained by giving U the Skorohod topology) we may determine a regular conditional distribution for u given any sub σ-algebra $\mathcal{B}$, of $\mathcal{B}(0,\infty) \times \mathcal{F}^0$. That is, there is a function $\bar{Q}(\Gamma, \omega)$ defined for ω in Ω and Γ in $\mathcal{F}^0$ with these properties:

a) for all ω, $\bar{Q}(\cdot, \omega)$ is a probability measure on $\mathcal{F}^0$

b) for all Γ, $\bar{Q}(\Gamma, \cdot)$ is $\mathcal{B}$ measurable

c) $Q\big(\bar{Q}(\Gamma, \cdot); \Delta\big) = Q\big(I_\Gamma(u); \Delta\big)$ for all Γ in $\mathcal{F}^0$ and Δ in $\mathcal{B}$.

For the rest of this section we will take $\mathcal{B}$ to be the intersection over all strictly positive t of $\sigma\{Z_s; s \le t\}$.

We may assume that for each ω the measure $\bar{Q}(\cdot, \omega)$ is carried on $\{u|u(0) = b$ and $\sigma(u) > 0\}$ since for almost all (Q) ω the u coordinate has this property. Define a family $\nu(\cdot, \omega)$ of measures on $(U, \mathcal{F}^0)$ by

$$\nu(\Gamma, \omega) = \bar{Q}\big(\Gamma/(1 - e^{-\sigma}), \omega\big).$$

Itô's representation of η as a mixture of extreme characteristic measures is contained in the following statement.

(6.4) Theorem. *For almost all (relative to Q) ω the following hold:*

a) $\nu(\cdot, \omega)$ is a characteristic measure

b) $\nu(\{u|u(0) \ne b\}, \omega) = 0$, $\nu(1 - e^{-\sigma}, \omega) = 1$

c) $\nu(\cdot,\omega)$ is extreme.

Furthermore $\eta = \int_\Omega \nu(\cdot,\omega)Q(d\omega)$.

Proof. An immediate consequence of the definition of Q is that for any positive $\mathcal{F}^0$ measurable function g on U we have $Q\big(g(u)/(1 - e^{-\sigma(u)})\big) = \eta(g)$. The assertion that η is a mixture of the measures $\nu(\cdot,\omega)$ follows immediately.

The first part of (b) is the statement we have made already about $\bar{Q}(\cdot,\omega)$. The second part follows from the fact that the $\bar{Q}$ measures are probability measures so that $\nu(1 - e^{-\sigma},\omega)$ is equal to $\bar{Q}(1,\omega)$ which equals 1 also.

Coming to the less trivial assertions let us suppose that (a) has been established and attend to (c). Let f be a positive bounded $\mathcal{E}$-measurable function. First we will establish that there is a set Λ_f in $\mathcal{B}$ with $Q(\Lambda_f) = 0$ and such that for $\omega \notin \Lambda_f$ the limit $\hat{H}^1 f\big(u(t_n)\big)$ exists and is constant almost everywhere relative to $\nu(\cdot,\omega)$. Here $\{t_n\}$ is any fixed sequence approaching 0. The existence of the limit follows from the martingale argument following (6.2) together with the observation that with Q probability one $Z_t = u(t)$ for all small t. Let us call the limit A_f. It is bounded and depends on ω only through its u coordinate, and it is $\mathcal{B}$ measurable. Applying Jensen's inequality for conditional expectations we have

$$A_f^2 = E_Q(A_f^2|\mathcal{B}) \geq \left(E_Q(A_f|\mathcal{B})\right)^2 = A_f^2$$

and so, in an obvious notation

$$\int A_f^2(u)\bar{Q}(du,\omega) = \left(\int A_f(u)\bar{Q}(du,\omega)\right)^2,$$

so the limit is constant. To establish the extremality of $\nu(\cdot,\omega)$ take a countable set $\{f_n\}$ of bounded continuous functions on E, and let $\Lambda = \bigcup_n \Lambda_{f_n}$. Then $Q(\Lambda) = 0$. Suppose ω is not in Λ. If $\nu(\cdot,\omega)$ is written as a convex combination of characteristic measures η_1 and η_2 then by the remarks following the proof of (6.1) we have

$$\nu\left(\int_0^\sigma e^{-t}f_n\big(u(t)\big)dt,\omega\right) = \eta_i\left(\int_0^\sigma e^{-t}f_n\big(u(t)\big)dt\right)$$

for all n and $i = 1, 2$. Obviously we can choose the family $\{f_n\}$ so that this sort of equality between two characteristic measures implies their identity and so $\nu(\cdot,\omega)$ must be extreme for all ω not in Λ.

Coming to assertion (a) we must prove that there is a set Λ with $Q(\Lambda) = 0$ and such that if ω is not in Λ then for all $t > 0$, bounded $\mathcal{F}^0$ measurable g and f of the form $\Pi f_i(u(t_i))$ with $0 \leq t_1 < \cdots < t_n \leq t$ and f_i bounded and $\mathcal{E}$ measurable we have

$$(6.6) \qquad \nu\big((g \circ \theta_t) f I_{t < \sigma}, \omega\big) = \nu\left((\bar{P}^{u(t)} g) f I_{t < \sigma}, \omega\right).$$

Suppose (6.6) has been established for a fixed t and ω and arbitrary g and f as described. Take $r > 0$ and consider an expression F of the form $F = \overset{m}{\underset{1}{\Pi}} h_i(s_i)$ with $0 < s_1 < \cdots < s_m \leq r$ so that functions of the form $f \cdot F \circ \theta_t$ generate $\sigma\{u(s); s \leq r + t\}$ and F is in $\sigma\{u(s); s \leq r\}$. The Markov property of the $(\bar{X}, \bar{P}^x)$ process implies that

$$\bar{P}^x\big((g \circ \theta_r) F I_{r < \sigma}\big)$$
$$\bar{P}^x\big((\bar{P}^{u(r)} g) F I_{r < \sigma}\big)$$

and applying this and the presumed validity of (6.6) for the given t and ω, and writing ν for $\nu(\cdot, \omega)$ we have

$$\nu\big((g \circ \theta_{t+r}) f (F \circ \theta_t) I_{t+r < \sigma}\big)$$
$$= \nu\left(\big\{(g \circ \theta_r) F I_{r < \sigma}\big\} \circ \theta_t I_{t < \sigma} f\right)$$
$$= \nu\left(\big\{\bar{P}^{u(t)}(g \circ \theta_r) F I_{r < \sigma}\big\} f I_{t < \sigma}\right)$$
$$= \nu\left(\bar{P}^{u(t)}\big\{(\bar{P}^{u(r)} g) F I_{r < \sigma}\big\} f I_{t < \sigma}\right)$$
$$= \nu\left(\big(\bar{P}^{u(r)}(g) F I_{r < \sigma}\big) \circ \theta_t I_{t < \sigma} f\right)$$
$$= \nu\big(\bar{P}^{u(r+t)}(g)(F \circ \theta_t) f I_{t+r < \sigma}\big).$$

In short when (6.6) is established for a particular t and ω it follows for that ω and all larger t. Consequently the exceptional set Λ may be allowed to depend on t. The fact that both sides of (6.6) are measures in g and f means that it is necessary to establish the equality only for a generating family; and since there is a countable generating family we need only establish (6.6) when the exceptional set Λ is allowed to depend on all the quantities involved. And finally in the representation of f as a product we may assume t_1 is strictly positive since $u(0) = b$ almost sure $\nu(\cdot, \omega)$ for almost all ω.

So turning to this verification we must show that if Λ is any set in $\mathcal{B}$ then

$$
\begin{aligned}
(6.7) \qquad & Q\big((g \circ \theta_t)f I_{t<\sigma}/(1 - e^{-\sigma}); \Lambda\big) \\
& = Q\big((\bar{P}^{u(t)}g)f I_{t<\sigma}/(1 - e^{-\sigma}); \Lambda\big).
\end{aligned}
$$

Let δ be strictly positive and strictly less than t_1 in the definition of f and set $f_\delta = \Pi f_i\big(u(t_i - \delta)\big)$ so that $g \circ \theta_t f$ is equal to $(g \circ \theta_{t-\delta} f_\delta) \circ \theta_\delta$ and $f \bar{P}^{u(t)}g$ is equal to $(f_\delta \bar{P}^{u(t-\delta)}g) \circ \theta_\delta$. The left hand side of (6.7) is equal to the limit as δ approaches 0 of

$$
(6.8) \qquad Q\bigg(\big\{ (g \circ \theta_{t-\delta} f_\delta I_{t-\delta<\sigma}) \circ \theta_\delta I_{\delta<\tau} \big\} / (1 - e^{-\sigma \circ \theta_\delta}); \Lambda \bigg)
$$

because the integrand in (6.8) approaches that in (6.7) and is dominated by a constant times $(1 - e^{-(t-\delta)})^{-1}$. The set Λ is in $\sigma\{Z_r ; r \leq \delta\}$ and so, by our conditional probability computations involving Q, (6.8) is equal to

$$
Q\bigg(I_{\delta<\tau} \bar{P}^{u(\delta)}\big((g \circ \theta_{t-\delta})f_\delta I_{t-\delta<\sigma}\big)/V^1 1\big(u(\delta)\big); \Lambda \bigg).
$$

If we note that f_δ is in $\sigma\{u(s); s \leq t - \delta\}$ and apply the Markov property of $(\bar{X}, \bar{P}^x)$ this last expression becomes

$$
Q\bigg(I_{\delta<\tau} \bar{P}^{u(\delta)}\big(\bar{P}^{u(t-\delta)}(g)f_\delta I_{t-\delta<\sigma}\big)/V^1 1\big(u(\delta)\big); \Lambda \bigg).
$$

Another application of the conditional probability computation shows that this last displayed expression is equal to

$$
\begin{aligned}
& Q\bigg(I_{\delta<\tau}\big(\bar{P}^{u(t-\delta)}(g)f_\delta I_{t-\delta<\sigma}/(1 - e^{-\sigma})\big) \circ \theta_\delta; \Lambda \bigg) \\
& = Q\big(I_{\delta<\tau}(\bar{P}^{u(t)}g)f I_{t<\sigma}/(1 - e^{-\sigma \circ \theta_\delta}); \Lambda\big),
\end{aligned}
$$

and as δ approaches 0 this approaches the right side of (6.7). Thus the proof of (6.4) is complete.

VI. Excursions and Local Time

1. Introduction.

For many processes each point in the state space is regular for itself, and so for each point x we have the notions of local time at x and of excursions away from x. Brownian motion in R and most one dimensional diffusion processes are important examples. There has been a great deal of attention focused on the question of how in such cases the local time $\{L(t,x); t \geq 0\}$ at x varies with x and on the use of excursion theory, not only to study this question but also to make other important constructions. In this chapter we will give some applications of excursion theory to such matters. We will be considering local time and excursions at several points simultaneously, but the basic notion is still excursions away from a single point, so that no new notions appear. This should be contrasted with the situation in the next chapter. There we will be considering processes essentially in higher dimensional state spaces, where the relevant notion is that of excursions of the path away from a rather general set and where much less is known about specific formulas and constructions.

2. Ray's local time theorem.

(a) Excursions and local time. We will have several occasions to use the fact that local time at a point can be recovered from the behavior of the excursions away from that point. A typical and important example of this appeared in II–3 where we obtained the local time at 0 for reflecting Brownian motion as a suitably scaled limit as $\varepsilon \to 0$, of the number of excursions away from 0 that reach a height ε. The following general theorem, (2.1) below, is adequate for our purposes. It puts in more general form an argument that we have used several times before. We will assume that X is a standard process and b is a point in the state space regular for itself and for $\{b\}^c$ so that local time $\{L_t; t \geq 0\}$ at b exists and the corresponding excursion measure $\hat{P}$ has infinite mass. Let $\{A_\varepsilon; \varepsilon > 0\}$ be a family of $\mathcal{F}^0$ sets in path space U such that $A_\varepsilon \supset A_\delta$ for $\varepsilon < \delta$, $\hat{P}(A_\varepsilon) < \infty$ for each ε and $\hat{P}(A_\varepsilon)$ approaches ∞ as $\varepsilon \to 0$. For any set A in $\mathcal{F}^0$ we will let N_t^A denote the number of points s in G (the left end points of the excursion intervals) which are less than t and are such that the excursion starting at s lies in A.

(2.1) Theorem. *There is a sequence $\{\varepsilon_n\}$ approaching zero such that almost surely P^b*

$$\lim_{n \to \infty} N_t^{A_{\varepsilon_n}} / \hat{P}(A_{\varepsilon_n}) = L_t$$

for all $t \geq 0$.

Proof. If $\tau_t = \inf\{s \geq 0 | L_s > t\}$ then $N_{\tau_t}^A$ has a Poisson distribution with expected value $t\hat{P}(A)$. It follows immediately that as $\varepsilon \to 0$ for each t, $N_{\tau_t}^{A_\varepsilon} / \hat{P}(A_\varepsilon)$ converges in probability to t. Thus we can take a sequence $\{\varepsilon_n\}$ approaching 0 such that along ε_n the convergence takes place with probability one for each rational t, from which it follows that the convergence takes place at all t by the monotoneity and regularity of the expressions involved. Now it is clear that $N_{\tau_{L_s}}^A$ equals N_s^A and so substituting in L_t for t the assertion of the theorem follows.

(2.2) Remark. The reader will find a more general version of this theorem, which avoids also the passage to subsequences, in a paper by Maisonneuve [Ma, 2].

One application we will make of this is to the case where X is Brownian motion or reflecting Brownian motion, $b = 0$ and A_ε is the set of paths (starting at 0) that reach ε before returning to 0. Thus $N_t^{A_\varepsilon}$ is just the number of downcrossings of the interval $[0, \varepsilon]$ by the path of the process viewed in the time interval $[0, t]$, perhaps increased by 1 to account for an excursion interval straddling t. We have, by IV–1.1, that $\hat{P}(A_\varepsilon) = (2^{3/2}\varepsilon)^{-1}$ if X is Brownian motion and twice that if X is reflecting Brownian motion. Thus if as in II–3 we let $n_\varepsilon(t)$ denote the number of times the path $s \to X_s$ crosses from ε down to 0 as s ranges over the interval $[0, t]$ then $\varepsilon_n n_{\varepsilon_n}(t)$ converges to a multiple of local time at 0 as $\varepsilon_n \to 0$. We will need this, as we did in Chapter II, to establish some measurability assertions.

We will need in addition to the above the following variant of (2.1) in which one starts with a characteristic measure rather than a process: using the notation and definitions of V–1 let $\hat{P}$ be a characteristic measure compatible with a given minimal process, $\overline{X}$, let Y be a Poisson point process having $\hat{P}$ as characteristic measure, and let $\{\tau(t); t \geq 0\}$ denote the subordinator

$$\tau(t) = mt + \sum_{\substack{r \in D_Y \\ r \leq t}} \sigma(Y_r),$$

where $m = 1 - \hat{P}(1 - e^{-\sigma})$. Now let $\{x_t; t \geq 0\}$ denote the process obtained by linking together the excursions of Y according to Itô's prescription. We will assume already known that the x process is a recurrent extension of $\overline{X}$ either because the minimal semigroup satisfies the hypotheses leading to V–2.10 or because we have made some special argument to cover the specific case at hand. And we will assume established also that the given $\hat{P}$ is indeed the excursion measure for excursions of x away from b. (Our application will be to Brownian motion, where these assumptions hold.) Let φ denote the right continuous inverse of τ.

(2.3) Theorem. $\{\varphi(t); t \geq 0\}$ *is local time at b for the process* $\{x_t; t \geq 0\}$.

Proof. Recall the basic assumption that m is strictly positive if $\hat{P}$ is a finite measure. This implies that in any case τ is strictly increasing and so φ is continuous. A review of Itô's construction procedure shows that the set of points of increase of φ contains $\{t | x_t = b\}$, and the two sets differ by at most a countable set so that the support of $d\varphi$ is the closure of the set of b values as required. The rest of the argument is essentially the same as that in II–3 which found the local time for the sawtooth processes. To be specific if $\hat{P}$ is finite so that m is strictly positive then the recurrent extension is of holding and jumping type, and obviously

$$\varphi(t) = m^{-1} \int_0^t I_{\{b\}}(x_s)ds.$$

This makes it clear that $\{\varphi(t); t \geq 0\}$ is a continuous additive functional with the right support. The fact that φ has the proper normalization follows from the calculations in V–2.

In the rest of the proof when discussing additivity properties it is to be understood that notation such as $f \circ \theta_t$ means a quantity obtained by taking a funtion f of the path $s \to x_s$ and applying it instead to the path $s \to x_{s+t}$. (Properly one should set up transformations on the probability space underlying the point process Y but we will leave that to a sufficiently determined reader.) Coming to the case where $\hat{P}$ is an infinite measure let $n_\epsilon(t)$ be the number of points s not exceeding t and such that $\tau(s) - \tau(s-)$ strictly exceeds ϵ. If $\lambda_\epsilon = \hat{P}(\sigma > \epsilon)$ then λ_ϵ approaches infinity as ϵ approaches 0 and so by the usual argument using the fact that $n_\epsilon(t)$ is Poisson with expectation $t\lambda_\epsilon$ we may take a sequence ϵ_n approaching 0 such that with probability one for all t

$$n_{\varepsilon_n}(t)/\lambda_{\varepsilon_n} \to t, \qquad n \to \infty.$$

(The probability space referred to is the one over which the original point process is defined.) Let G stand for the left ends of the excursion intervals away from b for the process $\{x_t; t \geq 0\}$. Let $M_\varepsilon(t)$ denote the number of points s in G such that $s+\varepsilon < t$ and $\sigma \circ \theta_s > \varepsilon$. Clearly $M_\varepsilon(t)$ is measurable relative to the appropriate completion of $\sigma\{x_r; r \leq t\}$, and $M_\varepsilon(t)$ can differ from $n_\varepsilon\big(\varphi(t)\big)$ by at most one, so that

$$M_{\varepsilon_n}(t)/\lambda_{\varepsilon_n} \to \varphi(t) \qquad n \to \infty.$$

This establishes the necessary measurability property of $\varphi(t)$. Also it is clear that $M_\varepsilon(t + s)$ differs from $M_\varepsilon(t) + M_\varepsilon(s) \circ \theta_t$ by at most one and from this the additivity property of φ follows. The fact that φ is normalized properly can be checked using III–3.30. This completes the proof.

(b) Some local time formulas. We will assume throughout this paragraph that X is Brownian motion in R. Then every point is regular for itself and so for every x there is a local time $\{l(t,x); t \geq 0\}$ at x. That is $l(\cdot, x)$ is a continuous additive functional which grows exactly when X_t is at x. Of course these requirements do not uniquely determine l; anything of the form $f(x)l(\cdot, x)$ will have the same properties if f takes on values only in $(0, \infty)$. But it is in fact possible to choose l so that it is jointly continuous in the pair (t, x) and is also an *occupation time density* relative to lebesgue measure; that is with probability one relative to any P^x we have

$$\int_A l(t, x)dx = \text{leb meas}\{s \leq t | X_s \in A\}$$

for all $t \geq 0$ and Borel sets A. This follows from a theorem of Trotter [T, 1] later improved and generalized by Boylan [Bo, 1]. It is fundamental for the general development of one dimensional diffusion processes. As we will elaborate on in a moment the above properties essentially determine l uniquely. We will denote this particular choice by $\{L(t, x); t \geq 0, x \in R\}$. As to the uniqueness, we know that local time at a fixed point x is unique up to multiplication by a constant. We always have chosen the normalization so as to have

$$E^x \int_0^\infty e^{-t}d_t l(t, x) = 1$$

for all x. Let us call that choice *normalized local time*. To compare l with L set

$$\Phi(x) = E^x \int_0^\infty e^{-t} d_t L(t, x)$$

and note that for any point b

$$E^b \int_0^\infty e^{-t} d_t L(t, x) = E^b e^{-\sigma_x} \Phi(x)$$

$$= e^{-\sqrt{2}|x-b|} \Phi(x).$$

Then using the presumed fact that L is an occupation time density we have for any positive Borel function f

$$\int_{-\infty}^\infty \frac{1}{\sqrt{2}} e^{-\sqrt{2}|x-b|} f(x) dx = E^b \int_0^\infty e^{-s} f(X_s) ds$$

$$= E^b \int_0^\infty e^{-t} \{ \int_0^t f(X_s) ds \} dt$$

$$= E^b \int_0^\infty e^{-t} \{ \int_{-\infty}^\infty f(x) L(t, x) dx \} dt$$

$$= \int_{-\infty}^\infty f(x) e^{-\sqrt{2}|x-b|} \Phi(x) dx,$$

and so $\Phi(x) = (1/\sqrt{2})$. Thus the occupation time density local time is normalized local time multiplied by $(1/\sqrt{2})$. What Itô and McKean call *standard* Brownian *local time* is $(1/2)L$. Denote this by $T(t, x), t \geq 0, x \in R$. It is for each fixed x a local time in t at x, jointly continuous in (t, x), an occupation time density relative to $2dx$ and satisfies

$$E^x \int_0^\infty e^{-t} d_t T(t, x) = 2^{-3/2}.$$

(The choice of $2dx$ as basic measure conforms to the usage of general diffusion theory.)

Obviously there is room for confusion. We will need to keep track of normalizations but we will not need the fact that local time at x can be chosen to be continuous or even measurable in x, but only that it exists for each fixed x and has the appropriate properties as a function of t. In the rest of section 2 we will use $\{L(t, x); x \in R, t \geq 0\}$ to denote a family of functionals which is for each x local time at x and satisfies $E^x \int_0^\infty e^{-t} d_t L(t, x) = 1/\sqrt{2}$. Aside from that we are not making any assumptions about how L varies with x.

Another definition that often is given for local time, at 0 let us say, for Brownian motion or reflecting Brownian motion is

$$l'_t = \lim_{\varepsilon \to 0} \frac{1}{2\varepsilon} \int_0^t I^\varepsilon(X_s)\,ds$$

where I^ε is the indicator of $[-\varepsilon, \varepsilon]$ if X is Brownian motion, and of $[0, \varepsilon]$ if X is reflecting Brownian motion. The existence has to be established of course, but once it is we have in either case

$$E^0 \int_0^\infty e^{-t} dl'_t = \lim_{\varepsilon \to 0} \frac{1}{2\varepsilon} \int_{-\varepsilon}^\varepsilon \frac{1}{\sqrt{2}} e^{-\sqrt{2}|x|} dx = 1/\sqrt{2},$$

so l' is normalized local time multiplied by $1/\sqrt{2}$.

There are a number of other important distributional facts about local time that appear frequently. One of the simplest but most important is the following:

(2.4) Theorem. *For any point x, $L(\sigma_x, 0)$ has, relative to P^0, the exponential distribution with rate $1/2|x|$. $(\sigma_x = \inf\{t > 0 | X_t = x\}.)$*

Proof. This is a Brownian motion result so it is understood that P^0 refers to Brownian motion starting at 0. For the proof of (2.4) let T_t denote the time when $L(\cdot, 0)$ reaches the level t. Then T_t is a stopping time and the events $\{L(\sigma_x, 0) > t\}$ and $\{T_t < \sigma_x\}$ are for each t almost surely the same. Now X_{T_t} must be 0 and T_{t+s} is equal to $T_t + T_s \circ \theta_{T_t}$, while σ_x is equal to $T_t + \sigma_x \circ \theta_{T_t}$ if $\sigma_x > T_t$. Thus

$$P^0(\sigma_x > T_{t+s}) = P^0(\sigma_x \circ \theta_{T_t} > T_s \circ \theta_{T_t}; T_t < \sigma_x)$$
$$= P^0(\sigma_x > T_s) P^0(\sigma_x > T_t)$$

and so

$$P^0\big(L(\sigma_x, 0) > t + s\big) = P^0\big(L(\sigma_x, 0) > s\big) P^0\big(L(\sigma_x, 0) > t\big).$$

This shows that $L(\sigma_x, 0)$ has an exponential distribution. The expected value of $L(\sigma_x, 0)$ is the reciprocal of the rate. To calculate this use the excursion formula

$$(2.5) \qquad E^0 \sum_{s \in G} Z_s f \circ \theta_s = \hat{P}(\sigma_x < \sigma_0) E^0 \int Z_s\, dL_s$$

where Z_s is the indicator of $\{s < \sigma_x\}$, f is the indicator of $\{\sigma_x < \sigma_0\}$ and, as they must be to make the formula correct, $\hat{P}$ is the excursion measure for Brownian motion and L is normalized local time at 0. The sum on the left of (2.5) is 1 as it records only the first excursion reaching x before returning to 0. According to IV–1.1 we have $\hat{P}(\sigma_x < \sigma_0) = (2^{3/2}|x|)^{-1}$ (recall $\hat{P}$ there is for reflecting Brownian motion), and $\int_0^\infty Z_s \, dL_s = L_{\sigma_x} = \sqrt{2} L(\sigma_x, 0)$. Putting these into (2.5) we get

$$E^0 L(\sigma_x, 0) = 2|x|,$$

which completes the proof.

In the discussion of how the local time $L(\cdot, x)$ varies with x the Bessel processes play an important role. We introduced them in Chapter II. The n-dimensional Bessel process is the radial part of n-dimensional Brownian motion. It turns out to be a (time homogeneous) Markov process, and hence a diffusion process, because of symmetries in the Brownian motion probabilities. One can obtain the transition density relative to lebesgue measure on $[0, \infty)$ by doing a polar coordinates integration on the transition density for n-dimensional Brownian motion. If $g(t, x, y)$ denotes the resulting transition density then for $t > 0, x > 0, y > 0$

$$g(t, x, y) = t^{-1} e^{-(x^2 + y^2)/2t} (xy)^{1-n/2} I_{\frac{n}{2}-1}\left(\frac{xy}{t}\right) y^{n-1}$$

where I_ρ denotes the Bessel function

$$I_\rho(x) = \sum_{k=0}^{\infty} \frac{1}{k! \Gamma(k + \rho + 1)} \left(\frac{x}{2}\right)^{2k+\rho} \qquad x > 0.$$

This is given in Itô and McKean on page 60.

In fact the processes directly relevant for our purposes are the squares of the Bessel processes in dimensions 2 and 4. These also are diffusion processes of course. One obtains the transition densities by replacing x and y by their square roots and multiplying by the Jacobian $(1/2)y^{-1/2}$. Thus if $\{r_n(t); t \geq 0\}$ denotes an n-dimensional Bessel process then the conditional density in y for $r_n^2(t)$ given $r_n^2(0) = x$ is in case $n = 2$

$$(2.6) \qquad \sum_{k=0}^{\infty} e^{-x/2t} \frac{(x/2t)^k}{k!} e^{-y/2t} \frac{(y/2t)^k}{\Gamma(k+1)} \frac{1}{2t}.$$

In case $n = 4$ the expression is exactly the same except that $(y/2t)^k$ is replaced with $(y/2t)^{k+1}$ and $\Gamma(k+1)$ is replaced with $\Gamma(k+2)$. The expressions are easy to interpret probabilistically: if S_k represents the sum of k independent random variables each having an exponential distribution with rate 1 then the conditional distribution of $r_2^2(t)$ given $r_2^2(0) = x$ is just that of $(2t)S_{N+1}$ where N denotes a random variable independent of the process $\{S_k; k \geq 1\}$ and having the Poisson distribution with parameter $x/2t$. In dimension 4 the conditional distribution is that of $(2t)S_{N+2}$ where otherwise the symbols have the same meaning. The variable S_k has Laplace transform $E\big(\exp(-\alpha S_k)\big) = (1+\alpha)^{-k}$ and so the conditional Laplace transforms are

$$
(2.7) \qquad
\begin{aligned}
&(1 + 2t\alpha)^{-1} \exp(-\alpha x/(1 + 2t\alpha)) & n = 2 \\
&(1 + 2t\alpha)^{-2} \exp(-\alpha x/(1 + 2t\alpha)) & n = 4
\end{aligned}
$$

respectively. The point is that these formulas are easy to recognize when they appear, as they will shortly in our discussion of the dependence of $L(\cdot, x)$ upon the variable point x.

(c) Ray's theorem.   Ray [Ra, 1] proved the following remarkable theorem.

(2.8) Theorem. *Under the probability law of Brownian motion starting at 0*

(a) *the process $\{L(\sigma_1, x); -\infty < x < \infty\}$ is a Markov process,*

(b) *the processes $\{L(\sigma_1, x); 0 \leq x \leq 1\}$ and $\{r_2^2(1-x); 0 \leq x \leq 1\}$ are equivalent,*

(c) *the processes $\{L(\sigma_1, x); x \leq 0\}$ and $\{(1-x)^2 r_4^2(\frac{1}{1-x} - \beta^*); x \leq 0\}$ are equivalent, where the Bessel processes r_2 and r_4 start at 0 and β^* is independent of r_4 and is uniformly distributed on $[0,1]$. In (c) we set $r_4(s) = 0$ if $s < 0$.*

We will give a proof of Ray's theorem which uses a truly clever argument given by David Williams [W, 3] based on excursion theory. Note that the assertion in (2.8) involves only the joint distributions of $L(\sigma_1, x)$ for finitely many x.

To state Williams' theorem, which is the basic fact in the excursion theory proof of Ray's theorem, let us use the notation and results of section

5 from Chapter IV where we got a reflecting Brownian motion $\{y_t; t \geq 0\}$ by making a random time change in a Brownian motion $\{x_t; t \geq 0\}$. This Brownian motion was obtained from synthesizing excursions, and the y process was identified as x^+, that is what one gets from x by using only positive excursions. We will use $L(t,d)$ to denote the choice of Brownian local time at d from paragraph (b) and let φ^+ denote normalized local time at 0 for y. By (2.3) this usage of φ^+ agrees with the one in IV-5. Probabilities and expectations are for Brownian motion starting at 0.

(2.9) Theorem. *For each $t > 0$, $\theta > 0$ and $b < 0$ we have*

$$E\{e^{-\theta L(\rho(t),b)}|y_t; t \geq 0\} = e^{-\sqrt{2\theta}\varphi^+(t)/1-2b\theta}.$$

Proof. By IV-5 we may assume that x is Brownian motion constructed from excursions and that y is in fact x^+. Quantities based on domain points D^- are independent of those based on D^+ as the corresponding paths lie in disjoint sets. Since $\rho(t) = t + \tau^-(\varphi^+(t))$ the excursion intervals corresponding to points in D^- that appear prior to the excursion interval (necessarily on the positive side) that straddles $\rho(t)$ arise exactly from those points r in D^- which are less than $\varphi^+(t)$. These are the only quantities contributing to $L(\rho(t),b)$, and so $L(\rho(t),b)$ emerges as a sum $\sum_1^N e_k$ where the number of summands N is the number of domain points r in D^- with r less than $\varphi^+(t)$ and $\sigma_b(X_r)$ less than $\sigma_0(X_r)$, an empty sum being 0, where the e_k are independent quantities distributed as $L(\sigma_0,b)$ for a Brownian motion starting at b. By (2.4) then each e_k is exponentially distributed with rate $-1/2b$ and hence has Laplace transform $(1 - 2b\theta)^{-1}$. The quantity N has the Poisson distribution with parameter $(\sqrt{2}|b|)^{-1}$ (the proper data is that for reflecting Brownian motion since we are focusing only on excursions to the negative side) and so we obtain the conditional Laplace transform as

$$E((1 - 2b\theta)^{-N}) = \sum_{k=0}^{\infty}(1 - 2b\theta)^{-k}e^{-\Delta}(\Delta)^k/k!$$

with Δ equal to $(\sqrt{2}|b|)^{-1}\varphi^+(t)$. This is exactly what is asserted in (2.9).

To apply (2.9) note that by a simple measure theory argument we may replace the fixed time t in the equality there by any positive function measurable relative to $\sigma\{y_t; t \geq 0\}$. In particular replace t by the time, σ_1^y, when the y-process first reaches 1. It is clear that $\rho(\sigma_1^y)$ is simply σ_1 (for the process x) and that replacement on the left side of (2.9) yields $L(\sigma_1, b)$. As to the right side first of all we assert that $\varphi^+(\sigma_1^y)$ is almost surely equal to $(1/\sqrt{2})L(\sigma_1, 0)$. Indeeed Theorem (2.1), with $A_\varepsilon = \{\sigma_\varepsilon < \sigma_0\}$, asserts that as ε approaches 0 through a subsequence we have almost surely

$$\varepsilon\sqrt{2}(N_{\sigma_1}^\varepsilon)^y \to \varphi^+(\sigma_1^y)$$

$$\varepsilon(2^{3/2})N_{\sigma_1}^\varepsilon \to \varphi(\sigma_1)$$

where N_t^ε denotes the number of downcrossings of $[0, \varepsilon]$ before time t, and the superscript y means the quantity is defined in term of the y process. Clearly $N_{\sigma_1}^\varepsilon$ and $(N_{\sigma_1}^\varepsilon)^y$ are equal and it follows that $\varphi^+(\sigma_1^y)$ and $L(\sigma_1, 0)$ differ by a multiplicative constant. The fact that $1/\sqrt{2}$ is the correct choice follows from the definitions in paragraph (b). Also exactly the same argument yields the fact that $L(\sigma_1, x)$ is y measurable for all x in $[0, 1]$. Thus, making these substitutions and taking conditional expectation relative to $\sigma\{L(\sigma_1, x); 0 \le x \le 1\}$ we obtain the following corollary:

(2.10) Corollary. *For each $\theta > 0$ and $b < 0$ we have*

$$E\{e^{-\theta L(\sigma_1, b)} | L(\sigma_1, x); x \ge 0\} = e^{-\theta L(\sigma_1, 0)/(1 - 2b\theta)}.$$

This result essentially contains the Markovian properties asserted by Ray's theorem and also the formulas necessary to relate $\{L(\sigma_1, x); x \le 1\}$ to the relevant Bessel processes. To justify this we will use (2.10) to find the conditional distribution of $L(\sigma_1, c)$ given $\sigma\{L(\sigma_1, x); d \le x \le 1\}$ for all choices of $c \le d \le 1$, breaking things up into cases. To ease the typography we will set

$$\psi(\theta, a, x) = \exp(-\theta x/(1 + 2a\theta)),$$

with θ, x and a positive, and we will write Z_x for $L(\sigma_1, x)$. Thus the right side in (2.10) reads $\psi(\theta, -b, Z_0)$.

(1) $0 \le c \le d$. Since we are starting the Brownian motion at 0 we have $Z_x = Z_x \circ \theta_{\sigma_d}$ if x exceeds d, whereas

$$Z_c = L(\sigma_d, c) + Z_c \circ \theta_{\sigma_d},$$

the first quantity on the right being exponentially distributed with rate $(2(d-c))^{-1}$ and the summands independent, since x_{σ_d} is constant. Thus making an obvious translation of (2.10) we obtain

$$(2.11) \qquad E\{e^{-\alpha Z_c}|Z_x; d \leq x \leq 1\} = \frac{1}{2\alpha(d-c)+1}\psi(\alpha, d-c, Z_d)$$

The distribution of Z_d itself is exponential with rate $(2(1-d))^{-1}$, which is the same as that of $r_2^2(1-d)$ started at 0, and the conditional Laplace transform in (2.11) with Z_d and $d-c$ replaced by x and t is exactly what appear in (2.7) as the transition data of $\{r_2^2(t); t \geq 0\}$. This establishes $\{Z_{1-x}; 0 \leq x \leq 1\}$ as equal in law to $\{r_2^2(x); 0 \leq x \leq 1\}$ with $r_2(0) = 0$.

(2) $c < 0 \leq d$. If f is measurable relative to $\sigma\{Z_x; d \leq x \leq 1\}$ then $f = f \circ \theta_{\sigma_d}$ and we have

$$E_0(e^{-\alpha Z_c}f) = E_0(e^{-\alpha L(\sigma_d, c)})E_d(e^{-\alpha Z_c}f).$$

In the first factor, $L(\sigma_d, c)$ has an exponential distribution provided we condition on the event $\{\sigma_c < \sigma_d\}$, which has probability $d/(d-c)$. Since a Brownian motion started at d is just a translate of Brownian motion starting at 0 the second factor can be read off from (2.10) and we obtain

$$E\{e^{-\alpha Z_c}|Z_x; x \geq d\} = \frac{1-2\alpha c}{1+2\alpha(d-c)}\psi(\alpha, d-c, Z_d).$$

(3) $c < d < 0$. The only complicating feature in this case is that with positive probability the Brownian motion starting at the origin reaches 1 before reaching d. This is equivalent to $Z_d = 0$ and of course implies that $Z_c = 0$. But if σ_d is less than σ_1 then starting at time σ_d the situation is identical to that of (2.10) translated from d to 0 and we conclude easily that in this case

$$(2.12) \qquad E\{e^{-\alpha Z_c}|Z_x; d \leq x \leq 1\} = \psi(\alpha, d-c, Z_d).$$

Note that now we have established the Markov property of $\{Z_x; x \leq 1\}$ at least if we read time from right to left. By the symmetry of past and future in the Markov property this gives the Markov property with "time" read in the usual direction. All that is left is the identification with the r_4 process for negative values of x. In the supposed equivalence (2.8c) the appearance of β^* will reflect the fact that Z_c is 0 if c is less than $\min_{s \leq \sigma_1} x_s$. Thus Z_c is 0 if and only if σ_c exceeds σ_1, an event of probability $-c/(1-c)$ which is indeed the probability that β^* exceeds $1/(1-c)$. The Laplace transform of $(1-c)^2 r_4^2 \, (\dfrac{1}{1-c} - \beta^*)$ on the set $\beta^* < 1/(1-c)$ is by paragraph (b) simply

$$\int_0^{1/(1-c)} \frac{dt}{1 + 2\alpha(1-c)^2 t}.$$

This is easy to cacluate. So is

$$E(e^{-dZ_c}; Z_c > 0) = \int_0^{\infty} \psi(\alpha, -c, x)\frac{1}{2}e^{-x/2}dx - \frac{1}{1-c}$$

and the results agree. Thus Z_c and the corresponding Bessel process expression have the same distribution. Knowing that both processes are Markovian with the same absolute distributions now it is necessary only to check the transition function. The Bessel data is given with time running one way and (2.12) gives the local time data with time running the other way. Since we now know the absolute distributions of Z_c and Z_d the conditional density of Z_d given Z_c can be obtained from that of Z_c given Z_d by multiplying and dividing by the marginal densities. The algebra is easy and one finds that (2.8c) is correct. We will leave the details to the reader.

(2.13) Remark. There is a lot of interesting mathematics associated with Ray's theorem that the forgoing presentation does not touch on. For example Williams' path decomposition, a most interesting topic in Markov process theory, arose (so he says) from an attempt to understand Ray's theorem. Readers who work through Williams' paper [W, 2] on path decomposition will find themselves challenged but enlightened as to general diffusion theory, time reversal and splitting times, which are important parts of general Markov process theory. Other approaches to Ray's theorem use martingale techniques, notably Tanaka's formula, which provide a calculus for identifying the component Bessel processes, (see [JY, 1].)

(2.14) Exercise. (a) Use the relationship $\rho(t) = t + \tau^-(\varphi^+(t))$ from the proof of (2.9) and the fact that $Ee^{-\alpha\tau^-(t)} = \exp(-t\sqrt{\alpha})$ to conclude that

$$E(e^{-\alpha\rho(t)}|y_t; t \geq 0) = e^{-\alpha t - \varphi^+(t)\sqrt{\alpha}}.$$

(b) Take expectation on both sides and make an explicit calculation on the resulting right side to calculate $Ee^{-\alpha\rho(t)}$ and conclude Lévy's arc-sine law, that is the time spent on the positive side by $\{x_t; 0 \leq t \leq 1\}$ has the arc-sine density, $(\pi\sqrt{x(1-x)})^{-1}$ for $0 < x < 1$. The calculations can be made by noting that $\varphi^+(t)$ is equal in law to $\sqrt{2}\max_{s\leq t} x_s$ and using the fact that $\exp -\sqrt{2\alpha}$ is the Laplace transform of a probability density of the form $kx^{-3/2}\exp(-1/2x)$, $x > 0$.

3. Trotter's theorem.

(a) **Occupation time density.** Trotter's theorem establishes $L(t, x)$ as an occupation time density, and one having joint continuity properties so that one can form expressions such as

$$A(t) = \int L(t, x) m(dx)$$

m being a measure on R, which are vital in setting up general diffusion theory. We do not intend to prove Trotter's theorem, which involves some difficult estimates, but we will need an accurate statement of the result as well as some steps in the proof.

He proceeds as follows: let $\{x_t; t \geq 0\}$ be Brownian motion with continuous paths defined over a probability space Ω. It is adequate for our purposes to take Ω to be the space of continuous functions on $[0, \infty)$ and x_t to be the coordinate function. Trotter defines

$$\mu(t, A, \omega) = \text{leb meas}\{s \leq t | x_s(\omega) \in A\}$$

with A a Borel set in R, and then he sets

$$F_n(t, x, \omega) = 2^n \mu(t, I_{k,n}, \omega) \qquad x \in I_{k,n}$$

where $I_{k,n} = [k/2^n, (k+1)/2^n)$. Then F_n satisfies

$$\int_A F_n(t, x, \omega) dx = \mu(t, A, \omega),$$

for all t and ω provided A is a union of intervals of the form $I_{k,m}$ with $m \leq n$. For each n the function $t \to F_n(t, x, \omega)$ is a continuous additive functional in t and it satisfies the additivity property

$$F_n(t + s, x, \omega) = F_n(t, x, \omega) + F_n(s, x, \theta_t \omega)$$

as an identity in t, s, ω. It is for fixed t and ω continuous in x except for possible discontinuities at the points $k/2^n$. Trotter makes a strong estimate on the probabilistic behavior of the maximum, for fixed t, of these discontinuities and then applies it cleverly to obtain the following fact.

(3.1) Theorem. *Let Λ denote the set of ω such that as $n \to \infty$ F_n converges uniformly on every set of the form $[0, t] \times R$ to a function which is jointly continuous. Then $P^x(\Lambda^c) = 0$ for all x.*

We will call this result Trotter's theorem. Define L by

$$L(t, x, \omega) = \lim_{n \to \infty} F_n(t, x, \omega) \qquad \omega \in \Lambda$$
$$= 0 \qquad\qquad\qquad \omega \notin \Lambda.$$

Clearly if ω is in Λ then L is an occupation time density as defined in section 2. Also from the additivity property of the F_n it follows that if ω is in Λ so is $\theta_t \omega$, and so we have

$$L(t + s, x, \omega) = L(t, x, \omega) + L(s, x, \theta_t \omega)$$

for all t, s, x if ω is in Λ. Clearly for each t and x, $F_n(t, x)$ is measurable in ω relative to $\sigma\{x_s; s \leq t\}$; and so $L(t, x)$ is measurable relative to $\mathcal{F}_t$, the usual completion. Thus for each x, $\{L(t, x); t \geq 0\}$ is a continuous additive functional of the Brownian motion. Clearly for each x, $L(t, x)$ should be a version of local time at x. This can be checked by computing a potential: indeed for real numbers x and y set

$$\Phi_x(y) = E^y \int_0^\infty e^{-t} d_t L(t, x) = E^y \int_0^\infty e^{-t} L(t, x) dt.$$

Then if f is bounded and continuous we have, using the fact that $L(t, x)$ is in x an occupation time density,

$$\int_R \Phi_x(y) f(x) dx = E^y \int_0^\infty e^{-t} \left(\int_0^t f(x_s) ds \right) dt$$
$$= E^y \int_0^\infty e^{-t} f(x_t) dt = \int_R \frac{1}{\sqrt{2}} e^{-\sqrt{2}|x-y|} f(x) dx$$

from which we have $\Phi_x(y) = (1/\sqrt{2}) \exp(-\sqrt{2}|x - y|)$. By the uniqueness theorem for continuous additive functionals $\{L(t, x); t \geq 0\}$ is normalized local time at x multiplied by $(1/\sqrt{2})$.

Note that the set Λ in (3.1) is in the σ-algebra generated by the coordinate functions since membership in Λ is determined by $F_n(t, x, \omega)$ for t and x rational. We will need this remark in a moment.

(b) **Excursion local time.** Let W be the subset of $\Omega(= \mathcal{C}([0, \infty), R))$ consisting of those paths which represent Brownian excursions away from 0. That is W is the set of continuous paths ω such that $\omega(0) = 0$,

$0 < \sigma(\omega) < \infty$ and $\omega(t) = 0$ for all t greater than $\sigma(\omega)$. Here $\sigma = \sigma_0$. Let $\hat{P}$ be Brownian excursion measure for excursions away from 0. By *excursion local time* we will mean a real valued function $\ell_t^x(\omega)$ defined for x in R, t positive and ω in W with these properties: (i) for each x and t, ℓ_t^x is measurable relative to the $\hat{P}$ completion of $\sigma(x_s; s \leq t)$, (ii) for each ω in W, $\ell_t^x(\omega)$ is jointly continuous in (t,x), increasing in t and constant on the interval $\sigma(\omega) \leq t$, and (iii) for almost all ω relative to $\hat{P}$

$$(3.2) \qquad \int_0^{t \wedge \sigma(\omega)} \varphi(\omega(s))ds = \int_R \ell_t^x(\omega)\varphi(x)dx$$

for all t positive and φ positive and Borel measurable on R. It seems clear that we can obtain such an object by setting $\ell_t^x(\omega)$ equal to $L(t \wedge \sigma(\omega), x, \omega)$ where L is the local time from paragraph (a). We will set this out as a formal statement. The mere fact that a jointly continuous occupation time density for Brownian motion exists does not seem to be quite adequate because the set W has P^x measure 0 for the Brownian motion measure.

(3.3) Theorem. *The function ℓ defined by*

$$\ell_t^x(\omega) = L(t \wedge \sigma(\omega), x, \omega)$$

is excursion local time.

Proof. Consider the set Λ from the proof of (3.1). We already have noted that if $\theta_s\omega$ is not in Λ then ω is not in Λ and so $\sum_{s \in G} e^{-s} I_{\Lambda^c} \circ \theta_s$ is non zero only on Λ^c. Thus

$$\hat{P}(\Lambda^c) = E^0 \sum_{s \in G} e^{-s} I_{\Lambda^c} \circ \theta_s$$
$$\leq P^0(\Lambda^c) = 0$$

and so almost surely relative to $\hat{P}$ the approximate densities F_n converge uniformly on $[0, \sigma] \times R$ to L. Now the fact that ℓ has the required properties follows as in the remarks following (3.1).

We need to carry over one more property from L to ℓ. Suppose $x_0(\omega) = 0$ and $0 < \epsilon < \delta$. For n large enough that 2^{-n} is less then $\delta - \epsilon$ it is clear that $\hat{F}_n(t - \sigma_\epsilon(\omega), x, \theta_{\sigma_\epsilon}\omega)$ is equal to $F_n(t, x, \omega)$ if x exceeds δ. Since for a set Γ in $\sigma\{x_s; s \geq 0\}$

$$\hat{P}(I_\Gamma \circ \theta_{\sigma_\epsilon}; \sigma_\epsilon < \sigma_0) = P^\epsilon(\Gamma)\hat{P}(\sigma_\epsilon < \sigma_0)$$

growth properties of $t \to L(t \wedge \sigma, x)$ which are almost sure relative to P^ϵ will be almost sure relative to $\hat{P}$. From our original discussion of local time at a point we know that relative to P^ϵ almost surely the two sets $\{t|x_t = x\}$ and $\{t|L(\cdot, x)$ is strictly increasing at $t\}$ are the same. It follows that almost surely relative to $\hat{P}$ these two sets are the same.

We will need this observation in a moment, and we will need excursion local time for an application in section 4.

(c) **The excursion filtration.** Williams' proof of Ray's theorem indicates the profit to be made from analyzing excursions in terms of their behavior above and below an arbitrary level and then regarding the variable level as a time parameter in some scheme that exhibits Markovian properties. This means introducing a filtration, indexed by the levels, on the space of excursion paths. As in paragraph (b) we will let W denote the set of excursion paths away from 0 for Brownian motion. Given a number $a \geq 0$ and a function ω in W set

$$A_a^\omega(t) = \text{leb meas}\{s \leq t|\omega(s) \leq a\}.$$

This is continuous and increasing in t and approaches infinity with t since $\omega(s)$ is 0 for all large s. Distinguishing $\sigma_{(a)}$ from the familiar σ_a set

$$\sigma_{(a)}^\omega(u) = \inf\{t|A_a^\omega(t) > u\},$$

so that $\sigma_{(a)}$ as a function of u is the right continuous inverse of A_a as a function of t. The function $\sigma_{(a)}$ is strictly increasing and right continuous, and the composition $\omega \circ \sigma_{(a)}^\omega$ is a continuous function because if $\sigma_{(a)}^\omega(t)$ strictly exceeds $\sigma_{(a)}^\omega(t-)$ then $\omega(\sigma_{(a)}^\omega(t-)) = \omega(\sigma_{(a)}^\omega(t)) = a$. The mapping ρ_a from W to W is defined by

$$\rho_a\omega(t) = \omega(\sigma_{(a)}^\omega(t)) \qquad t \geq 0.$$

Of course the effect of this reparametrization simply is to lift out the intervals where the path ω lies strictly above the level a. Thus ρ_0 is the operation from IV–5 and VI–2 that transforms Brownian motion into the negative of a reflecting Brownian motion. (And those facts about ρ_a that we will use all are based on the ideas in the proof of (2.9) from the present chapter.)

If we let $\mathcal{F}^0$ denote the σ-algebra in W generated by the coordinate functions then ρ_a as a function from $(W, \mathcal{F}^0)$ to itself is measurable. We will let $\mathcal{E}_a^0$ denote the σ-algebra $\rho_a^{-1}(\mathcal{F}^0)$ augmented by those sets Γ such that $\Gamma \subset \Delta$ with Δ in $\mathcal{F}^0$ and $\hat{P}(\Delta) = 0$, $\hat{P}$ being excursion measure for Brownian motion. It is fairly clear (see the exercises) that $\rho_a \circ \rho_b = \rho_a$ if a is less than b and so $\{\mathcal{E}_a^0; a \geq 0\}$ is indeed a filtration. It is called the *excursion filtration*.

For introducing Markovian properties the past up to time (or level) a will be represented by $\mathcal{E}_a^0$. The future is represented by *excursions above the level a* where we define this concept as follows: given ω in W call a *raw excursion of ω above the level a* the function ω regarded in the interval $[\alpha, \beta]$ where $\omega(\alpha) = \omega(\beta) = a$ and $\omega(t) > a$ for $\alpha < t < \beta$. For each ω there are at most countably many raw excursions above a, and no interval $[\alpha, \beta]$ with $\beta > \sigma(\omega)$ exists. Let us label the raw excursion intervals $[\alpha_i, \beta_i]$ with i ranging over an indexing set I_ω. Associate with each i the element ω_i of W defined by

$$\omega_i(t) = \omega((\alpha_i + t) \wedge \beta_i) - a$$

and then define the point function Y_ω by saying that the domain points are the numbers $\ell_{\alpha_i}^a(\omega)$ (ℓ is excursion local time) and associated with the domain point $\ell_{\alpha_i}^a(\omega)$ is the path ω_i. Recall from our discussion of excursion local time that almost surely $\hat{P}$ the function $t \to \ell_t^a(\omega)$ is strictly increasing at each α_i so the domain points indeed are distinct. We will call the point process Y the *future beyond level a*. Note that the definition of Y involves local time at a which also involves $\mathcal{E}_a^0$ events, and so Y and $\mathcal{E}_a^0$ are not independent.

The following theorem contains the basic probabilistic fact we will need for applying these notions. We will use the notation from IV–5. Let $\{x_t; t \geq 0\}$ be Brownian motion starting at 0 and let y^- be the reflecting Brownian motion on $(-\infty, 0]$ obtained by reparametrizing x in terms of the time it spends below 0, so that the σ-algebra of y^- is just $\mathcal{E}_0^0$, the past up to level 0, except defined over Ω rather than over W. Let L be normalized local time at 0 for x. Let X be the Poisson point process of excursions away from 0 of x (using normalized local time) and let X^+ be the domain restriction of X to the positive excursions and X^- the restriction to the negative excursions. In the terminology we have just been using the domain

points of X^+ are the values L_α as α ranges over the left ends of the raw
excursion intervals above the level 0, and the path associated with L_α is
just the corresponding raw excursion. So X^+ is what we just called the
future beyond level 0 except for the unimportant but temporarily annoying
fact that we are using normalized local time at 0 rather than $L(t,0)$ from
Trotter's theorem. Let $\hat{p}_t$ denote the product measure of lebesgue measure
restricted to $[0,t]$ with Brownian excursion measure restricted to the set of
positive excursions. And finally let d be a negative number and $T = L_{\sigma_d}$.

(3.4) Theorem. *The conditional distribution of $\alpha_T X^+$ given $\sigma\{y^-\}$ is
that of a Poisson random measure on $[0,\infty) \times W$ with intensity measure
$\hat{p}_T$.*

Proof. We first remark that one consequence of this theorem is the equality

$$(3.5) \qquad E^0(e^{-\sum_{s \in G, s < \sigma_d} \psi(L_s, x' \circ \theta_s)}; \Lambda) = E^0(\exp\{\int (e^{-\psi} - 1)d\hat{p}_T\}; \Lambda)$$

where Λ is in $\sigma\{y^-\}$, ψ is positive and measurable on $[0,\infty) \times W$ and $\psi(s, \omega)$
vanishes if ω is a negative excursion, G is the left ends of the excursion in-
tervals, and $x'_t = x_{t \wedge \sigma}$. Of course (3.5) will follow from (3.4) by applying
the exponential formula for Poisson random measures. The proof of (3.4) is
the same as the proof of (2.9). The notation will be that of IV–5. Specifically we may assume that X is a Poisson point process whose characteristic
measure is $\hat{P}$, Brownian excursion measure, and that the Brownian motion
x is obtained from X by synthesis. Then X is indeed the point process of
excursions of x away from 0, φ is normalized local time at 0, and, by the
discussion in IV–5, y^- is X^- measurable. Let t be a fixed time and consider the domain restriction of X^+ to $[0, \varphi(\rho^-(t))]$. Any event dependent
only on this point process is defined in terms of paths X_r corresponding to
points r in D^+ that appear prior to the domain point, necessarily in D^-,
such that the corresponding excursion interval straddles $\rho^-(t)$. By IV–5
these are the points r in D^+ with $r < \varphi^-(t)$. The function $\varphi^-(t)$, which is
equal to $\varphi(\rho^-(t))$, is $\sigma\{y^-\}$ measurable and hence is X^- measurable. The
processes X^+ and X^- are independent (positive and negative excursions)
so in expectations involving $\alpha_{\varphi(\rho^-(t))} X^+$ taken over sets in $\sigma\{y^-\}$ we may
treat $\varphi(\rho^-(t))$ as if it were constant; and so we conclude that (3.4) is valid
when T is replaced by $\varphi(\rho^-(t))$. In this statement we may replace the

fixed time t with any random time that is $\sigma\{y^-\}$ measurable. In particular replace t with $\sigma_d^{y^-}$, the time when y^- first reaches d, and then note that $\rho(\sigma_d^{y^-})$ is equal to σ_d. The conclusion of (3.4) follows.

The effect on (3.4) and (3.5) of a different normalization for local time is easy to see: if we set $L^1 = c^{-1}L$ and $T^1 = L^1(\sigma_d)$ then T^1 equals $c^{-1}T$ and (3.4) and (3.5) remain valid with L^1 and T^1 provided we replace $\hat{p}_t$ with $c\hat{p}_t$; that is, naturally enough, we use the normalization of the excursion law that makes the excursion equation III-3.27 valid under the new normalization of local time.

Finally we will need a slightly different form of (3.5) in which the basic measure P^0 is replaced with excursion measure $\hat{P}$. That is suppose a is strictly positive, ω is in W and $G_a^+(= G_a^+(\omega))$ denotes the left ends of the raw excursion intervals above the level a. And for s in G_a^+ let $x^a \circ \theta_s$ denote the corresponding raw excursion $(x_{t \wedge \sigma_a}) \circ \theta_s - a$. Then if ψ is positive and measurable on $[0, \infty) \times W$ and Λ is in $\mathcal{E}_a^0$ we have

$$(3.6) \quad \hat{P}(e^{-\sum_{s \in G_a^+, s < \sigma_0} \psi(\ell_s^a, x^a \circ \theta_s)}; \Lambda) = \hat{P}(\exp\{\sqrt{2} \int (e^{-\psi} - 1)d\hat{p}_T\}; \Lambda).$$

Here ℓ_s^a is excursion local time, $T = \ell_{\sigma_0}^a$, $\hat{p}_t$ has the same meaning as in (3.5) and the factor $\sqrt{2}$ reflects the normalization of excursion local time. Equation (3.6) follows immediately from (3.5): indeed it is obvious that the event $\{\sigma_a < \sigma_0\}$ is $\mathcal{E}_a^0$ measurable. On the set $\{\sigma_a \geq \sigma_0\}$ the integrands on each side of (3.6) are equal to 1 so that the equality in (3.6) with Λ replaced by $\Lambda \cap \{\sigma_a \geq \sigma_0\}$ is clear. The typical event $\Lambda \cap \{\sigma_a < \sigma_0\}$ may be written $F \cdot G \circ \theta_{\sigma_a}$ with F in $\mathcal{F}_{\sigma_a}$ and G in $\mathcal{F}^0$, and if the product is $\mathcal{E}_a^0$ measurable so is the second factor as it is clear that $F \cap \{\sigma_a < \sigma_0\}$ is $\mathcal{E}_a^0$ measurable. Also on $\{\sigma_a < \sigma_0\}$ the integrands on each side of (3.6) are unchanged by composition with θ_{σ_a}, and now it is clear that (3.6) follows by first using the strong Markov property at the time σ_a and then applying (3.4) with d equal to $-a$ and then making the space translation $x \to x + a$.

(3.7) Exercise. Write

$$A_a^{\rho_b \omega}(t) = \int_0^t I_{[0,a]}(\omega \circ \sigma_{(b)}^\omega)(s))ds$$

and make a change of variables in the integral to conclude that for $a < b$

$$A_a^{\rho_b \omega}(t) = A_a^\omega(\sigma_{(b)}^\omega(t)).$$

Conclude that $\rho_a \circ \rho_b = \rho_a$ for $a < b$.

4. Super Brownian motion.

In a recent paper Le Gall [Le, 1] has made a strong and imaginative application of excursion theory to the construction of what is called *super Brownian motion*. This is a diffusion process $\{Z_t; t \geq 0\}$ whose values are measures on R^1 and which essentially is characterized by the property that if Z_0 is unit mass at a point z then

$$\text{(4.1)} \qquad E e^{-\langle Z_t, \varphi \rangle} = e^{-V_t \varphi(z)}$$

where $\{V_t; t \geq 0\}$ is a semigroup of operators on continuous functions whose infinitesimal generator is given by

$$\text{(4.2)} \qquad \lim_{t \to 0} t^{-1}(V_t \varphi(z) - \varphi(z)) = (1/2)\varphi''(z) - 2\varphi^2(z),$$

so that one obtains a probabilistic structure for use in studying the non-linear operator on the right of (4.2). Here we are writing $\langle \mu, \varphi \rangle$ for $\int \varphi d\mu$. Of course suitable restrictions must be placed on the function φ and so forth.

The usual way to manufacture processes for the study of non-linear operators is to introduce what is called a branching diffusion. One starts with the rules governing a diffusion process in R^1 or R^n, Brownian motion in our case. A particle starts at a point z following these rules. At a time having an exponential distribution and independent of the spatial motion the particle vanishes or splits in two, each with probability $1/2$, and in the latter case the two particles proceed independently from the splitting point until exponentially distributed times, and so forth. Thus the state of the system at time t is specified by saying how many particles are in existence and where they are located, that is by the measure that puts unit mass at each location. But to obtain a process associated with the operator in (4.2) it is necessary to have a branching mechanism where branching occurs infinitely often in a finite time interval. It has been customary to treat this by taking a limit of a sequence of branching schemes where the rate of splitting and the number of particles extant at time 0 increase. One obtains limiting analytic expressions, but loses the rather concrete description of how the branching evolves.

Le Gall avoids the limit passage by a clever device. Let f denote a typical excursion path for excursions away from 0 of reflecting Brownian motion; that is f is a continuous function from $[0, \infty)$ to $[0, \infty)$ with

$f(0) = 0$, with $\tau(f) = \inf\{t > 0 | f(t) = 0\}$ strictly positive and finite and with $f(t) = 0$ for $t \geq \tau$. (We write τ instead of the familiar $\sigma_0(f)$ to preserve Le Gall's notation.) Assume also that f has some other properties typical of Brownian paths, namely that the times at which f has local minima are dense in $[0, \tau(f)]$ and that no two local minima have the same height. Le Gall uses the heights of these local minima to label the "times" at which branching is to occur, so that in the final set-up the time variable for the superprocess will be the space variable for the excursion path. This is a situation strongly reminiscent of what is going on in Ray's local time theorem, which turns out to be no accident. Of course it is necessary also to describe the spatial movement of the particles so that they follow the rules of the underlying diffusion and have independence properties compatible with the branching procedure. Le Gall does this by introducing for each excursion path f an appropriate collection of simultaneously running diffusion processes. Then he puts a measure on the entire collection of objects by randomizing f according to excursion measure for reflecting Brownian motion. This measure is taken as the intensity measure of a point process (or random measure) which then is used to define the measure valued diffusion of (4.1) and (4.2).

We will describe his development, focusing on the parts which make the most use of excursion theory and only sketching the rest. Our presentation just scratches the surface: we must refer the reader to all of Le Gall's difficult paper and the references for more information about this important and currently popular subject and for illustrations of the advantages of his construction.

Le Gall's development allows the underlying spatial process to be any well-behaved diffusion in R^n. In that case the operator $(1/2)\varphi''$ on the right of (4.2) gets replaced by the infinitesimal generator of the diffusion. Sticking with Brownian motion, as we do, simplifies the presentation. Also in this case all the random variables involved have joint normal distributions so that probability assertions can be checked by computing means and covariances. On the other hand we have avoided using simplifications special to the Brownian motion case that obscure the generality of Le Gall's technique.

(a) **The tree of excursions.** Let f denote a continuous function from

$[0, \infty)$ to $[0, \infty)$. We will assume that $0 < \tau(f) < \infty$ and that $f(t) = 0$ if t exceeds $\tau(f)$. We will say the f has a *local minimum* at t if $0 < t < \tau$ and $f(s) \geq f(t)$ throughout some interval containing t. The phrase "local maximum at t" has a similar definition. We will define $\mathcal{L}_f$ to be the set of all t such that f has a local minimum at t and we will agree also that 0 is in $\mathcal{L}_f$. We will assume that if t and s are distinct points in $\mathcal{L}_f$ then $f(t) \neq f(s)$. We will assume also that if t and s are distinct points in $(0, \tau)$ where local maxima occur then $f(t) \neq f(s)$. According to the exercises in II-2 almost all Brownian motion paths have local extrema with these properties and it follows that so do almost all paths relative to excursion measure. The hypothesis about uniqueness of local maxima is not needed, but it saves some complication in an already complicated subject.

A *raw excursion* of f will be an object $(x, (a, b))$ with $x \geq 0$ and (a, b) a non-empty subinterval of $(0, \tau)$ such that $f(a) = f(b) = x$, while $f(t) > x$ for $a < t < b$. For an interval (a, b) let $m(a, b) = \max\{f(x)|a \leq x \leq b\}$. Define an equivalence relation on the set of raw excursions by setting

$$(x, (a, b)) \sim (x', (a', b'))$$

if

$$x \leq x', \quad (a, b) \supset (a', b')$$

and

$$m(a, b) = m(a', b')$$

or if the situation with the primed and unprimed letters is reversed. This is an equivalence relation, the transitivity being guaranteed by the equality of the maxima. Denote by E_f the set of equivalence classes. We will call the equivalence classes *excursions*. Denote by M_f the set of all local minima so that $M_f = f(\mathcal{L}_f)$. The next two lemmas are basic. The reader can get a picture adequate for understanding them by drawing f piecewise linear and with at least six local minima.

(4.3) Lemma. *Let e be an excursion. Then there is a unique representative, $(x(e), (a(e), b(e)))$, of e having the property that for any raw excursion $(x, (a, b))$ in e we have $x(e) \leq x$, (a, b) contained in $(a(e), b(e))$. Further either $a(e)$ or $b(e)$ is in $\mathcal{L}_f$ and the range of the one-to-one mapping $e \to x(e)$ is all of M_f.*

Proof. If $(x,(a,b))$ and $(x',(a',b'))$ are equivalent then $m(a,b) = m(a',b')$ and so the expression $y(e) = m(a,b)$, where $(x,(a,b))$ is an element of e, defines a function on E_f. One notable raw excursion is $(0,(0,\tau))$. We will denote its class by e_0; obviously $x(e_0) = 0$, $y(e_0) = \|f\|$ and $(0,(0,\tau))$ is the representative whose existence is required. Now take a raw excursion $(x,(a,b))$ not equivalent to e_0. We have $m = m(a,b) < \|f\|$, $f(a) = f(b) = x > 0$ and $0 < a < b < \tau$. Let $[A,B]$ denote the largest interval containing (a,b), contained in $[0,\tau]$ and such that $f(t) \leq m$ for all t in $[A,B]$. Since $m < \|f\|$ we cannot have $A = 0$, $B = \tau$. Let us suppose $A > 0$ and $B < \tau$, leaving the other cases as exercises. We have $f(A) = f(B) = m$ and so $A < a$, $b < B$. Let α and β denote the infimum of $f(t)$ over $[A,a]$ and $[b,B]$ respectively and pick t_0 in $(A,a]$ and t_1 in $[b,B)$ such that $f(t_0) = \alpha$, $f(t_1) = \beta$. Both t_0 and t_1 are in $\mathcal{L}_f$ and so α and β are distinct. Suppose that β is the larger. Note that α and β are both no greater than x, and then let t_2 be the supremum of those $t \leq a$ such that $f(t) = \beta$. Quite clearly $(\beta,(t_2,t_1))$ is a raw excursion equivalent to $(x,(a,b))$. We assert that this is the desired minimal representative. To see this suppose $(x',(a',b'))$ also is equivalent to $(x,(a,b))$. We cannot have $x' < \beta$ for the component of $f^{-1}(x',\infty)$ containing b contains B also and so $m(a',b') > m$. If $x' > x$ then $(a',b') \subset (a,b) \subset (t_2,t_1)$ as required. The remaining case is $\beta \leq x' \leq x$. If $(a',b') \subset (t_2,t_1)$ we have again the required relationship between the raw excursions. The interval (a',b') cannot straddle either t_2 or t_1 as it is a raw excursion interval and $f(a') = f(b') = x' \geq f(t_i) = \beta$. So (a',b') lies to the left of t_2 or to the right of t_1 neither of which events would allow comparison with $(\beta,(t_2,t_1))$. Note that by construction t_1 is in $\mathcal{L}_f$ as the statement also requires. The last statement in the lemma requires us to show that any point t in $\mathcal{L}_f$ is equal to $a(e)$ or $b(e)$ for some equivalence class e. If $t = 0$ we take $e = e_0$ of course. Suppose $f(t) = x > 0$. If $K = f^{-1}(x)$ then K is a compact subset of $(0,\tau)$ containing points strictly less and strictly greater than t and t is isolated in K. Let a and b denote respectively the supremum (infimum) of those points in K which are strictly less (greater) than t. The numbers $m(a,t)$ and $m(t,b)$ are distinct. Assuming the first is smaller the reader will have no difficulty showing that $(x,(a,t))$ is the desired representative of a class e with $b(e) = t$.

Next we define an order on E_f, the set of all excursions, by setting $e <$

e' if $(a(e), b(e)) \supset (a(e'), b(e'))$. This implies that $x(e) \le x(e')$, $y(e) \ge y(e')$ and hence $h(e) \ge h(e')$ where the "height" h is $h(e) = y(e) - x(e)$. We set $e \ll e'$ if $e < e'$, $e \ne e'$ and also $e < e'' < e'$ implies that either $e'' = e$ or $e'' = e'$.

(4.4) Lemma. *If e is in E_f then there is a positive integer n and a finite sequence $(e_0, e_1, \ldots, e_n)$ of elements of E_f such that*

$$e_0 \ll e_1 \ll \cdots \ll e_{n-1} \ll e_n = e.$$

Moreover $e' < e$ if and only if $e' = e_i$ for some $i \le n$. The integer n and sequence of e's are unique.

Proof. As to existence, given excursions $e' < e''$ let $\mathcal{E}(e', e'')$ denote the set of excursions strictly between e' and e'' in the order $<$. If $\mathcal{E}(e_0, e)$ is empty we have the desired sequence by taking $e_1 = e$. Otherwise pick e' in $\mathcal{E}(e_0, e)$ and apply the same procedure to the pairs $e_0 < e'$ and $e' < e$. The reader should have no difficulty in arguing that this process cannot continue indefinitely, and the existence follows. If e' and e'' are distinct excursions their representative maximal intervals must be disjoint or else be comparable. So if $e' < e$ we have, in an obvious notation,

$$(a_{i+1}, b_{i+1}) \subset (a', b') \subset (a_i, b_i)$$

for some i, so e' equals e_i or e_{i+1}. Uniqueness follows for if $e'_1, \cdots, e'_m$ were another sequence we would have $e'_i < e$ and hence $e'_i = e_j$ for some j.

We will refer to the finite sequence in (4.4) as the *list* for e. The excursion e_{n-1} will be called the *predecessor* of e, so that for $e \ne e_0$ the predecessor is the unique e' with $e' \ll e$.

For use in the following exercise, which is important for the next subsection, define a function Φ_f from $\mathcal{L}_f$ to E_f by taking $\Phi_f(u)$ to be the excursion e (whose existence and uniqueness follow from (4.3)) such that $x(e) = f(u)$.

(4.5) Exercise. Suppose $u < u'$ are strictly positive elements of $\mathcal{L}_f$, and u'' is the unique number (necessarily in $\mathcal{L}_f$) such that $u \le u'' \le u'$ and

$$f(u'') = \inf\{f(v) | u \le v \le u'\}.$$

Let e, e', and e'' be defined by $\Phi_f(u) = e$, and so forth, and let $\tilde{e}$ be the predecessor of e''.

(a) Show that either $e'' < e$ or $e'' < e'$. (Hint: recall that one but not both of $\{a(e), b(e)\}$ are in $\mathcal{L}_f$.)

(b) Show that unless e and e' are comparable only one of the inequalities in (a) can hold.

(c) Show that both $\tilde{e} < e$ and $\tilde{e} < e'$ hold. (Hint: $x(\tilde{e})$ is strictly less than $x(e)$.)

(b) **Indexed families of stopped Brownian motions.** Let f be a function satisfying the conditions from the previous paragraph. Let $\{B^e; e \in E_f\}$ denote a mutually independent family of Brownian motions starting at 0 defined over some basic probability space. The terms "probability" and "expectation" will refer to this underlying space. We will define a family $\{\gamma^e(t); t \geq 0\}$ of stochastic processes indexed by e running over E_f as follows: if $e = e_0$ set $\gamma^{e_0}(t) = 0$ for all t. If $e \neq e_0$ take the list $(e_0, e_1, \cdots, e_n)$ for e and define

$$\gamma^e(t) = \sum_{j=0}^{i-1} B^{e_j}(x(e_{j+1}) - x(e_j)) + B^{e_i}(t - x(e_i)),$$

if $x(e_i) \leq t < x(e_{i+1})$ and $i < n$. If $t \geq x(e_n)$ set

$$\gamma^e(t) = \sum_{j=0}^{n-1} B^{e_j}(x(e_{j+1}) - x(e_j)).$$

An empty sum $\sum_{0}^{-1}$ is taken to be 0. A few simple facts are obvious:

(i) each $\{\gamma^e(t); t \geq 0\}$ is Brownian motion with initial position 0 and held fixed (that is, stopped) at time $x(e)$,

(ii) the process γ^e is measurable relative to the σ algebra of the processes $B^{e_0}, \ldots, B^{e_{n-1}}$.

(iii) for any $q \leq n$ the process $\{\gamma^e(t + x(e_q)) - \gamma^e(x(e_q)); t \geq 0\}$ is measurable relative to $\sigma\{B^{e_j}; q \leq j < n\}$.

Next define a family $\{\Gamma_u(t); t \geq 0\}$ indexed by u running over $\mathcal{L}_f$ by setting

$$\Gamma_u(t) = \gamma^{\Phi_f(u)}(t) \quad u \in \mathcal{L}_f, t \geq 0.$$

This is just a change of notation in the previous definition.

The most important property of the Γ family is expressed as follows:

(4.6) Theorem. *(i) For each u in $\mathcal{L}_f$, Γ_u is a Brownian motion with initial position 0 and stopped at time $f(u)$. (ii) if $u \leq u'$ are in $\mathcal{L}_f$ and u'' is the unique point of $\mathcal{L}_f$ such that $u \leq u'' \leq u'$ and $f(u'') = \inf\{f(v); u \leq v \leq u'\}$ then the two processes*

$$\{\Gamma_v(f(u'') + t) - \Gamma_v(f(u''));t \geq 0\}$$

that one obtains by setting $v = u$ and $v = u'$ are independent, and the pair is independent of the process $\Gamma_{u''}$. (iii) $\Gamma_u(t) = \Gamma_{u'}(t) = \Gamma_{u''}(t)$ for t in the interval $[0, f(u'')]$.

Proof. Statement (i) follows from the fact that if $e = \Phi_f(u)$ then $x(e) = f(u)$. For (ii) let us use the notation and facts from (4.5). Let $e = \Phi_f(u)$ with e' and e'' defined similarly. If e and e' are comparable, say $e < e'$, then $u = u''$; the independence assertions are then immediate consequences of the statements (ii) and (iii) about the γ family. So let us assume that e and e' are not comparable. We know from (4.5) that $e'' < e$ or $e'' < e'$ but not both. Assume $e'' < e$. By (4.5) $\tilde{e} < e$ and $\tilde{e} < e'$ so by (4.4) it appears in the list for each while e'' appears in the list for e but not in the list for e' and we have, in an obvious notation,

$$\tilde{e} \ll e'' = e_1 \ll \cdots \ll e_p = e$$

$$\tilde{e} \ll e'_1 \ll \cdots \ll e'_q = e'.$$

We cannot have $e_i = e'_j$ for some i and j or else $e'' < e'$. Also we must have $x(e'_1)$ larger (actually strictly larger) than $f(u'')$. The process $\{\Gamma_u(f(u'') + t) - \Gamma_u(f(u''));t \geq 0\}$ is defined in terms of $B^{e_1}, \cdots, B^{e_p}$. The process $\{\Gamma_{u'}(f(u'') + t) - \Gamma_{u'}(f(u''));t \geq 0\}$ involves $B^{e'_1}, \cdots, B^{e'_q}$ and also $B^{\tilde{e}}$ but it involves the latter process only through the differences $\{B^{\tilde{e}}(t + f(u'')) - B^{\tilde{e}}(f(u''))\}$. And finally $\Gamma_{u''}$ involves only processes B^g with $g < \tilde{e}$ with $B^{\tilde{e}}$ appearing only through variables $B^{\tilde{e}}(t)$ with $t \leq f(u'')$. The disjointness of $\{e_1, \cdots, e_p\}$, $\{e'_1, \cdots, e'_q\}$, $\{g : g < \tilde{e}, g \neq \tilde{e}\}$ and the independent increments property of the Brownian motion B^e then yields assertion (ii) of (4.6). The third assertion follows from the definition of the Γ processes and the description of the partial lists given in the proof of (ii).

If $\{B(t); t \geq 0\}$ is a Brownian motion then the covariance of $B(t)$ and $B(s)$ is $\min(s,t)$. From this and (4.6) it follows immediately that

$$(4.7) \qquad \mathrm{cov}(\Gamma_u(s), \Gamma_{u'}(t)) = \min(s, t, f(u''))$$

where $u \leq u'$ are in $\mathcal{L}_f$, u'' has the same meaning as in (4.6) and s and t are positive numbers.

(c) **The extended process.** We need to extend the definition of Γ_u from u in $\mathcal{L}_f$ to u ranging over all of $[0, \infty)$. Henceforth we will assume about f, in addition to the assumptions already made, that $\mathcal{L}_f$ is dense in $[0, \tau]$ and that f is Hölder continuous of some strictly positive order δ. Almost all paths of reflecting Brownian motion have these additional properties. From an argument like the one in section 3 it follows that almost all f relative to reflecting Brownian excursion measure have these properties as well.

To get started note first that for u in $\mathcal{L}_f$ $t \to \Gamma_u(t)$ defines with probability one a bounded continuous function (bounded because $\Gamma_u(t)$ is stopped at $f(u)$) on $[0, \infty)$. Let $\|\Gamma_u - \Gamma_{u'}\|$ denote the supremum over all positive t of $|\Gamma_u(t) - \Gamma_{u'}(t)|$. Suppose $u \leq u'' \leq u'$ are in $\mathcal{L}_f$ and u'' is as usual the place where f taken on its minimum over $[u, u']$. If v denotes either u or u' then by (iii) of (4.6) $\Gamma_v(t) = \Gamma_{u''}(t)$ for $t \leq f(u'')$ and thereafter the difference behaves like a Brownian motion initially at 0 and stopped at $f(v) - f(u'')$. Thus the law of $\|\Gamma_v - \Gamma_{u''}\|$ is that of the absolute maximum of a Brownian motion $\{B(t); t \geq 0\}$ over $[0, f(v) - f(u'')]$. From the standard relationship between the distribution of the maximum over an initial interval and that of the value of $B(t)$ at the right end (see II-2) we have for any $p > 0$

$$E\|\Gamma_v - \Gamma_{u''}\|^p \leq K_p E|B(\eta)|^p$$
$$= J_p \eta^{p/2}$$

where $\eta = f(v) - f(u'')$. Routine manipulations, the triangle inequality and an application of the Hölder continuity then yields

$$(4.8) \qquad E\|\Gamma_u - \Gamma_{u'}\|^p \leq C_p |u - u'|^{\delta p/2}$$

for u, u' in $\mathcal{L}_f$. This is the setting in which Kolmogorov's theorem (see the exercises in Chapter II) applies: Given δ we take p large enough that $\delta p/2$ exceeds 1. After a slight modification to take account of the fact that the random point Γ_u of the space of bounded continuous functions is defined only for a dense set of u we conclude that we can extend the Γ family to obtain $\{\Gamma_u; 0 \le u \le \tau\}$ varying continuously in the supremum metric. Obviously for the extended process (4.7) still holds for all u, u' and hence so do the conclusions of Theorem 4.6. (When the spatial motion is something other than Brownian motion this last assertion requires deeper arguments based on the branching structure.) Finally we set $\Gamma_u(t) = 0$ for all t if $u \ge \tau$.

Next we need some notation to accommodate what has been set up. Let C^* denote the set of continuous functions from $[0, \infty)$ to $[0, \infty)$ having the properties that $f(0) = 0$, $0 < \tau(f) < \infty$, $f(t) = 0$ for $t \ge \tau$, $\mathcal{L}_f$ is dense in $(0, \tau)$ with the local minima distinct, and f is Hölder continuous. Given $f \in C^*$ we have constructed a family $\{\Gamma_u; u \ge 0\}$ of stopped Brownian motions varying continuously in u when $t \to \Gamma_u(t)$ is regarded as an element of $C_b([0, \infty), R)$ with supremum metric. Let Ω denote the sample space for Γ; that is $\Omega = C([0, \infty), C([0, \infty), R))$ so that $\omega(u)(t)$ stands for $\Gamma_u(t)$ evaluated at a generic point of its underlying probability space. We will let Q_0^f denote the distribution of Γ in Ω so that the process $\{\Gamma_u(t); u \ge 0, t \ge 0\}$ is equal in law to $\{\omega(u)(t); u \ge 0, t \ge 0\}$ under Q_0^f. The subscript 0 stands for the fact that the Brownian motions Γ_u start at 0. If z is in R then under the mapping $\omega \to \omega + z$ we obtain Brownian motion starting at z but with all the other properties unchanged. We will denote by Q_z^f the corresponding probability law on Ω. (When the spatial process is a diffusion other than Brownian motion the initial point must be introduced at the beginning). Let Θ denote the product space $C^* \times \Omega$. These various spaces will be fitted out with the σ-algebras of the coordinate functions. Shortly we will need analytic manipulations such as integration of Q_z^f in f. We will leave to the reader the necessary (and not always easy) measurability arguments, while we simply will make freely whatever manipulations the notation allows.

Finally we will need to recall the components of the excursion filtration from section 3. Fix f in C^* and the corresponding random element Γ of Ω. Fix a number $a > 0$ and recall the reparametrization $f \circ \sigma_{(a)}^f$ and the raw excursions f_i of f above $[\alpha_i, \beta_i]$, $i \in I$, defined by

$$f_i(t) = f((\alpha_i + t) \wedge \beta_i) - a.$$

The function $f \circ \sigma^f_{(a)}$ is continuous even though $\sigma^f_{(a)}$ is not because $f(\alpha_i) = f(\beta_i) = a$. A slightly less obvious fact is that $\Gamma_{\alpha_i}(t) = \Gamma_{\beta_i}(t)$ for all t. This can be checked with a covariance calculation from (4.7), which is valid for all values of the arguments. It follows that if we define $\Gamma^{(a)}$ by

$$\Gamma_u^{(a)} = \Gamma_{\sigma^f_{(a)}(u)}$$

then the stopped Brownian motions $\{\Gamma_u^{(a)}(t); t \geq 0\}$ vary continuously with u. The function $f \circ \sigma^f_{(a)}$ is an element of C^* but has unique local maxima only for those maxima strictly less than a, an unimportant technicality. Define processes $\{\Gamma_u^i : u \geq 0\}$ for each raw excursion interval $[\alpha_i, \beta_i]$ of f above a by

$$\Gamma_u^i(t) = \Gamma_{(\alpha_i+u)\wedge\beta_i}(a+t).$$

The probabilistic structure of these objects is described as follows:

(4.9) Theorem. *The law of* $\{\Gamma^{(a)}\}$ *is* $Q_0^{\rho_a^f}$. *Conditionally given* $\{\Gamma_{\alpha_i}(a); i \in I\}$ *the indexed family* $\{\Gamma_u^i; u \geq 0\}$ *is an independent family and the family is independent of* $\{\Gamma_u^{(a)}; u \geq 0\}$. *The conditional distribution of* Γ^j *given* $\{\Gamma_{\alpha_i}(a); i \in I\}$ *is* $Q_{\Gamma_{\alpha_j}(a)}^{f_j}$.

Proof. Given the equality $\Gamma_{\alpha_i} = \Gamma_{\beta_i}$ the first assertion is quite obvious. The validity of the others can be checked from the covariance facts of (4.7). A better argument follows from a consideration of which of the Brownian motions B^e enter into the construction of $\Gamma^{(a)}$ and the various Γ^i exactly as in the argument for Theorem 4.6.

Everything we have done so far involves only a single path f. In this sense we have not yet used any excursion theory. Now we will bring in excursion theory. We will let η denote excursion measure on C^* for reflecting Brownian motion normalized (in keeping with Le Gall's choice) so that $\eta\{\max f > \epsilon\} = 1/2\epsilon$. Also let $\ell_t^a(f)$ denote excursion local time at the point $a > 0$ and time $t > 0$ on the path f. Define for z in R a measure M_z on Θ by

$$M_z(df, d\omega) = \eta(df) Q_z^f(d\omega).$$

The measure η is only σ-finite so that M_z is σ-finite. Its restriction to $\{f \mid \max f \geq \epsilon\}$ has total mass $1/2\epsilon$ of course. For $a > 0$ and $(f, \omega) \in \Theta$ define a measure $Y_a(f, \omega)$ on the Borel sets of R by

$$\langle Y_a(f, \omega), \varphi \rangle = \int_0^\infty d_s \ell_s^a(f) \varphi(\omega(s)(a)),$$

φ being a positive Borel function on R. The total mass, $\|Y_a\|$, is $\ell_\tau^a(f)$. This is 0 if $\max f < a$. At any rate by (2.4) of this chapter ℓ_τ^a is, relative to Brownian motion starting at a, exponential with rate $1/2a$; and so

$$\int \|Y_a\| dM_z = (1/2a) E^a \ell_\tau^a = 1$$

where, just for the next few lines, E^a refers to expectation for Brownian motion starting at a. More generally for a positive function φ we have

$$\int Q_z^f(d\omega) \varphi(\omega(s)(a)) = P_{a \wedge f(s)} \varphi(z),$$

with $P_t \varphi(z)$ being the transition operator for Brownian motion, and so

$$\begin{aligned}
(4.10) \qquad \int \langle Y_a, \varphi \rangle dM_z &= \int \eta(df) \int_0^\infty d_s \ell_s^a(f) P_{a \wedge f(s)} \varphi(z) \\
&= P_a \varphi(z)
\end{aligned}$$

since $d_s \ell_s^a(f)$ puts all its mass on $\{s \mid f(s) = a\}$. For use later on we have also

$$(4.11) \qquad \int \|Y_a\|^2 dM_z = (1/2a) E^a (\ell_\tau^a)^2 = 4a.$$

For $u > 0$ and $y \in R$ define a transformation $V_u \varphi(y)$ of positive functions φ by

$$V_u \varphi(y) = \int (1 - e^{-\langle Y_u, \varphi \rangle}) dM_y.$$

The fundamental fact in defining the superprocess and establishing its properties is the following:

(4.12) Theorem. *For $0 < a < b$ and $\Delta \in \sigma\{Y_u; u \leq a\}$ we have*

$$\int_\Delta e^{-\langle Y_b, \varphi \rangle} dM_z = \int_\Delta e^{-\langle Y_a, V_{b-a} \varphi \rangle} dM_z.$$

Proof. The statement would be phrased in conditional probability language if M_z were a probability measure. The conditioning σ-algebra means the one generated by the real-valued functions $\langle Y_u, \varphi \rangle$, $u \le a$ and φ ranging over the positive, or positive and continuous, functions. Let $\mathcal{E}_a$ and $\mathcal{E}_a^0$ denote the σ-algebras generated by the functions $(f, \omega) \to (f \circ \sigma_{(a)}^f, \omega \circ \sigma_{(a)}^f)$ and $(f, \omega) \to f \circ \sigma_{(a)}^f$ respectively. We must include sets of M_z measure 0 to account for the fact that only almost surely is $\omega \circ \sigma_{(a)}^f$ continuous. Now Y_a is $\mathcal{E}_a$ measurable. To see this let $\lambda_f^a(ds)$ be the measure

$$\lambda_f^a([0, t]) = \ell_{\sigma_{(a)}^f(t)}^a(f).$$

Use the continuity and density properties of excursion local time and a change of variables to write

$$\lambda_f^a[0, t] = \lim_{\epsilon \downarrow 0} \frac{1}{\epsilon} \int_0^t I_{[a-\epsilon, a]}(f \circ \sigma_{(a)}^f(s)) ds$$

so that $\lambda_f^a(ds)$ depends on f through $f \circ \sigma_{(a)}^f$; and then note that

$$\langle Y_a(f, \omega), \varphi \rangle = \int_0^\infty \lambda_f^a(ds) \varphi(\omega \circ \sigma_{(a)}^f(s)(a)),$$

so that dependence in both f and ω is through composition with $\sigma_{(a)}^f$ as required. Thus we need only to check the equality in (4.12) when Δ is replaced by a function $F(f \circ \sigma_{(a)}^f)G(\omega \circ \sigma_{(a)}^f)$ with F on C^* and G on Ω positive and measurable. With this replacement the left side of (4.12) becomes

$$\int \eta(df) F(f \circ \sigma_{(a)}^f) \int Q_z^f(d\omega) e^{-\int_0^\infty d_s \ell_s^b(f) \varphi(\omega(s)(b))} G(\omega \circ \sigma_{(a)}^f).$$

In calculating the inner integral we regard f as fixed. The only contributions to the exponential integral come from the raw excursions of f above a that reach b (of which there are only finitely many.) Each of these makes a contribution

$$\int_{\alpha_i}^{\beta_i} d_s \ell_s^b(f) \varphi(\omega(s)(b))$$

which because of the definition of local time as a derivative is equal to

$$\int_0^\infty d_s \ell_s^{b-a}(f_i) \varphi(\omega_i(s)(b-a))$$

where ω_i is the path of Γ^i, that is

$$\omega_i(u)(t) = \omega((\alpha_i + u) \wedge \beta_i)(a + t).$$

Thus the inner integral reads

$$(4.13) \qquad \int Q_z^f(d\omega) e^{-\sum_i \int_0^\infty d_s \ell_s^{b-a}(f_i) \varphi(\omega_i(s)(b-a))} G(\omega \circ \sigma_{(a)}^f).$$

This is an expression to which the considerations of (4.9) apply: relative to the measure Q_z^f the summands in the exponential, once the $\omega_i(\alpha_i)(a)$ are given, are independent and independent of $\omega \circ \sigma_{(a)}^f$ and the ith one follows the law $Q_{\omega(\alpha_i)(a)}^{f_i}$. Thus (4.13) can be written as

$$(4.14) \qquad \int Q_z^f(d\omega) G(\omega \circ \sigma_{(a)}^f) \prod_i H(f_i, \omega(\alpha_i)(a))$$

where

$$H(f, y) = \int \{\exp - \int_0^\infty d_s \ell_s^{b-a}(f) \varphi(\omega'(s)(b-a))\} Q_y^f(d\omega').$$

Of course in the product only finitely many terms are not equal to one. Next let

$$\tau_{(a)}^f(u) = \inf\{t | \ell_t^a(f) > u\}$$

$$\gamma_{(a)}^f(u) = \inf\{t | \lambda_f^a[0, t] > u\},$$

so that

$$\omega \circ \tau_{(a)}^f = \omega \circ \sigma_{(a)}^f \circ \gamma_{(a)}^f$$

and

$$\omega(\alpha_i)(a) = \omega \circ \tau_{(a)}^f(\ell_i)$$

where $\ell_i = \ell_{\alpha_i}^a(f)$. Since $\omega \circ \sigma_{(a)}^f$ is $\mathcal{E}_a$ measurable and $\gamma_{(a)}^f$ is $\mathcal{E}_a^0$ measurable it follows that $\omega \circ \tau_{(a)}^f$ is $\mathcal{E}_a$ measurable. With this notation the product in (4.14) reads

$$e^{\sum \log H(f_i, \omega \circ \tau_{(a)}^f(\ell_i))}.$$

Rewriting the product this way and using the fact that $\omega \circ \tau = \omega \circ \sigma \circ \gamma$ we see that the integrand in (4.14) depends on ω through $\omega \circ \sigma_{(a)}^f$. According to the first assertion in (4.9) an integral $\int Q_z^f(d\omega)h(\omega \circ \sigma_{(a)}^f)$ is equal to $\int Q_z^{f \circ \sigma_{(a)}^f}(d\omega)h(\omega)$ and so the left side of (4.12) can be written as

$$(4.15) \quad \int \eta(df) \int Q_z^{f \circ \sigma_{(a)}^f}(d\omega) e^{\sum \log H(f_i, \omega \circ \gamma_{(a)}^f(\ell_i))} F(f \circ \sigma_{(a)}^f)G(\omega \circ \sigma_{(a)}^f).$$

In (4.15) think of ω as being fixed so that the inner integral in its dependence on $\{f_i, \ell_i\}$ is a function of excursions above the level a whereas the rest of the expression (including the $\gamma_{(a)}^f$ part) depends on f through $f \circ \sigma_{(a)}^f$, that is, it is $\mathcal{E}_a^0$ measurable. Thus when we integrate against η and apply (3.6) we obtain

$$\int \eta(df) \int Q^{f \circ \sigma_{(a)}^f}(d\omega) \exp\{ \int [H(g, \omega \circ \gamma_{(a)}^f(\ell)) - 1]\eta(dg)d\ell \} F(f \circ \sigma_{(a)}^f)G(\omega)$$

where the integral in braces is on g running over the set of all positive excursions and $\ell \leq \ell_\tau^a(f)$. The reader will argue without much difficulty that the change in order of integration involved in making this step is permissible. Also one should check that the normalization of Brownian excursion measure appearing in the exponential part of (3.6) and the one appearing here are the same. Upon reversing the change of variables that preceded (4.15) this last display becomes

$$(4.16) \quad \begin{aligned} &\int \eta(df)F(f \circ \sigma_{(a)}^f) \int Q_z^f(d\omega)G(\omega \circ \sigma_{(a)}^f) \cdot \\ &\qquad \cdot \exp\{- \int_{\ell \leq \ell_\tau^a(f)} [1 - H(g, \omega \circ \tau_{(a)}^f(\ell))]\eta(dg)d\ell\}. \end{aligned}$$

In the integral in braces we make the change of variable $\tau_{(a)}^f(\ell) = s$, which as is easily seen, changes it to

$$\int \{ \int_0^\infty d_s \ell_s^a(f)(1 - H(g, \omega(s)(a))) \}\eta(dg)$$

which is the integral against the measure $Y_a(f, \omega)$ of the function ψ defined by

$$\psi(y) = \int (1 - H(g, y))\eta(dg).$$

One sees immediately that $\psi(y)$ is simply $V_{b-a}\varphi(y)$, and substituting this into (4.16) we obtain the right side of (4.12), so the proof is complete.

Now we are ready to define the superprocess and check its properties. Let z be a real number and construct a Poisson random measure whose intensity measure is M_z; that is we set up a probability space $(X, \mathcal{G}, P)$ and associate with each x in X a countable subset D_x of Θ so that for each measurable subset A of Θ, the number of points $N_A(x)$ of D_x which lie in A is $\mathcal{G}$ measurable and has a Poisson distribution with parameter $M_z(A)$ and such that the independence properties of III-2.4 hold. Then define a measure valued process $\{Z_t; t \geq 0\}$ over X by setting

$$Z_0(x) = \epsilon_z$$
$$Z_t(x) = \sum_{(f,\omega) \in D_x} Y_t(f, \omega) \quad t > 0,$$

where $Y_t(f, \omega)$ is the measure from (4.12). The exponential formula for Poisson measures will let us reduce questions about Z to those involving Y. Specifically we have the following.

(4.17) Theorem. *$\{Z_t; t \geq 0\}$ is the superprocess with initial position ϵ_z and satisfying (4.1) with the operator V_t defined after (4.11).*

Proof. For the Markovian property and (4.1) we have to show that given numbers $0 < a_1 < \cdots < a_{n+1}$ and positive Borel functions $\varphi_1, \cdots, \varphi_{n+1}$ on R we have

$$(4.18) \qquad E\left(e^{-\sum_1^{n+1}(Z_{a_k}, \varphi_k)}\right) = E\left(e^{-\sum_1^n (Z_{a_k}, \varphi_k)} e^{-(Z_{a_n}, V_\delta \varphi_{n+1})}\right),$$

$\delta = a_{n+1} - a_n$, where E refers to expectation with respect to P. The exponential formula applied to the left side changes it to

$$\exp \int \left(e^{-\sum_1^{n+1}(Y_{a_k}, \varphi_k)} - 1\right) dM_z$$

and similarly the right side is changed to the exponential of an integral dM_z. Upon then applying (4.12) the equality at (4.18) follows immediately. (Note that in this last display the integrand is non-zero only on $\{\ell_r^a(f) > 0\}$. That set has finite M_z measure so the subtraction of 1 in the integrand causes no problems.) This gives the major part of what is needed to establish $\{Z_t; t \geq 0\}$ as super Brownian motion. We have not yet shown that Z_t varies continuously with t nor have we established (4.2). The continuity follows from that of Y_t which in turn follows from the continuity in the space variable, t, of excursion local time ℓ_s^t. To render (4.2) at least plausible make the crude approximation

$$\int (1 - e^{-\langle Y_t, \varphi \rangle}) dM_y = \int \langle Y_t, \varphi \rangle dM_y - \frac{1}{2} \int \langle Y_t, \varphi \rangle^2 dM_y + \varepsilon$$

and

$$\int \langle Y_t, \varphi \rangle^2 dM_y = \varphi^2(y) \int \|Y_t\|^2 dM_y + \varepsilon'$$

where ε and ε' are error terms which Le Gall shows are $o(t)$. Applying (4.10) and (4.11) to these expressions we have

$$V_t \varphi(y) = P_t \varphi(y) - 2t\varphi^2(y) + o(t)$$

and (4.2) follows by subtracting $\varphi(y)$ and using the fact that $\varphi''/2$ is the infinitesimal operator for Brownian motion. This completes our presentation.

VII Excursions Away From a Set

1. Introduction.

Part of the theory of excursions away from a point can be generalized as follows. Let $\{X_t; t \geq 0\}$ be a standard process and let V be a subset of the state space E. We will assume that V is closed and that every point of V is regular for V, that is $P^x(\sigma = 0) = 1$ for all x in V where $\sigma = \sigma_V = \inf\{t > 0 | X_t \in V\}$. As in Chapter III let $G = G(\omega)$ denote the strictly positive left ends of the open intervals making up the complement of the closure of $\{t | X_t(\omega) \in V\}$. Then there is a continuous additive functional $\{L_t; t \geq 0\}$ which grows exactly on $\{t | X_t \in V\}$ and a family $\{\hat{P}^x; x \in V\}$ of excursion measures compatible with the minimal process X_t killed at time σ, such that the "excursion formula"

$$(1.1) \qquad E^x \sum_{s \in G} Z_s f \circ \theta_s = E^x \int_0^\infty Z_t \hat{P}^{X_t}(f) dL_t$$

holds for all x, all positive previsible Z and all positive measurable f on path space, exactly as in III-3. In this chapter we will describe, following Maisonneuve [Ma 1] and Motoo [Mo 1], the theory of excursions away from a set; and we will say what we can about the synthesis problem, that is the possibility of constructing a process based on a minimal process and excursion data.

One obvious difference between (1.1) and III-3.26 is that in the latter case $\hat{P}^{X_t}(f) \equiv \hat{P}(f)$ and hence it factors through the integrals resulting in much simplification and a wealth of specific formulas. But even in the present situation (1.1) and related facts allow us to say a good bit about the structure of X by looking at its excursions away from V. The major difference between the general case and the special case $V = \{b\}$ is that in the general case we do not know of a relatively simple substitute for Itô's synthesis theorem and thus have no general way of constructing interesting recurrent extensions based on the minimal process and some sort of analytic excursion data such as a family of characteristic measures. Another aspect of this is that while one may define the point process Y of excursions away from V exactly as in III-3, the future $\theta_t Y$ is independent of the past $\alpha_t Y$ only conditionally given $X_{L_t^{-1}}$. Since this quantity is not constant the Y process is not a Poisson point process but only a "point process of Markov type". There does not seem to be at present a useful general theory of such processes.

Our treatment of excursions away from a set will omit many details. We will focus on techniques and facts which are basic to the specific theory, have the most probabilistic content, or are particularly puzzling. A complete treatment requires at some points attention to measure-theoretic issues associated with the advanced "general theory of stochastic processes." For these points we only will give references. One thing we will need from the general theory is the proper definition of previsible and of optional processes. In Chapter III we loosely defined the *previsible* processes $\{Z_s; s \geq 0\}$ to be those obtainable by linear and monotone operations from processes of the form $Z_s = I_{[0,R]}(s)$, R ranging over the stopping times. And we defined the *optional* processes to be those obtainable in the same way but starting with those of the form $I_{[0,R)}(s)$. Since $I_{[0,R]} = \lim_n I_{[0,R+1/n)}$ any previsible process is optional; but there is no reason to think that $I_{[0,R)}(s)$ is previsible unless there is a sequence R_n of stopping times increasing to R from strictly below. According to the general theory of processes (see Dellacherie-Meyer [DM, 1]) the class of previsible processes includes all the processes which are $\{\mathcal{F}_t\}$ *adapted* (that is for all t Z_t is $\mathcal{F}_t$ measurable) and are such that for all ω, $s \to Z_s(\omega)$ is left continuous and has everywhere right hand limits. The class of optional processes contains all those which are $\{\mathcal{F}_t\}$ adapted and are such that $s \to Z_s$ is right continuous with left limits.

2. Additive functionals and Lévy systems.

(a) Potentials.

In Chapter III we gave some of the theory of additive functionals. Now we need a bit more of this theory.

We will make the basic assumptions of III-3 (a) about X and $\{\mathcal{F}_t\}$. Suppose $\{A_t; t \geq 0\}$ is a continuous additive functional, that λ is positive and that the λ-potential $E^x \int_0^\infty e^{-\lambda t} dA_t = f(x)$ is finite for all x. Then we know that f is λ-excessive. Now it is a fact from the general theory of additive functionals that a continuous additive functional satisfies a strong additivity property: namely if T is an $\{\mathcal{F}_t\}$ stopping time then almost surely

$$A_{t+T} = A_T + A_t \circ \theta_T.$$

(Then by the right continuity for each T almost surely this equality holds for all t). From this it follows immediately that the λ-potential f satisfies

$$P_T^\lambda f(x) = E^x \int_T^\infty e^{-\lambda t} dA_t.$$

Thus if $\{T_n\}$ is an increasing sequence of stopping times with limit T then

$$(2.1) \qquad\qquad P_{T_n}^\lambda f(x) \to P_T^\lambda f(x) \qquad n \to \infty.$$

Any finite λ-excessive function for which (2.1) holds for all increasing sequences $\{T_n\}$ is called a *regular λ-potential*. We have just noted that the λ-potential of A is a regular λ-potential provided it is finite. We will have need of the remarkable and important fact, due to M. G. Šur, that the converse of this assertion is true.

(2.2) Theorem. *If f is a regular λ-potential then there is a unique continuous additive functional A such that*

$$f(x) = E^x \int_0^\infty e^{-\lambda t} dA_t \qquad x \in E.$$

Proof. Šur's proof is a fine piece of probabilistic reasoning. Given $\epsilon > 0$ let

$$B_n = \{f - P_{1/n}^\lambda f > \epsilon\}$$

and let $T_n = \sigma_{B_n}$. Pick a number $h > 0$ and set $T'_n = T_n + h$. If $1/n$ is less than h then

$$
\begin{aligned}
P^\lambda_{T'_n} f(x) &= E^x(e^{-\lambda(T_n+h)} f(X_{T_n+h}); T_n < \infty) \\
&= E^x(e^{-\lambda T_n} P^\lambda_h f(X_{T_n}); T_n < \infty) \\
&\leq E^x(e^{-\lambda T_n} f(X_{T_n})) - \epsilon E^x(e^{-\lambda T_n}; T_n < \infty).
\end{aligned}
$$

Since B_n decreases, T_n increases to a limit T and T'_n increases to $T + h$, and applying the regularity hypothesis twice we have for all $h > 0$

$$
P^\lambda_{T+h} f(x) \leq P^\lambda_T f(x) - \epsilon E^x(e^{-\lambda T}; T < \infty).
$$

The left side of this display is $P^\lambda_T P^\lambda_h f(x)$. This increases to $P^\lambda_T f(x)$ as h decreases to zero, and so we conclude that T is infinite almost surely. Now the function f need not be uniformly λ-excessive - if it were we would get our desired additive functional from III-3.7; but the fact that T_n approaches infinity says essentially that the path X_t does not see in finite time the failure of uniform excessiveness. In particular the argument following III-3.11 can be carried out to show that the additive functionals A_n defined there converge to a limiting continuous additive functional whose λ-potential is f. The argument for uniqueness is essentially the same as that given in III-3.7. We refer the reader to IV-3.8 of [BG, 1] for the details.

The following notation is common when dealing with a continuous additive functional:

(i) gA for the continuous additive functional

$$
\int_0^t g(X_s) dA_s.
$$

Here g is assumed to be positive and $\mathcal{E}^*$ measurable; and also we must impose some condition to guarantee continuity in t; boundedness of g would suffice.

(ii) $u^\lambda_A(x)$ for the λ-potential $E^x \int_0^\infty e^{-\lambda t} dA_t$

(iii) $U^\lambda_A g(x)$ for $E^x \int_0^\infty e^{-\lambda t} g(X_t) dA_t$ provided g is $\mathcal{E}^*$ measurable and positive. Clearly $U^\lambda_A g$ is λ-excessive if $U^\lambda_A g$ is finite. Of course $u^\lambda_{gA} = U^\lambda_A g$.

(b) Local time and time changes.

As an example of Theorem (2.2) let V be a closed subset of the state space and assume that every point in V is regular for V. Then the function φ defined by

$$\varphi(x) = E^x e^{-\sigma_V}$$

is 1-excessive. In fact it is a regular 1-potential. Indeed if T is any stopping time then $P_T^1 \varphi(x)$ is equal to $E^x(e^{-(T+\sigma_V \circ \theta_T)})$; thus if $\{T_n\}$ is an increasing sequence of stopping times with limit T then $T_n + \sigma_V \circ \theta_{T_n}$ is equal to $\inf\{t > T_n | X_t \in V\}$ and this increases to a stopping time R with $T \leq R \leq T + \sigma_V \circ \theta_T$. Since V is closed we have $X_{T_n + \sigma_V \circ \theta_{T_n}}$ in V if $T_n + \sigma_V \circ \theta_{T_n}$ is finite. If R is finite then X_R is in V either because R equals $T_n + \sigma_V \circ \theta_{T_n}$ for some n or by quasi-left-continuity. Since each point of V is regular for V we have $\sigma_V \circ \theta_R = 0$ almost surely. Thus R equals $T + \sigma_V \circ \theta_T$ and it follows that $P_{T_n}^1 \varphi$ approaches $P_T^1 \varphi$.

Let $\{L_t; t \geq 0\}$ denote the continuous additive functional, whose existence is guaranteed by (2.2), with

$$E^x e^{-\sigma_V} = E^x \int_0^\infty e^{-t} dL_t.$$

We will call L *local time on* V. We need to justify this name by showing that L grows exactly when X_t is in V. To do this set

$$R = \inf\{t | L_t > 0\} = \sup\{t | L_t = 0\}.$$

Then an argument essentially identical to the one given in III-3 (c) shows that $R = \sigma_V$ almost surely and then that $L^R \subset \{t | X_t \in V\} \subset L^I$ almost surely, where L^I and L^R are the points of increase and of right increase of L. It follows that almost surely the measure dL is carried by $\{t | X_t \in V\}$ and that the two additive functionals L and $I_V L$ are the same.

Now let $\{A_t; t \geq 0\}$ be a continuous additive functional and consider the *right continuous inverse* $\{\beta_t; t \geq 0\}$ defined by

$$\beta_t(\omega) = \inf\{u | A_u(\omega) > t\}.$$

The reader will see immediately that each β_t is a stopping time, the mapping $t \to \beta_t$ is right continuous and

$$\beta_{t+u} = \beta_u + \beta_t \circ \theta_{\beta_u} \quad 0 \leq u, t.$$

From the strong Markov property for X it follows quickly that the process $\widetilde{X}$ defined by

$$\widetilde{X}_t = X_{\beta_t}$$

is, relative to the measures P^x for X, a right continuous strong Markov process. It is called the *time-changed process* (associated with the additive functional A). (To be precise we must set $\beta_t = \infty$ if $t \geq A_\infty$ and then set $\widetilde{X}_t$ equal to δ, an adjoined cemetary point, for such t.) The translation operators for $\widetilde{X}$ are the operators $\widetilde{\theta}_t = \theta_{\beta_t}$, where θ_t denotes the translation operator for X. (This shows the reason for axiomatizing these objects rather than relying always on a function space set up.) In the particular case where A is the local-time-on-V additive functional from the previous example we have β_t in L^R for $t < L_\infty$ and thus almost surely $\widetilde{X}_t = X_{\beta_t}$ is in V for all such t. The process $\widetilde{X}$ is called X *restricted to* V. In our application of these ideas V will denote the boundary of a dense open subset of the state space. Then $\widetilde{X}$ is called the *process on the boundary*. In all our work so far V is the single point b and the process on the boundary is trivial: $X_t = b$ for all t (or for all t less than L_∞ if we are in a non-recurrent case as discussed in III-3 (g).) But in the general case $\widetilde{X}$ will be an important part of the excursion description of X.

There are a number of technicalities associated with $\widetilde{X}$ that we have not dealt with. For example if we want to know that $\widetilde{X}$ is a standard process so that we can use all the facts developed for such processes then we need to know that $E^x f(\widetilde{X}_t)$ defines an $\mathcal{E}$ measurable function of x when f is $\mathcal{E}$ measurable and that $\widetilde{X}$ is quasi-left-continuous up to its lifetime. In the case of the process on the boundary with V satisfying our additional conditions these additonal facts can be established. See V-2.11 in [BG, 1] for more discussion.

(c) Absolute continuity.

Let A and B be continuous additive functionals of X. We will say that A is *absolutely continuous with respect to* B if there is a positive universally measurable function f on E such that $A = fB$. We need to develop a simple test for this in the spirit of the Radon-Nikodym theorem. The test is the following:

(2.3) Theorem. *Let A and B be continuous additive functionals such* that

$$E^x(A_t + B_t) < \infty$$

for all x and t. Suppose that whenever g is positive and universally measurable $gB = 0$ implies $gA = 0$. Then A is absolutely continuous with respect to B. (Note that for a continuous additive functional C, $u_C^\lambda < \infty$ for some positive λ implies $E^x C_t < \infty$ for all x and t; and if $E^x C_t < \infty$ for all x and t then $gC = 0$ is equivalent to $E^x \int_0^\infty e^{-\lambda t} g(X_t) dC_t = 0$ for all x and some positive λ.)

Proof. Let us assume first of all that we are given continuous additive functionals A and B with $A_t \le B_t$ for all t and B_t strictly increasing in t. Let $\{\beta_t; t \ge 0\}$ be the inverse to B and let $\widetilde{X}$ be the time-changed process $\widetilde{X}_t = X_{\beta_t}$. If $\widetilde{A}_t = A_{\beta_t}$ and $\widetilde{B}_t = B_{\beta_t}$ then one checks easily that $\widetilde{A}$ and $\widetilde{B}$ are continuous additive functionals (the σ-algebras are $\mathcal{F}_{\beta_t}$). Obviously $\widetilde{B}_t = t$ and we have $\widetilde{A}_t \le t$. Let

$$Z_t = \varliminf_{s \downarrow 0}(\widetilde{A}(t+s) - \widetilde{A}(t))/s$$

and let $f(x) = E^x Z_0$. From the additive functional property we have $Z_t = Z_0 \circ \widetilde{\theta}_t$ and then from the Markov property of X at the stopping time β_t we have $Z_t = f(\widetilde{X}_t)$ almost surely. Since (use additivity and right continuity) almost surely $A_{t+s} - A_t \le B_{t+s} - B_t$ the measure $d_t \widetilde{A}_t$ is absolutely continuous with respect to lebesgue measure and hence is the integral of its derivative. Thus we have

$$\widetilde{A}_t = \int_0^t Z_s\, ds.$$

A technical argument, which the reader can supply by using the proof of (2.2) and the right continuity of excessive functions composed with the process, shows that $f(\widetilde{X}_t(\omega))$ can be taken to be jointly measurable in (t, ω); and then applying Fubini's theorem we may assert that for almost all ω (P^μ for any μ) $Z_t(\omega) = f(\widetilde{X}_t(\omega))$ for almost all (lebesgue measure) t. Thus we have

$$\widetilde{A}_t = \int_0^t f(\widetilde{X}_s)\, ds.$$

We can make the change of variables $\beta_s = r$ on the right side and we obtain almost surely

$$\widetilde{A}_t = A_{\beta_t} = \int_0^{\beta_t} f(X_r)\, dB_r$$

for all t. As t ranges over $[0, \infty)$ β_t covers the interval $[0, B_\infty)$ and so $A = fB$ follows in the present special case.

Coming to the general case let A and B be as in the hypotheses of (2.3) and let C be the continuous strictly increasing additive functional $C_t = A_t + B_t + t$. Applying to the pairs A, C and B, C the previous paragraph we obtain positive universally measurable functions h and k such that $A = kC$, $B = hC$. If g is the indicator of $\{h = 0\}$ then $gB = 0$ and so by hypothesis $gA = 0$. Thus $\{h = 0\}$ is A null and we may write

$$A = (k/h)hC = (k/h)B,$$

so $A = fB$ with $f = k/h$. This argument and the one in the next paragraph are taken from Benveniste and Jacod [BJ, 1].

(d) Lévy systems.

Let $\mathcal{P}(x, A)$, $x \in E$, $A \in \mathcal{E}$ be a kernel on E; that is $\mathcal{P}(x, A)$ is a measure in A, and universally measurable in x. For a positive function f from $E \times E$ to R which is $\mathcal{E} \times \mathcal{E}$ measurable we will write $\mathcal{P}f(x)$ for the expression $\int \mathcal{P}(x, dy)f(x, y)$. Write $(\mathcal{E}^2)^0_+$ for the set of all such f which also satisfy $f(x, x) = 0$ for all x in E. A *Lévy system* for X is a pair $(\mathcal{P}, \mathcal{L})$ where $\mathcal{P}$ is a kernel and $\mathcal{L}$ is a continuous additive functional such that, with $Z_t = e^{-t}$, we have

$$(2.4) \qquad E^x \sum Z_t f(X_{t-}, X_t) = E^x \int_0^\infty Z_s \mathcal{P}f(X_s)d\mathcal{L}_s$$

for all x in E and f in $(\mathcal{E}^2)^0_+$. First of all, because $f(x, x) \equiv 0$ the sum on the left of (2.4) is really only over those t such that $X_{t-} \neq X_t$. Secondly, granted the existence of $(\mathcal{P}, \mathcal{L})$ it follows by the same arguments as those in III-3 that (2.4) holds whenever $\{Z_t; t \geq 0\}$ is a positive previsible process.

We will sketch a proof that the Lévy system always exists. First of all a fact which we will use several times in the sequel is that $\{t | X_{t-} \neq X_t\}$ can be embedded in a countable family of stopping times: to be specific let ρ be a metric on E compatible with the topology, and set

$$T_1^k = \inf\{t | \rho(X_{t-}, X_t) > 1/k\}.$$

So T_1^k is the first time at which the path jumps by more than $1/k$. It is not hard to show that T_1^k is a stopping time, and then $T_{n+1}^k = T_n^k + T_1^k \circ \theta_{T_n^k}$

defines the successive times at which such jumps occur. Every t such that $X_{t-} \neq X_t$ is of the form T_n^k for all large k and some n depending on k. To get started on the construction of $(\mathcal{P}, \mathcal{L})$, we need to argue that there is a strictly positive $\mathcal{E}^*$ measurable function h such that for all x

$$(2.5) \qquad E^x \sum_{X_{s-} \neq X_s} e^{-s} h(X_s) \leq 1.$$

Indeed let j and k be positive integers, set

$$A_{jk} = \{x | E^x e^{-T_1^k} \leq (j-1)/j\}$$

and let $S_1, S_2, \ldots$ denote the succesive values of $\{T_n^k, n \geq 1\}$ such that $X_{T_n^k}$ lies in A_{jk}. We have

$$E^x(e^{-S_{n+1}}) \leq E^x(e^{-S_n} e^{-T_1^k \circ \theta_{S_n}})$$
$$\leq (j-1)/j E^x(e^{-S_n})$$

and so $E^x \sum_n e^{-S_n} \leq \sum((j-1)/j)^n = j$. Since T_1^k is almost surely strictly positive we have $E = \bigcup_j A_{jk}$ so if $h_k = \sum_j \frac{1}{2^{k+j} j} I_{A_{jk}}$ then h_k is strictly positive and

$$E^x \sum_n e^{-T_n^k} h_k(X_{T_n^k}) \leq 2^{-k}.$$

Clearly then $h = \sum h_k$ defines a function for which (2.5) holds. Now let g be a positive bounded $\mathcal{E}$ measurable function and set

$$(2.6) \qquad \psi_g(x) = E^x \sum_{X_{s-} \neq X_s} e^{-s} g(X_s) h(X_s).$$

If T is a stopping time then $P_T^1 \psi_g(x)$ is the sum in (2.6) but taken only over those s with $s > T$. It follows immediately that ψ_g is 1-excessive. In fact ψ_g is a regular 1-potential. To see this let T_n be stopping times increasing to T. Then the difference between $\lim P_{T_n}^1 \psi_g$ and $P_T^1 \psi_g$ is the one term

$$E^x(e^{-T} g(X_T) h(X_T); X_{T-} \neq X_T, T_n < T \text{ for all } n).$$

By quasi-left-continuity X_{T_n} approaches X_T almost surely so this last expectation is zero. Thus there is by Šur's theorem a continuous additive functional $\mathcal{L}_g$ such that $\psi_g(x) = E^x \int_0^\infty e^{-s} d\mathcal{L}_g(s)$. Let us denote $\mathcal{L}_1$ simply by $\mathcal{L}$. Since $\mathcal{L}_g + \mathcal{L}_{1-g} = \mathcal{L}$ we have $\mathcal{L}_g$ absolutely continuous with respect to $\mathcal{L}$ and so there is an $\mathcal{E}^*$ measurable function Pg such that $\mathcal{L}_g = Pg\mathcal{L}$. Let us call a set Γ in $\mathcal{E}^*$ $\mathcal{L}$-*null* if $E^x \int_0^\infty I_\Gamma(X_t)d\mathcal{L}_t$ is zero for all x; and then say that $\mathcal{E}^*$ measurable functions h and k are *equal* $\mathcal{L}$-a.e. if the set on which they differ is $\mathcal{L}$-null. The uniqueness assertion in (2.2) says that Pg is uniquely determined up to equality $\mathcal{L}$-a.e. If we regard as equal two functions which are equal $\mathcal{L}$ a.e. then $g \to Pg$ is a positive linear mapping and because of the nice structure of E we may assume it is given by a kernel $Pg(x) = \int P(x, dy)g(y)$, where $P(x, A)$ is a measure on $\mathcal{E}$ in A and is $\mathcal{E}^*$ measurable in x. Define the kernel $\mathcal{P}$ by

$$\mathcal{P}g = P(g/h).$$

Then we have shown that

$$E^x \sum_{X_{s-} \neq X_s} e^{-s}g(X_s) = E^x \int_0^\infty e^{-s}\mathcal{P}g(X_s)d\mathcal{L}_s$$

for any positive $\mathcal{E}$ measurable g; and by the argument from III-3 we may in this equality replace e^{-s} with any positive previsible process Z_s. In particular let g' be a positive bounded continuous function on E and take $Z_s = e^{-s}g'(X_{s-})$. This process is left continuous, hence previsible, and we have

$$E^x \sum_{X_{s-} \neq X_s} e^{-s}g'(X_{s-})g(X_s) = E^x \int_0^\infty e^{-s}g'(X_{s-})\mathcal{P}g(X_s)d\mathcal{L}_s.$$

On the right side of this equation we may replace X_{s-} with X_s since these differ only countably often and $\mathcal{L}$, being continuous, charges no countable set. This equality now can be written

$$E^x \sum_{X_{s-} \neq X_s} e^{-s}f(X_{s-}, X_s) = E^x \int_0^\infty e^{-s}\{\int \mathcal{P}(X_s, dy)f(X_s, y)\}d\mathcal{L}_s$$

with $f(x, y) = g'(x)g(y)$. Monotone class reasoning implies that this equality holds for any positive $\mathcal{E} \times \mathcal{E}$ measurable function f. If we take f to be the indicator of the diagonal in $E \times E$ then the left side is 0 and so we have $\mathcal{P}(x, \{x\})$ equal to zero for $\mathcal{L}$-a.e. x. We set this expression equal to 0 for all x. For a function in $(\mathcal{E}^2)^0_+$ the restriction $X_{s-} \neq X_s$ is superfluous so (2.4) holds and the existence of the Lévy system is established.

From the uniqueness part of (2.2) it is clear that if $(\mathcal{P}, \mathcal{L})$ and $(\mathcal{P}', \mathcal{L}')$ are Lévy systems then for each f the continuous additive functionals $(\mathcal{P}f)\mathcal{L}$ and $(\mathcal{P}'f)\mathcal{L}'$ are the same.

A pair of examples will illustrate different types of Lévy systems. Suppose X is a Lévy process of the simplest type - a generalized Poison process. Then the process is constant except for jumps of size Y_n at times $J_1 + \cdots + J_n$, where the J's are independent exponentials with rate α, the Y's are independent with common probability law μ, and the J's and Y's are independent. The reader should have no difficulty in checking that $\mathcal{L}(t) = t$ and $\mathcal{P}f(x) = \int f(x, x + y)\alpha\mu(dy)$ is a Lévy system for X. Noting that $\alpha\mu$ is the Lévy measure for X leads one to conjecture that the same description holds for the most general process with stationary independent increments. We will leave this as an exercise. An apparently trivial, but in fact more interesting, example is obtained when X is a recurrent extension beyond $\sigma_{\{b\}}$ of a minimal process. Let us assume that the given minimal process is a diffusion so that the only discontinuities of the path occur at points s in G such that $X_s \neq b$. It follows then from the material in Chapter III that $\mathcal{P}(b, dy) = \hat{P}(X_0 \in dy)$ restricted to $E - \{b\}$ and $\mathcal{L} = L$ where L is local time at b for X and $\hat{P}$ is the excursion measure. The definition of $\mathcal{P}(x, dy)$ at $x \neq b$ is arbitrary. We will give more interesting examples in section 4 in discussing the Lévy system for the process on the boundary.

3. Exit systems.

(a) The kernel $\hat{P}$.

As in the introduction X will be a standard process and V will be a closed subset of E such that each point in V is regular for V. The hitting time σ_V will be denoted simply by σ; and G will denote the strictly positive left ends of the excursion intervals. We will assume that X is in fact the canonical right continuous realization. We will use $\{\overline{P}^x ; x \in E\}$ for the probabilities for the minimal process $X_{t \wedge \sigma}$, and as usual $P^0(t, x, A) = \overline{P}^x(X_t \in A, t < \sigma)$, $V^\lambda f(x) = \overline{E}^x \int_0^\sigma e^{-\lambda t} f(X_t) dt$; and P_V^λ will be the hitting kernel

$$P_V^\lambda f(x) = \overline{E}^x(e^{-\lambda \sigma} f(X_\sigma); \sigma < \infty)$$

for f in $\mathcal{B}_+(V)$, the positive functions on V measurable relative to the Borel sets of V.

Set

$$\varphi(x) = E^x e^{-\sigma}.$$

According to (b) of section 2 we have $\varphi(x)$ equal to $E^x \int_0^\infty e^{-t} dL_t$ where L, the local time on V, is a continuous additive functional growing exactly at those t such that X_t is in V.

Now take a function f on path space Ω, measurable with respect to the σ-algebra $\mathcal{F}^0$ of the coordinate functions and with $0 \le f \le 1$. Then we have

$$\begin{aligned}
\varphi(x) &= E^x \int_\sigma^\infty e^{-t} dt \\
&= E^x \sum_{s \in G} e^{-s}(1 - e^{-\sigma}) f \circ \theta_s \\
&\quad + E^x \sum_{s \in G} e^{-s}(1 - e^{-\sigma})(1 - f) \circ \theta_s \\
&\quad + E^x \int_\sigma^\infty e^{-t} I_V(X_t) dt = \varphi_1(x) + \varphi_2(x) + \varphi_3(x).
\end{aligned}$$

The functions φ_i are 1-excessive, and since their sum is a regular 1-potential, each of them must be regular also. Thus they are potentials of continuous additive functionals L_1, L_2, L_3, and by uniqueness $L_1 + L_2 + L_3 = L$. Thus the L_i are absolutely continuous with respect L so by (2.3) applied to L_1

and L_3 there are positive $\mathcal{E}^*$ measurable function $P_1^{\cdot} f$ and ℓ such that

$$E^x \sum_{s \in G} e^{-s}\{(1 - e^{-\sigma})f\} \circ \theta_s = E^x \int_0^\infty e^{-s} P_1^{X_s}(f) dL_s;$$

$$(3.1) \qquad E^x \int_0^\infty e^{-s} I_V(X_s) ds = E^x \int_\sigma^\infty e^{-s} I_V(X_s) ds$$

$$= E^x \int_0^\infty e^{-s} \ell(X_s) dL_s.$$

(The choices of the letter ℓ and of m later on are unfortunate since ℓ is analogous to the delay coefficient m from earlier chapters. But we are trying to preserve notation from the paper of Motoo to be discussed in section 4.) Now clearly two different choices of $P_1^{\cdot} f$ must be equal L-a.e. If we regard as equal two functions on E which agree L-a.e. then $f \to P_1^{\cdot} f$ is linear and preserves monotone limits. We need to know that it can be realized by a kernel on $\mathcal{E}^* \times \mathcal{F}$. In fact there is a function $P_1^x(\Gamma)$ defined for x in E and Γ in $\mathcal{F}^*$ (the universal completion of $\mathcal{F}^0$) which is $\mathcal{E}^*$ measurable in x and a measure in Γ of mass no greater than 1 and such that for each positive bounded $\mathcal{F}^0$ measurable f the integral $\int P_1^x(d\omega) f(\omega)$ as a function of x is a version of $P_1^{\cdot} f$. The argument is given in [Ma, 1]. We will take this fact as known.

Take $f = I_{\{\sigma=0\}}$. (Strictly speaking f is measurable only with respect to a completion of $\mathcal{F}^0$. We refer the reader to the arguments in III-3 to justify applying the forgoing material to such functions.) With this choice of f we have $E^x \sum_{s \in G} e^{-s}(1 - e^{-\sigma}) f \circ \theta_s = 0$ and since this equals $E^x(\int_0^\infty e^{-t} P_1^{X_t}(\sigma = 0) dL_t)$ we have

$$P_1^x(\sigma = 0) = 0 \qquad L\text{-a.e.}x.$$

We can set this equal to 0 for all x without destroying the other properties and will assume that done. As in Chapter III set

$$P_2^x(f) = P_1^x(f/(1 - e^{-\sigma}))$$
$$\hat{P}^x = P_2^x \circ \tau$$

where $\tau(X) = X'$ with $X_t' = X_{t \wedge \sigma}$. What we have established are the *excursion formulas*

$$(3.2a) \qquad E^x \sum_{s \in G} Z_s f \circ \theta_s = E^x \int_0^\infty Z_s P_2^{X_s}(f) dL_s$$

$$(3.2b) \qquad E^x \sum_{s \in G} Z_s f \circ \tau \circ \theta_s = E^x \int_0^\infty Z_s \hat{P}^{X_s}(f) dL_s$$

for $Z_s = e^{-s}$. By exactly the same argument as was given in III-3 we conclude that (3.2) is valid whenever Z is any positive previsible process. As usual it is the second of these formulas that we will use. Typically f is a function of the stopped process only and then we write f rather than $f \circ \tau$.

Before going on we have to compare some notation and definitions with those from Maisonneuve's paper [Ma, 1]. Maisonneuve gives a separate treatment of those points s in G such that X leaves V at s by means of a jump. That is he sets

$$G^r = \{s \in G | X_s \in V\}$$

$$G^i = \{s \in G | X_s \notin V\}.$$

Then by multiplying f in (3.2) by $I_{X_0 \in V}$ we obtain

$$(3.3) \qquad E^x \sum_{s \in G^r} Z_s f \circ \theta_s = E^x \int_0^\infty Z_s P_2^{X_s}(f; X_0 \in V) dL_s$$

with a similar replacement for the second equality in (3.2). It is an important fact that in (3.3) Z_s can be any positive optional process (not just a previsible one). To see this in (3.3) start with a stopping time R and $Z_s = I_{(0,R]}(s)$, for which (3.3) is known to hold. Replacing Z_s with $I_{(0,R)}(s)$ leaves the right side unchanged and changes the left side only through deletion of a term $E^x(f \circ \theta_R I_{R \in G^r})$. Since every point in V is regular for V we have for any stopping time R

$$P^x(R \in G, X_R \in V) \le P^x(P^{X_R}(\sigma > 0); X_R \in V) = 0,$$

so the deleted term is zero and (3.3) holds for $Z_s = I_{(0,R)}(s)$. Since optional process are obtained from those of the form $I_{(0,R)}$, the assertion is established. The points in G^i are discontinuity points of the path. Hence (refer back to section 2) they are embeddable in stopping times so that Maisonneuve can treat the sum $E^x \sum_{s \in G^i} Z_s f \circ \theta_s$ in a different way. Besides using only G^r, Maisonneuve deals mainly with the entire future path $\{X_t; t \ge s\}$ rather than the "excursion" $\{X_t; s \le t \le s + \sigma \circ \theta_s\}$. Thus his excursion formula is (3.3) rather than the second equality in (3.2) which is ours. We have chosen to lump G^r and G^i together and to focus on the excursions because this is more relevant to Itô's approach and to the problem of constructing extensions of a minimal process.

We will call the pair $(\hat{P}, L)$ the *previsible exit system.*

(b) Markov properties.

The Markov properties of the coordinate process relative to the measures P_2^x and $\hat{P}^x$ are summed up in the following theorem.

(3.4) Theorem. *For L-a.e. x whenever T is an $\{\mathcal{F}_{t+}^0\}$ stopping time with $T > 0$, g is a positive function which is $\mathcal{F}_{T+}^0$ measurable and F is a positive $\mathcal{F}^0$ measurable function we have*

$$(3.5) \qquad P_2^x(g \cdot F \circ \theta_T) = P_2^x(g P^{X_T}(F))$$

and

$$(3.6) \qquad \hat{P}^x(g \cdot F \circ \tau \circ \theta_T; T < \sigma) = \hat{P}^x(g \overline{P}^{X_T}(F); T < \sigma).$$

(Recall $\overline{P}^x(\Gamma) \equiv P^x(X' \in \Gamma)$ where $X'_t = X_{t \wedge \sigma}$. Note that the exceptional set is to be shown independent of g, F and T.)

Proof. We need establish only (3.5) as (3.6) follows automatically. Now exactly the same argument as that for Theorem 3.28 in Chapter III shows that for all x

$$E^x \int_0^\infty e^{-s} P_2^{X_\bullet}(g \cdot F \circ \theta_T) dL_s = E^x \int_0^\infty e^{-s} P_2^{X_\bullet}(g \cdot P^{X_T}(F)) dL_s,$$

and so by the uniqueness theorem for continuous additive functionals (3.5) holds for L-a.e. x but with the exceptional set depending perhaps on g, F, T. To deal with this matter, for each strictly positive rational number u let $\mathcal{G}_u$ denote a countable class of $\mathcal{F}_u^0$ sets with the property that a finite measure on $\mathcal{F}_u^0$ is determined by its values on $\mathcal{G}_u$ sets, and let $\mathcal{H}$ consist of all functions of the form $\lambda U^\lambda f$ where U^λ is the potential operator for X and λ and f range respectively over the strictly positive rationals and a countable collection $\mathcal{J}$ of bounded continuous functions on E large enough that a finite measure μ on $\mathcal{E}$ is determined by the integrals $\int f d\mu$ with $f \in \mathcal{J}$. The separability properties of E and right continuity of $t \to X_t$

makes it clear that such classes exist. If h is in $\mathcal{H}$, t is positive and k is the function $k(x) = P^x h(X_t)$ then k is λ excessive for some λ and so for each x in E almost surely P^x the mapping $t \to k(X_t)$ is right continuous (see (3b) of Chapter III). Now if $t \to k(X_t) \circ \theta_s$ fails to be right continuous then so does $t \to k(X_t)$. Thus if we set

$$\Lambda = \{\omega | t \to k(X_t(\omega)) \text{ is not right continuous}\}$$

then for any x in E

$$0 = P^x(\Lambda) \geq E^x\big(\sum_{s \in G} I_\Lambda \circ \theta_s\big)$$

$$= E^x\{\int_0^\infty P_2^{X_s}(\Lambda) dL_s\}$$

and it follows that for L-a.e. x

$$(3.7) \qquad P_2^x(r \to P^{X_r} h(X_t) \text{ is not right continuous}) = 0.$$

We will delete from E the set of points x where for some strictly positive rational u, $g \in \mathcal{G}_u$, F of the form $h(X_t)$ with h in $\mathcal{H}$ and t rational and $T \equiv u$ the equality (3.5) fails. According to the observation at the start of the proof we are deleting only an L-null set. Also we will delete the L-null set where (3.7) fails for some h in $\mathcal{H}$ and rational t. With these deletions understood let T be an $\{\mathcal{F}_{t+}^0\}$ stopping time with T strictly positive. We will in fact assume $T > \delta$ for some $\delta > 0$ passing from here to the general case by monotone convergence. Let T_n denote the usual approximant,

$$T_n = (k+1)/2^n \quad \text{if } k/2^n \leq T < (k+1)/2^n$$

Then $\{T_n = j/2^n\}$ is in $\mathcal{F}_{j/2^n}^0$ and if Λ is in $\mathcal{F}_{T+}^0$ then $\Lambda \cap \{T_n = j/2^n\}$ is in $\mathcal{F}_{j/2^n}^0$. If ϵ is less than $j/2^n$ we may regard $\{\sigma > \epsilon\}$ as being in $\mathcal{F}_{j/2^n}^0$ and then we have, with $F = h(X_t)$,

$$P_2^x(g \cdot F \circ \theta_{j/2^n}; \sigma > \epsilon) = P_2^x(g \cdot P^{X_{j/2^n}}(F); \sigma > \epsilon),$$

first for g the indicator of a $\mathcal{G}_{j/2^n}$ set and then for g any positive $\mathcal{F}_{j/2^n}^0$ measurable function, since each side is a finite measure in g. Then it follows that for $\epsilon < \delta$ we have

$$P_2^x(g \cdot h(X_{t+T_n}); \sigma > \epsilon) = P_2^x(g \cdot P^{X_{T_n}} h(X_t); \sigma > \epsilon)$$

for g in $\mathcal{F}_{T+}^0$ as in the argument from Chapter I for the strong Markov property. Now we let $n \to \infty$. By right continuity of the composition of h and of $P^{\cdot}h(X_t)$ with the path and the finiteness of $P_2^x(\sigma > \epsilon)$ we obtain

$$(3.8) \qquad P_2^x(g \cdot h(X_{t+T}); \sigma > \epsilon) = P_2^x(g \cdot P^{X_T}h(X_t); \sigma > \epsilon)$$

for g in $\mathcal{F}_{T+}^0$, h in $\mathcal{H}$ and t rational. We may replace h with any bounded $\mathcal{E}$ measurable function and t with any positive real number by first taking $h = \lambda U^\lambda f$, which converges boundedly to f as $\lambda \to \infty$ for f bounded and continuous, then using right continuity of the path and then using the fact that each side of (3.8) is a measure in h. We can now delete the qualification $\{\sigma > \epsilon\}$ in (3.8) by letting ϵ decrease to 0 and using the fact that $P_2^x(\sigma = 0)$ is 0 for L-a.e. x. Finally having established (3.5) for all F of the form $f(X_t)$ with f being $\mathcal{E}$ measurable we obtain it for any F in $\mathcal{F}^0$ by the usual iteration of conditional expectations; see (4.3) in Chapter I.

(3.9) Exercise. Define $\eta_s^x(A) = \hat{P}^x(X_s \in A, s < \sigma)$ for $s > 0$, A in $\mathcal{E}$ and $A \subset V^c$. Show that except for x in an L-null set $\{\eta_s^x; s > 0\}$ is an entrance law for the minimal semigroup $P^0(t, x, A)$.

(3.10) Exercise. If X has *no sojourn on* V, that is $P^x(\int_0^\infty I_V(X_t)dt) = 0$ for all x, then use the second part of (3.2) with $Z_s = e^{-s}$ and $f = 1 - e^{-\sigma}$ to conclude that for L-a.e. x we have $\hat{P}^x(1 - e^{-\sigma}) = 1$.

(3.11) Exercise. Show that the Markov properties (3.5) and (3.6) can be strengthened to allow $T \geq 0$ provided $\{T = 0\} \subset \{X_0 \notin V\}$. (Hint: practice on the case where V is single point, b. Do this case by repeating the argument for III-3.28, but now defining G_n^ϵ to be the nth smallest point $s \in G$ such that $\sigma_{O_\epsilon} \circ \theta_s < \sigma_b \circ \theta_s$, where $O_\epsilon = \{x \in E | \rho(x, b) > \epsilon\}$. Set $T_n^\epsilon = G_n^\epsilon + \sigma_{O_\epsilon} \circ \theta_{G_n^\epsilon}$ and note that every s in G is of the form G_n^ϵ for all small ϵ, that T_n^ϵ is a stopping time and that if s is in G and $X_s = X_0 \circ \theta_s$ is not equal to b then s is equal to some T_n^ϵ for all small ϵ. Now proceed to the general case.)

(3.12) Exercise. Use (3.11) to show that for L-a.e. x

$$\hat{P}^x(F; X_0 \notin V) = \overline{P}^{\mu_x}(F), \quad F \in \mathcal{F}^0$$

where $\mu_x(A) = \hat{P}^x(X_0 \in A)$, $A \in \mathcal{E}$ and $A \subset V^c$.

(c) The excursion data.

For the situation of Chapter V the description of X through excursions away from $\{b\}$ is based on the excursion measure $\hat{P}$. We have seen that there may be some advantage to decomposing $\hat{P}$ into its restrictions to $\{X_0 = b\}$ and $\{X_0 \neq b\}$ and to calling attention to sojourn at $\{b\}$ through the delay coefficient $m = 1 - \hat{P}(1 - e^{-\sigma})$. We have seen also that in the important special case where the ratio $V^1 f(x)/V^1 1(x)$ (f bounded continuous) has a continuous extension from $E - \{b\}$ to all of E the part of $\hat{P}$ not dependent on sojourn or discontinuous exit can be described explicitly in terms of that limiting ratio and hence entirely in terms of the minimal process. (See V-4.2). When we proceed from V being a single point to the general case we still have excursion measures, their decompositions in terms of exit behavior, and delay coefficients, except that now these vary from point to point. The new element that is needed is the process on the boundary. We will describe these objects and their relation to the original process.

The only situation of interest to us is the one in which V is the complement of a dense open subset D of E. Henceforth this assumption on V will be in force, and D will denote V^c.

Define functions m and n on V by

$$m(x) = \hat{P}^x(1 - e^{-\sigma}; X_0 \in V)$$

$$n(x) = \hat{P}^x(1 - e^{-\sigma}; X_0 \notin V).$$

(Caution: m is not the delay coefficient). As in (3.1) define a function ℓ on V by the requirement that

$$E^x \int_0^\infty e^{-t} I_V(X_t) dt = E^x \int_0^\infty e^{-s} \ell(X_s) dL_s.$$

Define a kernel $Q(x, A)$ for x in V and A in $\mathcal{E}$ with $A \subset D$ by

$$Q(x, A) = \hat{P}^x((1 - e^{-\sigma}); X_0 \in A).$$

For $\lambda > 0$, x in D and f bounded and $\mathcal{E}$ measurable set

$$\hat{H}^\lambda f(x) = V^\lambda f(x)/V^1 1(x).$$

If, for a particular f, $\hat{H}^\lambda f$ has an extension to all of E, that is for each x in V

$$\lim_{\substack{y \to x \\ y \in D}} \hat{H}^\lambda f(y)$$

exists, we will denote the extension by $\hat{H}^\lambda f$ and say "$\hat{H}^\lambda f$ is defined on all of E." We will call the following statement Motoo's *hypothesis*:

(3.13) for all $\lambda > 0$ and bounded continuous f on E, $\hat{H}^\lambda f$ is defined on all of E and defines a continuous function on E. Furthermore there is a kernel $\hat{H}^\lambda(x, A)$, $x \in E$, $A \subset \mathcal{E}$ such that

$$\hat{H}^\lambda f(x) = \int f(y)\hat{H}^\lambda(x, dy) \quad x \in E, f \in C_b^+(E).$$

Of course for $x \in D$ the kernel $\hat{H}^\lambda(x, A)$ simply is the ratio $V^\lambda I_A(x)/V^1 1(x)$. Motoo does not specifically assume the existence of a kernel but the existence follows automatically from his assumption that E has been compactified. In a bit we will derive an important consequence, (3.17), of Motoo's hypothesis.

First note the following simple facts.

(3.14) Proposition. For L-a.e. x we have $(\mu_x = \hat{P}^x(X_0 \in \cdot))$
    (a) $Q(x, D) = n(x)$,
    (b) $Q(x, A) = \int_A \mu_x(dy)V^1 1(y) \qquad A \in \mathcal{E}$,
    (c) $\ell + m + n = 1$

Proof. (a) and (b) simply are matters of notation. As for (c) we have for $x \in V$

$$1 = E^x \int_0^\infty e^{-t} I_E(X_t)dt = E^x \sum_{s \in G} e^{-s}(1 - e^{-\sigma}) \circ \theta_s$$

$$+ E^x \int_0^\infty e^{-t} I_V(X_t)dt$$

$$= E^x \int_0^\infty e^{-s}(\hat{P}^{X_s}(1 - e^{-\sigma}) + \ell(X_s))dL_s$$

and since $E^x \int_0^\infty e^{-s}dL_s = 1$ the assertion follows from the uniqueness for potentials.

Let U^α denote the α potential operator for X. If g is positive and $\mathcal{E}$-measurable then clearly we have for x in V

$$U^\alpha g(x) = E^x \int_0^\infty e^{-\alpha t}(gI_V + gI_D)(X_t)dt$$

$$(3.15) \qquad = E^x \int_0^\infty e^{-\alpha t}g\ell(X_t)dL_t$$

$$+ E^x \int_0^\infty e^{-\alpha t}\hat{P}^{X_t}\left(\int_0^\sigma e^{-\alpha r}g(X_r)dr\right)dL_t.$$

In the second of the summands on the right of (3.15) write $\hat{P}^{X_t}$ as the sum of its restrictions to $\{X_0 \in V\}$ and $\{X_0 \in D\}$ and note that by (3.12)

$$E^x \int_0^\infty e^{-\alpha t}\hat{P}^{X_t}\left(\int_0^\sigma e^{-\alpha r}g(X_r)dr; X_0 \notin V\right)dL_t$$

$$= E^x \int_0^\infty e^{-\alpha t}Q\hat{H}^\alpha g(X_t)dL_t.$$

Thus we may write

$$(3.16) \qquad U^\alpha g(x)$$

$$= E^x \int_0^\infty e^{-\alpha t}\{g\ell + \hat{P}'\left(\int_0^\sigma e^{-\alpha r}g(X_r)dr; X_0 \in V\right) + Q\hat{H}^\alpha g\}(X_t)dL_t$$

This formula is found in Motoo [Mo, 1] and in section 8 of [Ma, 1]. We will call it the Maisonneuve-Motoo *decomposition*.

When (3.13) is in force we can make a more specific statement slightly generalized as follows

(3.17) Theorem. *Suppose $\alpha > 0$, g is positive bounded and continuous and $\hat{H}^\alpha g$ exists on all of E and defines a continuous function. Then for x in V*

$$U^\alpha g(x) = E^x \int_0^\infty e^{-\alpha t}\{\ell + (m + Q)\hat{H}^\alpha\}g(X_t)dL_t.$$

Proof. Clearly all we need to do is show that

$$(3.18) \qquad E^x \int_0^\infty e^{-\alpha t}\hat{P}^{X_t}\left(\int_0^\sigma e^{-\alpha r}g(X_r)dr; X_0 \in V\right)dL_t$$

$$= E^x \int_0^\infty e^{-\alpha t}m\hat{H}^\alpha g(X_t)dL_t.$$

Note that $V^\alpha g = U^\alpha g - P_V^\alpha U^\alpha g$ and hence $V^\alpha g(X_t)$, as the difference of excessive functions composed with the process, is right continuous in t. The same thing applies to the ratio $(V^\alpha g/V^1 1)(X_t)$ except at t such that X_t is in V. But the ratio, now denoted $\hat{H}^\alpha g(X_t)$, is by hypothesis defined and right continuous at such points also. Let G_n^ϵ denote as usual the nth smallest element s of G with $\sigma \circ \theta_s > \epsilon$ and s in G^r. Let $T_n^\epsilon = G_n^\epsilon + \epsilon$. Then the left side of (3.18) is

$$\lim_{\epsilon \to 0} E^x \sum_{s \in G^r} e^{-\alpha s} \int_\epsilon^\sigma e^{-\alpha r} g(X_r) dr$$

$$= \lim_{\epsilon \to 0} E^x \sum_n e^{-\alpha T_n^\epsilon} V^\alpha g(X_{T_n^\epsilon})$$

$$= \lim_{\epsilon \to 0} E^x \sum_n e^{-\alpha T_n^\epsilon} \hat{H}^\alpha g(X_{T_n^\epsilon})(1 - e^{-\sigma}) \circ \theta_{T_n^\epsilon}$$

$$= E^x \sum_{s \in G^r} e^{-\alpha s} \hat{H}^\alpha g(X_s)\{(1 - e^{-\sigma}) I_{X_0 \in V}\} \circ \theta_s.$$

We may use the excursion formula on this last expression: admittedly the term $e^{-\alpha s} \hat{H}^\alpha g(X_s)$ is only optional, not previsible, but as we noted near the end of 3(a) this is enough when the sum is only over s in G^r. Thus the last line in the display is

$$E^x \int_0^\infty e^{-\alpha t} \hat{H}^\alpha g(X_t) \hat{P}^{X_t}(1 - e^{-\sigma}; X_0 \in V) dL_t,$$

and bringing in the definition of m we see that the proof is complete. We will refer to the equality asserted in (3.17) as Motoo's *formula*. Note that when (3.13) is in force then (3.17) holds for all positive $\mathcal{E}$ measurable g because each side is a measure in g.

4. Motoo Theory.

In the rest of this Chapter we will assume that Motoo's hypothesis, (3.13), holds. Note that this is a condition on the minimal process only. As noted in V-4 the condition holds for Brownian motion on $(0, \infty)$ but fails for Brownian motion on $R - \{0\}$. The failure in this case is remedied by keeping track of the (two) extremal modes of continuous entrance from $\{0\}$. No doubt something similar must be done in the general case if (3.13) fails; but, to our knowledge, this complicated task has not been carried out. An example rich enough to illustrate the general theory and in which (3.13) holds is Brownian motion in the upper half plane with V being the real axis. Shortly we will discuss this example in more detail.

For us the term "Motoo theory" will mean two things. Firstly, it will mean starting with an extension X of a given minimal process and seeing how the excursion data determine the law and some aspects of the path behavior of X. Secondly, it will mean starting with excursion data and attempting to construct an X whose excursion data is as given.

The hypotheses and notation will be those from the beginning of Chapter V except that the one point set $\{b\}$ will be replaced by the set V. The fact that X is an extension of $\overline{X}$ will be expressed as

$$(4.1) \qquad U^\alpha g(x) = V^\alpha g(x) + P_V^\alpha U^\alpha g(x)$$

where U^α is the α potential operator for X and $P_V^\alpha(x, A) = \overline{P}^x(X_\sigma \in A) = P^x(X_\sigma \in A)$, A a Borel set in V. Note that P_V^α is determined by the minimal process. The equalities

$$(4.2) \qquad \begin{aligned}
V^\alpha - V^\beta &= (\beta - \alpha)V^\alpha V^\beta & \alpha &> 0, \quad \beta > 0 \\
P_V^\alpha - P_V^\beta &= (\beta - \alpha)V^\beta P_V^\alpha & \alpha &\geq 0, \quad \beta > 0 \\
V^\alpha f(x) &= 0, \ P_V^\alpha f(x) = f(x) & \alpha &> 0, \quad x \in V \\
\hat{H}^\alpha - \hat{H}^\beta &= (\beta - \alpha)\hat{H}^\alpha V^\beta & \alpha &> 0, \quad \beta > 0
\end{aligned}$$

are the resolvent equation for the semigroup $\{P_t^0\}$ and easy consequences of the basic hypotheses. (Caution: in Motoo's paper U^α is G_α , V^α is G_α^0 and more dangerously P_V^α is H_α.) The process on the boundary, introduced in section 2, will be denoted by $\widetilde{X}$.

Any object definable in terms of the minimal process is regarded as basic data. In particular (3.17) expresses in terms of $\hat{H}^\alpha$ and hence in terms of the minimal process that part of the kernel $\hat{P}$ for an extension which is not associated with sojourn on V or jumping into D from V.

(a) The boundary system.

Following Motoo we will call $(\widetilde{X}, \ell, m, Q)$ from section 3(c) the *boundary system* for X. We should note right away that ℓ and m, and also Q since it is dependent on $\hat{P}^{\cdot}$ are determined only up to L-null sets. However, if Γ is in $\mathcal{E}$ and $\{\beta_t\}$ is the right continuous inverse of $\{L_t\}$ so that $\widetilde{X}_t = X_{\beta_t}$ then

$$E^x \int_0^\infty e^{-t} I_\Gamma(X_t) dL_t = E^x \int_0^\infty e^{-\beta_t} I_\Gamma(\widetilde{X}_t) dt,$$

which makes it clear that Γ is L-null if and only if it is of potential 0 for the process on the boundary. Since processes equal in law have the same sets of potential zero we may regard objects $(\widetilde{X}, \ell, m, Q)$ and $(\widetilde{X}', \ell', m', Q')$ as the same if $\widetilde{X}$ and $\widetilde{X}'$ are equal in law and $\ell = \ell', m = m'$ and $Q = Q'$ up to sets of potential 0.

We need to develop some properties of the boundary system. We will say that a relationship holds a.e.$(\widetilde{X})$ if the set on which it fails has $\widetilde{X}$ potential zero.

(a 1). $\qquad\qquad \ell(x) + m(x) + Q(x, D) = 1 \qquad$ a.e.$(\widetilde{X})$.

Proof.: This simply is (3.14)(c).

(a 2). $\qquad\qquad m\hat{H}^\alpha I_V = 0 \qquad$ a.e.$(\widetilde{X})$.

Proof.: Take $g = I_V$ in (3.17). The function ℓ was constructed so that ℓL and $I_V t$ are the same additive functional. Hence they have equal potentials so

$$E^x \int_0^\infty e^{-\alpha t} m\hat{H}^\alpha I_V(X_t) dL_t = 0$$

as asserted in (a 2).

The next property of the boundary system corresponds, to a certain extent at least, to the fact that except in the case of holding and jumping processes there are infinitely many excursions in a short time if we start at a point x in V.

(a 3) Let $J = \{x | \ell(x) + m(x) > 0\}$. Then almost surely P^x for all x

$$\int_0^t \{(\tfrac{1}{1-I_J} + Q(\tfrac{1}{V^1 1}))\}(\tilde{X}_s)ds = \infty$$

for all $t > 0$.

Proof. Of course $Q(1/V^1 1)(x)$ is simply $\hat{P}^x(X_0 \in D)$. In writing out the proof we will assume that $\ell + m \equiv 0$ since the difficulty in the assertion is associated with the Q term. Let

$$\rho = \inf \{t | \int_0^t \|\hat{P}^{X_\bullet}\| dL_s = \infty\}.$$

Let x in V be fixed. Unless $P^x(\rho = 0) = 1$, which is what we are trying to prove, we have $P^x(\rho > 0) = 1$. The mapping $t \to \int_0^t \|\hat{P}^{X_\bullet}\| dL_s$ is everywhere left continuous and is continuous and finite on $[0, \rho)$. Set

$$\rho_1 = \inf \{t | \int_0^t \|\hat{P}^{X_\bullet}\| dL_s \geq 1\}.$$

Then ρ_1 is a stopping time, strictly positive almost surely P^x if ρ is, and $\int_0^{\rho_1} \|\hat{P}^{X_\bullet}\| dL_s \leq 1$. We have

$$\alpha U^\alpha 1(x) = E^x \int_0^\infty e^{-\alpha t} \hat{P}^{X_t}(1 - e^{-\alpha\sigma})dL_t$$
$$= E^x \int_0^{\rho_1} e^{-\alpha t} \hat{P}^{X_t}(1 - e^{-\alpha\sigma})dL_t$$
$$+ E^x e^{-\alpha\rho_1}(\alpha U^\alpha 1(X_{\rho_1})).$$

Let α tend to infinity. Since $\alpha U^\alpha 1 \leq 1, \rho_1$ is strictly positive and we have the equality $\hat{P}^{X_t}(1 - e^{-\alpha\sigma}) \leq \|\hat{P}^{X_t}\|$ the right side approaches zero, while $\alpha U^\alpha 1(x)$ approaches 1. This shows $\rho = 0$, which clearly is equivalent to the assertion in (a 3).

(a 4) - The Lévy system for $\tilde{X}$. Let $\Psi(x, dy)$ denote the kernel $\hat{P}^x(X_\sigma \in dy)$ restricted to $V - \{x\}$. First of all we will argue that Ψ is determined by the minimal process and the boundary system. (Note that the lack of uniqueness in defining $\hat{P}^x$ is up to sets of $\widetilde{X}$ potential zero.) In the following x will denote a point such that the Markov properties of (3.6) and (3.11) as well as the conclusion of (3.12) hold, and also such that the equality

$$\hat{P}^x(\int_0^\sigma e^{-\alpha r}g(X_r)dr; X_0 \in V) = m(x)\hat{H}^\alpha g(x)$$

holds for all $\alpha > 0$ and $g \in \mathcal{E}_+$. The set of x for which one of these relationships fails has $\widetilde{X}$ potential zero. Let us use the displayed equation with $g = P_V^\beta h$ with h in $\mathcal{B}_+(V)$. A simple calculation using the Markov properties yields

$$(4.3) \qquad (\alpha - \beta)m(x)\hat{H}^\alpha P_V^\beta h(x) = \hat{P}^x((e^{-\beta\sigma} - e^{-\alpha\sigma})h(X_\sigma); X_0 \in V).$$

The companion equality

$$(4.4) \quad (\alpha - \beta)\int Q(x,dy)\hat{H}^\alpha P_V^\beta h(y) = \hat{P}^x(e^{-\beta\sigma} - e^{-\alpha\sigma})h(X_\sigma); X_0 \notin V)$$

follows from (3.12) and the Markov properties of the minimal process. The left sides are determined by the minimal process and the boundary system. Now letting $\beta \to 0$ and $\alpha \to \infty$ the assertion about Ψ follows.

Coming to the Lévy system itself let f and g be positive $\mathcal{E}$ measurable functions, g being continuous and bounded and apply the excursion formula with $Z_s = e^{-\alpha L_s}g(X_{s-})$, which is previsible. We have

$$E^x \sum_{s \in G} e^{-\alpha L_s}g(X_{s-})f(X_\sigma) \circ \theta_s$$

$$= E^x \int_0^\infty e^{-\alpha L_s}g(X_{s-})\hat{P}^{X_s}(f(X_\sigma))dL_s$$

$$\geq E^x \int_0^\infty e^{-\alpha L_s}g(X_s)\Psi(X_s, f)dL_s.$$

The replacement of $g(X_{s-})$ with $g(X_s)$ in the integral is valid since dL_s does not charge the countable set of s such that X_{s-} is not equal to X_s. The inequality rather than equality is due to the fact that $\Psi(x, \cdot)$ is restricted to $V - \{x\}$. Rewriting the left and right sides of this inequality by making the change of variables $L_s = t$ and noting that if $L_s = t$ and s is in G we have $\widetilde{X}_{s-} = X_{\beta_s-}$ we obtain for F positive and $\mathcal{E} \times \mathcal{E}$ measurable

$$(4.5) \qquad E^x \sum_{t=L_s, s \in G} e^{-\alpha t}F(\widetilde{X}_{t-}, \widetilde{X}_t) \geq E^x \int_0^\infty e^{-\alpha t}\Psi F(\widetilde{X}_t)dt.$$

Here we have appealed to monotone classes to replace $g(x)f(y)$ with $F(x,y)$, and of course the notation $\Psi F(x)$ stands for $\hat{P}^x(F(x, X_\sigma)\,;\, X_\sigma \neq x)$. The sum on the left side of (4.5) is over only some values of t and hence the pair consisting of the kernel Ψ and the additive functional $A_t \equiv t$ is dominated by the Lévy system $(\widetilde{P}, \widetilde{L})$ of $\widetilde{X}$ in the sense that

$$(4.5') \qquad E^x \int_0^\infty \widetilde{Z}_t \widetilde{P} F(\widetilde{X}_t) d\widetilde{L}_t \geq E^x \int_0^\infty \widetilde{Z}_t \Psi F(\widetilde{X}_t) dt$$

for any positive F and positive previsible process $\widetilde{Z}$ of $\widetilde{X}$.

In many cases the pair (Ψ, dt) is actually equal, in the sense of (2d), to the Lévy system of $\widetilde{X}$. Suppose for example that for L-a.e. x, $\hat{P}^x(X_\sigma = x)$ is equal to zero and that the process X_t has no discontinuities with X_{t-} and X_t both in V. Then (4.5) is an equality and, because of the growth properties of L the only discontinuities of $\widetilde{X}_t$ are at times $t = L_s$ with s in G. Thus in such a case (Ψ, dt) equals $(\widetilde{P}, \widetilde{L})$. The jump condition will hold of course whenever the process X is a diffusion, the case of greatest interest. The first condition will hold whenever the minimal process is such that the distribution of X_σ charges no points of V provided one starts from a point in D (see Exercise (4.6)). This is the usual situation as soon as V is anything other than a finite or countable set.

For use later on we need a substitute for Ψ that makes no mention of the excursion measures. Just for this paragraph let us make the assumption that for any $\beta > 0$ and bounded continuous function h on V, $P_V^\beta h$ defines a continuous function on E. (In most cases of interest this is satisfied). For x in D the equation

$$(\alpha - \beta)V^\alpha P_V^\beta h(x) = \overline{P}^x(e^{-\beta\sigma} - e^{-\alpha\sigma}; h(X_\sigma)),$$

which holds for h in $\mathcal{B}_+(V)$ as well, shows that the kernel $(\alpha - \beta)V^\alpha P_V^\beta$ increases in α and decreases in β. Thus the same behavior holds for the kernel $(\alpha - \beta)\hat{H}^\alpha P_V^\beta$. Define Θ to be the kernel

$$\Theta(x, dy) = \lim_{\alpha \to \infty} \lim_{\beta \to 0} (\alpha - \beta)\hat{H}^\alpha P_V^\beta(x, dy) \qquad dy \in V - \{x\}.$$

From (4.3) and (4.4) it follows that

$$(m + Q)\Theta = \Psi$$

and the left side depends only on m, Q and the minimal process.

(4.6) Exercise. Show that if $\overline{P}^x(X_\sigma = y) = 0$ for all x in D and y in V then for L-a.e. x, $\hat{P}^x(X_\sigma = x) = 0$. (Hint: use (4.3) and (4.4). In applying (4.3) it is necessary to use property (a2).)

(b) Feller Brownian motions in the upper half plane.

The following example is adequate for illustrating most of the general theory including the construction problem. Let E be the upper half plane $\{(x,y)|y \geq 0\}$ with V equal to the x axis. Take for minimal process Brownian motion held fixed upon reaching V. We will call any extension of this minimal process a Feller Brownian motion (in the upper half plane). The most obvious example is two-dimensional *reflecting* Brownian *motion*, which we define to be (B^1, B^2), with B^1 a Brownian motion on $(-\infty, \infty)$, B^2 a reflecting Brownian motion on $[0, \infty)$ and B^1 and B^2 independent.

For our minimal process Motoo's hypothesis holds. Indeed if P_t and P_t^0 denote the transition operators for one dimensional Brownian motion and Brownian motion killed at $\sigma_{\{0\}}$ then for a function f of the form $f(x,y) = h(x)k(y)$ with h and k bounded we have

$$V^\alpha f(x,y) = \int_0^\infty e^{-\alpha t} P_t h(x) P_t^0 k(y)dt,$$

and clearly $V^1 1(x,y) = 1 - e^{-y\sqrt{2}}$. Using the material developed in Chapter IV we find that if h and k are also continuous then as $(x,y) \to (x_0, 0)$ the ratio $V^\alpha f/V^1 1$ approaches

$$\int_0^\infty P_t h(x_0)\eta_t(k)e^{-\alpha t}dt$$

where $\{\eta_t; t > 0\}$ is the entrance law for reflecting Brownian motion. We will leave to the reader the extension from product functions to all bounded continuous ones. It follows almost immediately from this that if the Feller Brownian motion leaves V continuously and has no sojourn on V then we can represent the excursion measure for this process as

$$\hat{P}^{(x,0)}(d\omega_1 \times d\omega_2) = P^x(d\omega_1)\hat{P}(d\omega_2)$$

where $\Omega = \Omega_1 \times \Omega_2$, $\Omega_1 = \mathcal{C}([0, \infty), R)$, $\Omega_2 = \mathcal{C}([0, \infty), [0, \infty))$ and P^x is the measure on Ω_1 for Brownian motion starting at x while $\hat{P}$ on Ω_2 is excursion measure for reflecting Brownian motion on $[0, \infty)$. For reflecting Brownian motion in the upper half plane the process on the boundary is easy to find: in the above notation we can take local time to be $L_t^2(\omega_2)$ where L^2 means the local time at 0 for the reflecting Brownian motion B^2. Then the process on the boundary (ignoring the trivial second coordinate) is $B^1(\beta_t(\omega_2), \omega_1)$, that is Brownian motion run with an independent subordinator. This is itself a process with stationary independent increments (called *subordinate* to the process B^1.) Using the material from Chapter IV the reader should be able to show that $E^{(0,0)}e^{-\lambda B^1(\beta_t)} = \exp{-t|\lambda|/\sqrt{2}}$. The corresponding Lévy process is called the Cauchy *process* so this is the process on the boundary for reflecting Brownian motion in the upper half plane.

One obtains a rather trivial generalization of this by replacing the reflecting Brownian motion B^2 with any Feller Brownian motion on $[0, \infty)$ still independent of B^1. The local time is now local time at $\{0\}$ for B^2 and the process on the boundary will be the sum of three independent processes one a scaled Cauchy process, one a scaled Brownian motion and one a process subordinate to a Cauchy process. We leave the details to the reader.

A simple but more enlightening example is

$$X_t = (B_t^1 + aL_t^2, B_t^2)$$

where B^1, B^2, L^2 are from the first example and the "drift coefficient", a, is some number. It is a Feller Brownian motion because L_t^2 is constant when the second coordinate is not 0. Clearly X has continuous paths and has no sojourn on the real axis because the reflecting Brownian motion B^2 has no sojourn at $\{0\}$. The local time is simply L^2, and making a change of variables we obtain the process on the boundary as $C_t + at$ where C is the Cauchy process from the previous example. The excursion measures $\hat{P}^x$ of course are the same as those for the first example. Thus the process on the boundary will be a vital part of the excursion description of any extension of the minimal process. This process is said to have *constant drift on the boundary* or to have *oblique reflection at the boundary*.

A generalization is obtained by letting the drift function (or reflection angle) vary from point to point on the real axis but in a very regulated way. Specifically let λ be a function satisfying $|\lambda(x) - \lambda(y)| \leq K|x - y|$ for all real x and y with K constant, let B^1 denote a continuous real valued function on $[0, \infty)$ and L^2 denote a continuous non-decreasing function on $[0, \infty)$ with $L^2(0) = 0$. Then (Picard's method) the equation

$$X_t^\lambda = B_t^1 + \int_0^t \lambda(X_s^\lambda) dL_s^2$$

has a unique solution X^λ. If we let B^1 and L^2 be the Brownian motion and local time from our previous example then the process $\{(X_t^\lambda, B_t^2); t \geq 0\}$ is a Feller Brownian motion. When λ is constant we obtain the previous example.

Much more interesting is the case where the drift coefficient λ has a discontinuity. Suppose we consider the simplest case

$$\lambda(x) \quad \begin{aligned} &= a \quad x < 0 \\ &= b \quad x \geq 0. \end{aligned}$$

The interesting case is $a > b$. Approximating this λ with those from the previous example and using monotoneity properties in λ of the X^λ from the previous example it is easy to obtain a Feller Brownian motion X that agrees in law with the constant drift process $(B_t^1 + aL_t^2, B_t^2)$ until the time σ^+ when this process first meets the positive real axis and with $(B_t^1 + bL_t^2, B_t^2)$ until the time σ^- when it reaches the negative real axis. This description characterizes, informally at least, the probabilistic behavior of X in the interval between successive passages from one side of the real axis to the other. The paths in these time intervals can be linked together, but this will yield a probabilistic description of X only up to the time, σ, when X first meets the origin. It turns out that σ is finite almost surely, although this is not easy to prove. Continuing the process beyond σ then becomes a question about uniqueness subject to additional requirements on non-sojourn and continuity of departure from $(0,0)$, of extensions of the minimal process X killed when it hits $(0,0)$. As indicated by V-6.1, the issue turns into one of verifying the existence of the limit as x approaches 0 of $\widetilde{V}^\alpha f(x) / \widetilde{V}^1 1(x)$ where $\widetilde{V}^\alpha$ is the potential operator for the minimal process on the boundary (the Cauchy process with opposing drifts killed when it reaches 0). Establishing the existence of this limit for f on R bounded and continuous requires clever use of probabilistic and analytic techniques, which is what lends interest to the subject. The more general problem in which the half-plane is replaced by a wedge has received a good deal of attention. We refer the reader to the paper by Ruth Williams and Varadhan [VW, 1] and the attendant references. These papers use techniques of weak convergence, and martingale characterizations of the constructed process.

(c) Uniqueness.

The fact that the law of an extension of the minimal process is determined by the boundary system is summed up in the following statement.

(4.7) Theorem. *Suppose X and X' are extensions of the minimal process with boundary systems $(\widetilde{X}, \ell, m, Q)$ and $(\widetilde{X}', \ell', m', Q')$. If $\widetilde{X}$ and $\widetilde{X}'$ are equal in law and a.e. $(\widetilde{X})$-hence a.e. $(\widetilde{X}')$-we have $\ell = \ell'$, $m = m'$, and $Q = Q'$ then X and X' are equal in law.*

Proof. We will shorten the argument somewhat by assuming ℓ and ℓ' are zero a.e. $(\widetilde{X})$ and thus the extensions have no sojourn on V. For α and λ positive and $\alpha + \lambda > 0$ set, for $x \in V$, $f \in \mathcal{B}_+(V)$

$$(4.8) \qquad K_\alpha^\lambda f(x) = E^x \int_0^\infty e^{-\alpha t} e^{-\lambda L_t} f(X_t) dL_t$$

where the right side of (4.8) uses the data for the extension X of the minimal process. For g in $\mathcal{B}_+(V)$ we have

$$(4.9) \qquad \hat{P}^x \left(\int_0^\sigma e^{-t} g(X_t) dt \right) = (m + Q) \hat{H}^1 g(x)$$

so (see III-3.32) the excursion laws $\hat{P}^x$ are determined by the boundary system. Set $K^\lambda = K_0^\lambda$ and $K_\alpha = K_\alpha^0$. The transformations K^λ are the potential operators for $\widetilde{X}$, hence also are determined by the boundary system. On the other hand if $f(x) = \hat{P}^x(\int_0^\sigma e^{-\alpha t} g(X_t) dt)$ then

$$K_\alpha f(x) = E^x \int_0^\infty e^{-\alpha t} f(X_t) dL_t = U^\alpha g(x)$$

where U^α is potential operator for X. Since $\{U^\alpha g\}$ determines the law of X it will be enough to show that a knowledge of K^λ and $\hat{P}$ determines $K_\alpha f$ for all f of the form $\hat{P}^\bullet F$ with F positive $\mathcal{F}^0$ measurable and $\hat{P}^\bullet F$ bounded.

Let V_α denote the operator

$$V_\alpha k(x) = \hat{P}^x(1 - e^{-\alpha\sigma}; k(X_\sigma)) \qquad\qquad k \in \mathcal{B}(V), x \in V.$$

We will show first of all that if $f = \hat{P}^{\cdot} F$ as above then

$$(4.10) \qquad K^\lambda f - K_\alpha^\lambda f = K_\alpha^\lambda V_\alpha K^\lambda f.$$

Indeed, applying the excursion formula we have

$$V_\alpha K^\lambda f(x) = \hat{P}^x (1 - e^{-\alpha\sigma}; E^{X_\sigma} \sum_{r \in G} e^{-\lambda L_r} F \circ \theta_r)$$

and so

$$K_\alpha^\lambda V_\alpha K^\lambda f(x) = E^x \sum_{s \in G} e^{-\alpha s} e^{-\lambda L_s} (1 - e^{-\alpha\sigma \circ \theta_s}) E^{X_\sigma \circ \theta_s} (\sum_{r \in G} e^{-\lambda L_r} F \circ \theta_r).$$

Let $D_s = s + \sigma \circ \theta_s$ and use the fact that $L_s = L_{D_s}$ to rewrite this last sum as

$$(4.11) \qquad E^x \sum_{s \in G} (e^{-\alpha s} - e^{-\alpha D_s}) e^{-\lambda L_{D_s}} E^{X_{D_s}} (\sum_{r \in G} e^{-\lambda L_r} F \circ \theta_r).$$

If G_n^ϵ denotes as usual the left end of the nth excursion interval of length exceeding ϵ then $D_n^\epsilon = G_n^\epsilon + \sigma \circ \theta_{G_n^\epsilon}$ is a stopping time. Thus by a limit passage it is clear that we may treat the D_s in (4.11) as stopping times and rewrite (4.11), with $Z_s = (e^{-\alpha s} - e^{-\alpha D_s}) e^{-\lambda L_{D_s}}$, as

$$E^x \sum_{s \in G} Z_s \sum_{r \in G \circ \theta_{D_s}} e^{-\lambda L_r \circ \theta_{D_s}} F \circ \theta_{r + D_s}$$

$$= E^x \sum_{s \in G} (e^{-\alpha s} - e^{-\alpha D_s}) \sum_{\substack{r \in G \\ r > D_s}} e^{-\lambda L_r} F \circ \theta_r.$$

This last sum is the increasing limit, as ϵ decreases to zero, of

$$E^x \sum_n (e^{-\alpha G_n^\epsilon} - e^{-\alpha D_n^\epsilon}) \sum_{k > n} e^{-\lambda L_{G_k^\epsilon}} F \circ \theta_{G_k^\epsilon}.$$

A change of summation order and a little algebra changes this to

$$(4.12) \qquad E^x \sum_{k=2}^\infty e^{-\lambda L_{G_k^\epsilon}} F \circ \theta_{G_k^\epsilon} \{ (1 - e^{-\alpha G_k^\epsilon}) - \sum_{n=1}^k (e^{-\alpha D_{n-1}^\epsilon} - e^{-\alpha G_n^\epsilon}) \}.$$

We assert that as ϵ approaches zero the sum

$$(4.13) \qquad E^x \sum_{k=2}^\infty e^{-\lambda L_{G_k^\epsilon}} F \circ \theta_{G_k^\epsilon} (\sum_{n=1}^k e^{-\alpha D_{n-1}^\epsilon} - e^{-\alpha G_n^\epsilon})$$

approaches zero. This will prove (4.10) since the rest of the expression in (4.12) approaches

$$E^x \sum_{s \in G} e^{-\lambda L_s}(1 - e^{-\alpha s})F \circ \theta_s$$

which is the left side of (4.10). As to the sum in (4.13) we have

$$E^x \sum_{k=2}^{\infty} e^{-\lambda L_{G_k^\epsilon}} F \circ \theta_{G_k^\epsilon} \leq E^x \sum_{s \in G} e^{-\lambda L_s} F \circ \theta_s$$

$$= E^x \int_0^\infty e^{-\lambda L_s} \hat{P}^{X_s}(F)dL_s \leq \lambda^{-1} \sup\{\hat{P}^{\cdot} F\} < \infty$$

while the sum $\sum_{s \in G} e^{-\alpha D_{n-1}^\epsilon} - e^{-\alpha G_n^\epsilon}$ is always less than 1 and converges, as $\epsilon \to 0$, to $\int_0^\infty e^{-\alpha t} I_V(X_t)dt$ which, because of no sojourn on V, is P^x almost surely equal to zero.

Now we can complete the proof. Apply (4.10) repeatedly to $f = V_\alpha K^\lambda k$, legitimate since $V_\alpha h$ is of the form $\hat{P}^{\cdot} F$, and we obtain

$$(4.14) \qquad K_\alpha^\lambda f = \sum_{1}^{N}(-1)^{n-1}K^\lambda(V_\alpha K^\lambda)^{n-1}f + (-1)^N K^\lambda(V_\alpha K^\lambda)^N f.$$

If $\lambda > 1$ and $\alpha \leq 1$ we have for $h \in \mathcal{B}_+(V)$

$$V_\alpha K^\lambda h(x) \leq \sup h \hat{P}^x(1 - e^{-\alpha\sigma})\sup_y E^y \int_0^\infty e^{-\lambda L_t}dL_t \leq 1/\lambda(\sup h)$$

so the remainder term in (4.14) tends to zero as $N \to \infty$. Thus

$$K_\alpha^\lambda f = K^\lambda(\sum_{1}^{\infty}(-1)^{n-1}(V_\alpha K^\lambda)^{n-1}f).$$

So the knowledge of K^λ and of $\hat{P}$ (and hence V_α) determines $K_\alpha^\lambda f$ on functions $f = \hat{P}^{\cdot} F$ if $\alpha \leq 1$ and $\lambda > 1$. Now the resolvent equation

$$(4.15) \qquad\qquad K_\alpha^\lambda - K_\alpha^\mu = (\mu - \lambda)K_\alpha^\mu K_\alpha^\lambda$$

is easy to establish by the same argument that yields the ordinary resolvent equation. The usual manipulation with this yields a power series

$$K_\alpha^\mu = \sum_{n=1}^{\infty}(-1)^{n-1}(\mu - \lambda)^{n-1}(K_\alpha^\lambda)^n$$

convergent if $|\mu - \lambda| < \lambda$ and $\alpha \leq 1$, in particular for $0 < \mu < 2\lambda$. Thus $K_\alpha^\mu f$ is determined for $f = \hat{P}^\cdot F$, $\alpha \leq 1$ and $\mu > 0$. Now letting $\mu \to 0$ we have $K_\alpha f$ determined, which is all we need.

(d) Additive functional formulas.

As preparation for understanding Motoo's construction of a recurrent extension based on a minimal process and a given $(\tilde{X}, \ell, m, Q)$ we want to develop an expression for the operators K_α^λ from (4.8) that involves only the boundary system and the minimal process but not the local time, whose existence would be known only after the construction were complete.

(4.16) Lemma. *Let μ be a finite atomless measure on $\mathcal{B}(0, \infty)$. Let S be a countable subset of $(0, \infty)$ and suppose for each $s \in S$ we have a number α_s with $0 \leq \alpha_s < 1$. Set $a(t) = \prod_{s \leq t}(1 - \alpha_s)$ and let $T = \inf\{t | a(t) = 0\}$. Then*

$$\sum_S a(s-)\alpha_s \mu(s, \infty) = \int_0^T (1 - a(s))\mu(ds).$$

Proof. Suppose first that $S = \{s_1, \ldots, s_n\}$ and set $a(s) = 1$ for $s < s_1$. Then starting with the left side above we have

$$\sum_{k=1}^n a(s_k-)(\alpha_{s_k})\mu(s_k, \infty) = \sum_{k=1}^n a(s_k-)\alpha_{s_k} \sum_{j=k}^n \mu(s_j, s_{j+1}]$$

$$= \sum_{j=1}^n \{\sum_{k=1}^j a(s_k-)\alpha_{s_k}\}\mu(s_j, s_{j+1}].$$

(Here $s_{n+1} = \infty$.) The sum in braces is a collapsing sum whose value is $1 - a(s_j)$, the value of $1 - a(s)$ in the interval $[s_j, s_{j+1})$. Thus the assertion of (4.16) clearly is correct in this case, in which $T = \infty$. One proceeds from here to the general case by passing to larger and larger finite subsets of S. We will leave the details to the reader, who will discover the need for the upper limit T in the integral on the right.

We have two applications of (4.16). The first is to a formula, (4.17), that is used repeatedly in Motoo's construction. In Theorem (4.17) X will be a standard process, in practice an extension of a minimal process or a process $\widetilde{X}$ on V which is to be shown to be the process on the boundary for X. The pair $(\mathcal{P}, \mathcal{L})$ will be the Lévy system for X.

(4.17) Theorem. *Let α and β be measurable functions on the state space for X with values in $[0, 1)$ and with $\alpha(x, x) = \beta(x, x) = 0$ for all x. Let A, B, C be continuous additive functionals of X and let*

$$a(t) = \prod_{s \leq t}(1 - \alpha(X_{s-}, X_s))$$

$$b(t) = \prod_{s \leq t}(1 - \beta(X_{s-}, X_s)).$$

Suppose $E^x \int_0^\infty e^{-A_t} dC_t$ and $E^x \int_0^\infty e^{-B_t} dC_t$ are bounded in x and set

$$K_1 f(x) = E^x \int_0^\infty e^{-A_t} a(t) f(X_t) dC_t$$

$$K_2 f(x) = E^x \int_0^\infty e^{-B_t} b(t) f(X_t) dC_t$$

$$U_1 f(x) = \int \mathcal{P}(x, dy) \alpha(x, y) f(y)$$

$$U_2 f(x) = \int \mathcal{P}(x, dy) \beta(x, y) f(y)$$

and suppose that $E^x \int_0^\infty U_i 1(X_t) d\mathcal{L}_t$ is bounded in x for $i = 1, 2$. Then

$$K_1 f(x) - K_2 f(x)$$

$$= E^x \left(\int_0^\infty e^{-A_t} a(t) \{ K_2 f(X_t)(dB_t - dA_t) + (U_2 - U_1) K_2 f(X_t) d\mathcal{L}_t \} \right).$$

Proof. The only difficulty in the proof is connected with the product expressions $a(t)$ and $b(t)$. And the method of dealing with these is illustrated adequately by taking $A_t \equiv B_t$ and $\beta \equiv 0$, which is what we will do. Then the left side in (4.17) is

$$(4.18) \qquad E^x \int_0^\infty e^{-A_t} (a(t) - 1) f(X_t) dC_t.$$

For the right side, using the continuity of $\mathcal{L}$, the definition of a Lévy system and the fact that $e^{-A_t} a(t-)$ is previsible we obtain the negative of

$$E^x \int_0^\infty e^{-A_t} a(t-) U_1 K_2 f(X_t) d\mathcal{L}_t$$

$$= E^x \sum_s e^{-A_s} a(s-) \alpha(X_{s-}, X_s) E^{X_s} \int_0^\infty e^{-A_r} f(X_r) dC_r.$$

We may treat the right side of this equality via the usual strong Markov property expression because the times s appearing there are path discontinuity times and thus are embeddable in a countable family of stopping times. The resulting expression is

$$(4.19) \qquad E^x \sum_s a(s-)\alpha(X_{s-}, X_s) \int_0^\infty e^{-(A_s + A_r \circ \theta_s)} f(X_{r+s}) dC_r \circ \theta_s$$

$$= E^x \sum_s a(s-)\alpha(X_{s-}, X_s) \int_s^\infty e^{-A_r} f(X_r) dC_r.$$

We can apply (4.16) to this last expression. The measure $d\mu$ is of course $e^{-A_r} f(X_r) dC_r$ and $\alpha_s = \alpha(X_{s-}, X_s)$ with S being the discontinuity points. By hypothesis $E^x \int_0^\infty U_1 1(X_t) d\mathcal{L}_t$ which equals $E^x \sum_s \alpha(X_{s-}, X_s)$ is finite so that almost surely T in the lemma is infinite. Thus the right side of (4.19) is the negative of (4.18) as required.

The second application of (4.16) is to developing an alternative expression for the operators K_α^λ. Here we are assuming that X is an extension of the minimal process with local time L, excursion measures $\hat{P}^x$ and boundary system $(\widetilde{X}, \ell, m, Q)$. Make the following definitions:

$$v_\alpha(x) = \hat{P}^x(1 - e^{-\alpha\sigma}; X_\sigma = x)$$

$$(= (m + Q)\alpha\hat{H}^\alpha P_V(x, \{x\}) \qquad\qquad x \in V,$$

$$w_\alpha(x) = \alpha\ell(x) + v_\alpha(x),$$

$$W_\alpha(t) = \int_0^t w_\alpha(X_r) dL_r,$$

$$k_\alpha(x, y) = d\hat{P}^x(1 - e^{-\alpha\sigma}; X_\sigma \in dy - \{x\})/d\hat{P}^x(X_\sigma \in dy - \{x\}),$$

$$\mathcal{K}_\alpha(t) = \prod_{\substack{s \in G \\ s \le t}} k_\alpha(X_{s-}, X_{D_s}).$$

(4.20) Theorem. For every $f \in \mathcal{B}_+(V)$ and $x \in V$

$$K_\alpha^\lambda f(x) = E^x \int_0^\infty e^{-\lambda L_t} e^{-W_\alpha(t)} \mathcal{K}_\alpha(t) f(X_t) dL_t.$$

Remark. Our interest in (4.20) is as follows. Let us assume that X has no discontinuity times s with X_{s-} and X_s in V. Make the change of variables $r = L_t$ on the right of (4.20). Then it reads

$$E^x \int_0^\infty e^{-\lambda r} e^{-\widetilde{W}_\alpha(r)} \widetilde{\mathcal{K}}_\alpha(r) f(\widetilde{X}_r) dr$$

where

$$\widetilde{W}_\alpha(r) = \int_0^r w_\alpha(\tilde{X}_u)du$$

$$\tilde{K}_\alpha(r) = \prod_{u \leq r} k_\alpha(\tilde{X}_{u-}, \tilde{X}_u).$$

Thus the right side of (4.20) expresses the basic operators K_α^λ in terms of the boundary system and the minimal process only.

Proof of (4.20). Simply to avoid notational tangles we will carry out the proof under the assumption that $\ell = v_\alpha = 0$, so that X has no sojourn on V and almost no excursion leaves from and returns to the same point. This still leaves us the heart of the matter to deal with.

Denote the right side of (4.20) by $\overline{K}_\alpha^\lambda f$. When α is zero so is k_α, so $\overline{K}^\lambda = K^\lambda$. Let Q be a bounded continuous function on V and $q = K^\lambda Q$. Recall the operators V_α from paragraph (c) and note that

$$V_\alpha h(x) = \hat{P}^x(1 - e^{-\alpha\sigma}; h(X_\sigma)) = \hat{P}^x(k_\alpha(x, X_\sigma)h(X_\sigma)).$$

Apply (4.16) with $S = G$ and $\alpha_s = (1 - e^{-\alpha\sigma\circ\theta_s})$. Since there is no sojourn on V the equality

$$t = \sum_{\substack{s\in G \\ s \leq t}} \sigma \circ \theta_s$$

holds for L-a.e. t and so a.e. L we have

$$a(t) = \prod_{\substack{s \leq t \\ s \in G}} e^{-\alpha\sigma\circ\theta_s} = e^{-\alpha t}.$$

Then

$$K_\alpha^\lambda V_\alpha K^\lambda Q(x) = E^x \int_0^\infty e^{-\lambda L_t} a(t-)\hat{P}^{X_t}((1 - e^{-\alpha\sigma})q(X_\sigma))dL_t$$

$$= E^x \sum_{s\in G} e^{-\lambda L_s} a(s-)(1 - e^{-\alpha\sigma\circ\theta_s})E^{X_{D_s}} \int_0^\infty e^{-\lambda L_r}Q(X_r)dL_r.$$

Using as usual the fact that $L_s = L_{D_s}$ and applying a strong Markov property calculation in this last sum we see it can be written as

$$E^x \sum_{s\in G} a(s-)\alpha_s \int_s^\infty e^{-\lambda L_r}Q(X_r)dL_r$$

which by (4.16) equals

$$E^x \int_0^\infty (1 - a(s))e^{-\lambda L_s} Q(X_s) dL_s.$$

Since $a(s) = e^{-\alpha s}$ a.e.-L we have shown that

$$K_\alpha^\lambda Q + K_\alpha^\lambda V_\alpha K^\lambda Q = K^\lambda Q,$$

for continuous, and hence for all positive Q.

Next we will show that the same equality holds when K_α^λ is replaced with $\overline{K}_\alpha^\lambda$. Recall that in finding the Lévy system for $\widetilde{X}$ we established the fact that for $H \in \mathcal{B}_+(V \times V)$ with $H(x,x) \equiv 0$ and Z previsible we have

$$(4.21) \qquad E^x \sum_{s \in G} Z_s H(X_{s-}, X_{D_s}) = E^x \int_0^\infty Z_s \hat{P}^{X_s}(H(X_s, x_\sigma)) dL_s$$

where $\hat{P}^y H(y, x_\sigma)$ stands for $\int H(y,z)\hat{P}^y(X_\sigma \in dz)$. Apply this with

$$Z_s = e^{-\lambda L_s} K_\alpha(s-)$$
$$H(x,y) = k_\alpha(x,y)K^\lambda Q(y).$$

Because $\hat{P}^y H(y, X_\sigma)$ equals $\hat{P}^y((1 - e^{-\alpha\sigma})K^\lambda Q(X_\sigma))$ the right side of (4.21) is $\overline{K}_\alpha^\lambda V_\alpha K^\lambda Q(x)$ while on the left by the same manipulations used in the previous argument we obtain

$$(4.22) \qquad E^x \sum_s K_\alpha(s-)k_\alpha(X_{s-}, X_{D_s}) \int_s^\infty e^{-\lambda L_r} Q(X_r) dL_r.$$

Apply (4.16) again. The calculation

$$E^x \sum_{s \in G} e^{-\alpha s} k_\alpha(X_{s-}, X_{D_s}) = E^x \int_0^\infty e^{-\alpha s} \hat{P}^{X_s}(1 - e^{-\alpha\sigma}) dL_s \leq 1$$

shows that for all t, $\sum_{s \leq t} k_\alpha(X_{s-}, X_{D_s})$ is finite almost surely. Thus $K_\alpha(t)$ is never zero, that is in (4.16) $T = \infty$ so we may write (4.22) as

$$E^x \int_0^\infty e^{-\lambda L_s}(1 - K_\alpha(s))Q(X_s) dL_s,$$

so the replacement of K_α^λ with $\overline{K}_\alpha^\lambda$ is permisible.

Thus we have shown

$$\overline{K}_\alpha^\lambda(I + V_\alpha K^\lambda)Q = K_\alpha^\lambda(I + V_\alpha K^\lambda)Q = K^\lambda Q,$$

and so

$$\overline{K}_\alpha^\lambda = K^\lambda(I + V_\alpha K^\lambda)^{-1} = K_\alpha^\lambda$$

provided the operator $I + V_\alpha K^\lambda$ has an inverse. It will, obviously, if $\alpha\lambda > 1$ and so (4.20) is established for such pairs α, λ. The operators $\overline{K}_\alpha^\lambda$ satisfy a resolvent equation in λ as the reader will prove easily so the equality can be extended from $\lambda > \alpha^{-1}$ to all λ as in the conclusion of Theorem (4.7).

(e) Construction. Now we will construct a recurrent extension based on boundary data.

The following assumptions will be in force throughout the rest of this chapter. We will assume that E has been compactified in the usual way if it is not compact to start with, and that the transition probabilities have been extended to the point at infinity in the usual way (see I-7). Thus V is compact also. The hypothesis from previous chapters that $\overline{P}^x(\sigma < \infty) = 1$ for all x can be dispensed with-its only purpose was to ensure that the point process of excursions will be a genuine Poisson point process rather than an absorbed process. Besides this assumption on E we will assume the following smoothness conditions on the minimal process:

(h1) $V^\alpha f \in \mathcal{C}(E)$ for all $f \in \mathcal{C}(E)$, $\alpha > 0$,

(h2) $P_V^\alpha f \in \mathcal{C}(E)$ for all $f \in \mathcal{C}(V)$, $\alpha > 0$,

(h3) $\{\hat{H}^\alpha f | f \in \mathcal{C}(E)\}$ is for each $\alpha > 0$ dense in $\mathcal{C}(E)$.

We will assume we are given $(\widetilde{X}, \ell, m, Q)$ where $\widetilde{X}$ is a standard process with state space V, ℓ and m are positive $\mathcal{B}(V)$ measurable functions and $Q(x, A)$, $x \in V$, $A \in \mathcal{B}(E - V)$ is a kernel on $V \times (E - V)$ putting no mass at ∞; about these objects we will assume that properties (a1) through (a3) from paragraph (a) hold and that (a4) holds in the form that the Lévy system $(\widetilde{P}, \widetilde{\mathcal{L}})$ of $\widetilde{X}$ dominates the pair consisting of the kernel $(m + Q)\Theta$ and the additive functional $A_t \equiv t$ in the sense of (4.5)'; $(\Psi = (m + Q)\Theta.)$. Besides all this we will make the following smoothness assumptions on $(\widetilde{X}, \ell, m, Q)$, $(\widetilde{U}^\alpha$ will denote the α-potential operator for $\widetilde{X}$):

(a5) for each $f \in \mathcal{C}(V)$ and $\alpha > 0$ both $\widetilde{U}^\alpha f$ and $\widetilde{U}^\alpha(\{\ell + (m + Q)\hat{H}^\alpha\}f)$ are in $\mathcal{C}(V)$.

(4.23) Theorem. *Under the above conditions there is a unique (in law) extension of the minimal process whose boundary system is* $(\widetilde{X}, \ell, m, Q)$.

Comments on the theorem. The reader should argue that (h1) through (h3) are valid for Brownian motion in the upper half plane so that this substantial case is covered by (4.23). Also once the construction is made, the uniqueness follows from paragraph (c). The conditions (a1) through (a4), in spite of the somewhat mysterious appearance of (a2) and (a3), are necessary, as we showed earlier. The need for (a2) is clear enough from our discussion in V-4 on existence. Condition (a4) appears related to the task of finding an appropriate time scale for use in synthesizing a process from minimal process data. However (4.23) is not a synthesis theorem in the sense that the desired X is not constructed by hooking together paths.

Proof. Define operators C_α, U_α on $\mathcal{B}(V)$, a kernel V_α on $V \times \mathcal{B}(V)$ and functions v_α and w_α on V by

$$C_\alpha = \ell + (m + Q)\hat{H}^\alpha$$
$$U_\alpha = \alpha C_\alpha P_V$$
$$V_\alpha(x, A) = (m + Q)\alpha \hat{H}^\alpha P_V(x, A) \quad A \in \mathcal{B}(V - \{x\})$$
$$v_\alpha(x) = (m + Q)\alpha \hat{H}^\alpha P_V(x, \{x\})$$
$$w_\alpha = \alpha\ell + v_\alpha$$
$$\widetilde{W}_\alpha(t) = \int_0^t w_\alpha(\widetilde{X}_r)\,dr.$$

The easiest way to keep track of these objects and manipulations with them is to consider the case in which $(\widetilde{X}, \ell, m, Q)$ is known to be the boundary system of an extension X, and one is trying to reconstruct a probabilistic replica of X armed only with $(\widetilde{X}, \ell, m, Q)$ and the minimal process.

From the discussion following (a.4) we know that $V_\alpha \leq (m+Q)\hat{H}^\alpha P_V \leq (m + Q)\Theta$. By (a4) then $(V_\alpha, t) \leq (\widetilde{P}, \widetilde{L})$ in the sense of (4.5)' ($\Psi = V_\alpha$) and so by an argument quite similar to the ones from (2c) and (2d) there is a function k_α in $\mathcal{B}(V \times V)$ with values in $[0, 1)$ and $k_\alpha(x, x) \equiv 0$ such that (V_α, t) is equivalent to $(\widetilde{P}k_\alpha, \widetilde{L})$ in the sense that

$$\widetilde{E}^x \int_0^\infty \widetilde{Z}_t V_\alpha F(\widetilde{X}_t)\,dt = \widetilde{E}^x \int_0^\infty \widetilde{Z}_t \widetilde{P}(k_\alpha F)(\widetilde{X}_t)\,d\widetilde{L}_t$$

for every positive $\widetilde{X}$ previsible process $\widetilde{Z}$ and every F in $\mathcal{B}_+(V \times V)$ with $F(x, x) \equiv 0$. Define a family $\{K_\alpha^\lambda; \alpha \geq 0, \lambda \geq 0, \alpha + \lambda > 0\}$ of operators on $\mathcal{B}_b(V)$ by

$$(4.24) \qquad K_\alpha^\lambda f(x) = \widetilde{E}^x \int_0^\infty e^{-\lambda t - \widetilde{W}_\alpha(t)} \widetilde{\mathcal{K}}_\alpha(t) f(\widetilde{X}_t) dt$$

$(\widetilde{\mathcal{K}}_\alpha(t) = \prod_{s \leq t} 1 - k_\alpha(\widetilde{X}_{s-}, \widetilde{X}_s))$. As the notation suggests the motivation for defining the K_α^λ this way is Theorem (4.20). For $\alpha > 0$ define U^α on bounded $\mathcal{E}$ measurable functions by

$$(4.25) \qquad U^\alpha = V^\alpha + P_V^\alpha K_\alpha C_\alpha.$$

The first step in the proof of (4.23) will consist of showing that under an additional condition $\{U^\alpha; \alpha > 0\}$ is the resolvent associated with a transition function (suitably smooth) and that the corresponding Markov process is an extension of the minimal process and that its boundary system is equivalent to $(\widetilde{X}, \ell, m, Q)$. But before turning to this we need some remarks about the operators K_α^λ and similar operators to appear later on in the proof of (4.23).

The family $\{K_\alpha^\lambda\}$ defined in (4.24) satisfies the following conditions:

(k1) $K^\lambda = \widetilde{U}^\lambda$

(k2) K_α^λ is positive, bounded, and $\alpha K_\alpha^\lambda C_\alpha 1 \leq 1$

(k3) $K_\alpha^\lambda - K_\alpha^\mu = (\mu - \lambda) K_\alpha^\lambda K_\alpha^\mu$

(k4) $K_\alpha^\lambda - K_\beta^\lambda = K_\alpha^\lambda (U_\beta - U_\alpha) K_\beta^\lambda$.

Proof. Statement (k1) is immediate from the definition of K_α^λ and the fact that when $\alpha = 0$, $W_\alpha = 0$ and $k_\alpha = 0$ so that $\mathcal{K}_\alpha = 1$. The second assertion in (k2) is a bit more delicate: we will establish it assuming $Q = 0$, $w_\alpha = 0$ and $P_V 1 = 1$ as this illustrates the heart of the argument. Then, taking $\lambda > 0$ we have

$$\alpha K_\alpha^\lambda C_\alpha 1 = \alpha K_\alpha^\lambda V_\alpha 1 = \widetilde{E}^\cdot \int_0^\infty e^{-\lambda t} \widetilde{\mathcal{K}}_\alpha(t-) V_\alpha 1(\widetilde{X}_t) dt$$

$$= \widetilde{E}^\cdot \int_0^\infty e^{-\lambda t} \widetilde{K}_\alpha(t-) \widetilde{P} k_\alpha(\widetilde{X}_t) d\widetilde{L}_t$$

$$= \widetilde{E}^\cdot \sum e^{-\lambda t} \widetilde{K}_\alpha(t-) k_\alpha(\widetilde{X}_{t-}, \widetilde{X}_t)$$

$$\leq \widetilde{E}^\cdot \int_0^\infty (1 - \mathcal{K}_\alpha(t)) \lambda e^{-\lambda t} dt \leq 1$$

where the last step follows from (4.16). As λ decreases to zero the inequality is preserved so the case $\lambda = 0$ follows as well. Assertions (k3) and (k4) are straightforward consequences of (4.17).

Expanding on the foregoing call a family $\{K_\alpha^\lambda; \alpha, \lambda \geq 0, \alpha + \lambda > 0\}$ of bounded kernels on $V \times \mathcal{B}(V)$ (not necessarily the family defined in (4.24)) an (ℓ, m, Q) *system* if it satisfies the relationships (k3) and (k4) where the transformations U_α are those we have been using. Exactly as in the arguments for the uniqueness theorem (4.7) the operators in an (ℓ, m, Q) system are, for different values of λ and μ connected by power series expansions and it follows that in an (ℓ, m, Q) system a knowledge of K_α^λ for one pair (λ, α) determines the entire family.

We will need one more observation concerning transformations of the type K_α^λ. Let K be a bounded kernel on $V \times \mathcal{B}(V)$, let ℓ, m, ℓ', m' be elements of $\mathcal{B}_+(V)$ and let Q and Q' be bounded kernels on $V \times \mathcal{B}(D)$. Then if for some $\alpha > 0$

$$Km\hat{H}^\alpha I_V = 0 \quad \text{and} \quad Km'\hat{H}^\alpha I_V = 0$$

and

$$K(\ell + (m + Q)\hat{H}^\alpha)f = K(\ell' + (m' + Q')\hat{H}^\alpha)f \quad f \in \mathcal{C}(E)$$

then $K\ell g = K\ell' g$, $Kmg = Km'g$ and $KQh = KQ'h$ for all $g \in \mathcal{B}(V)$ and $h \in \mathcal{C}_b(D)$. We will leave the proof to the reader.

Now we are ready to return to the proof of Motoo's construction theorem (4.23).

Step 1. The resolvent $\{U^\alpha; \alpha > 0\}$.

The family U^α of positive operators satisfies the resolvent equation $U^\alpha - U^\beta = (\beta - \alpha)U^\alpha U^\beta$ and $\alpha U^\alpha 1 \leq 1$. The proof of this is easy if a bit tedious: it uses the resolvent equation for $\{V^\alpha; \alpha > 0\}$ associated with the killed minimal process, properties (k3) and (k4) and the relationships (4.2). But to utilize the resolvent operators we need some continuity and range properties as well. Introduce the following condition:

(a6) for some constant $p > 0$ we have $\ell \geq p$.

If we use the continuity assumptions (a5) and (h1) through (h3) then under the additional assumption (a6) it follows rather easily that each U^α

maps $\mathcal{C}(E)$ into itself and on $\mathcal{C}(E)$ αU^α converges strongly to the identity as $\alpha \to \infty$. Let us assume (a6) temporarily. (We will have to get rid of this condition later on since in the most interesting examples ℓ is zero.) Then by the Hille-Yosida theorem and the discussion from Chapter 1 there is a standard process X (with expectations E^x) such that

$$E^x \int_0^\infty e^{-\alpha t} f(X_t) dt = U^\alpha f(x) \quad x \in E, f \in \mathcal{E}_+.$$

Step 2. The process X.

Recall we are assuming (a6). We will show first of all that the process X is an extension of the minimal process. Suppose $f \in \mathcal{E}$ is bounded and vanishes off V. By (a2) $m\hat{H}^\alpha f$ is 0 relative to the operators K_α^λ with $\lambda > 0$ and hence relative to K_α also. Also $\hat{H}^\alpha f$ vanishes on D and so $Q\hat{H}^\alpha f$ is 0. Thus $C_\alpha f = \ell f$ and so writing

$$
\begin{aligned}
\text{(4.26)} \qquad U^\alpha f(x) &= V^\alpha f(x) + P_V^\alpha K_\alpha C_\alpha f \\
&= E^x \int_0^\sigma e^{-\alpha t} f(X_t) dt + E^x(e^{-\alpha\sigma} U^\alpha f(X_\sigma))
\end{aligned}
$$

we obtain

$$P_V^\alpha g(x) = E^x(e^{-\alpha\sigma} g(X_\sigma))$$

for any g of the form $K_\alpha \ell f$ and hence any g of the form $K_\alpha f$ since $\ell \geq p > 0$. From the fact that the range of U^α is dense in $\mathcal{C}(E)$ it follows that the restriction to V of $\{U^\alpha h | h \in \mathcal{C}(E)\}$ is dense in $\mathcal{C}(V)$. Thus the set of g of the form $K_\alpha f$ is large in $\mathcal{C}(V)$ and so the hitting operators $P_V^\alpha f(x)$ and $E^x(e^{-\alpha\sigma} f(X_\sigma))$ for the minimal process and for X are the same. By definition $U^\alpha h(x) = K_\alpha C_\alpha h(x)$ for x in V and any $h \in \mathcal{E}_+$ so for $h \in \mathcal{E}_+$ the second summands in (4.26) are the same and so

$$V^\alpha h(x) = E^x \int_0^\sigma e^{-\alpha t} h(X_t) dt.$$

Thus X agrees with $\overline{X}$ up to time σ as asserted.

Next we need to argue that the boundary system of X is the given $(\tilde{X}, \ell, m, Q)$. Actually we need a slightly more general fact: suppose our given $(\tilde{X}, \ell, m, Q)$ satisfies the weaker condition

(a1)′ $$\ell + m + Q \geq 1$$

but the hypotheses on $\widetilde{X}$ are as before. Define the transformations K_α^λ exactly as in (4.24). The equality $\ell + m + Q = 1$ was not needed - the inequality suffices - and one sees that $\{K_\alpha^\lambda\}$ is an (ℓ, m, Q) system. Clearly K^λ is equal to $\widetilde{U}^\lambda$. The argument in Step 1 shows that U^α defined by (4.25) is the α-potential operator for a standard process X, which by the argument at the start of Step 2 is an extension of the minimal process. Let (X^*, ℓ^*, m^*, Q^*) denote the boundary system for X and let $K_\alpha^{*\lambda}$ denote the transformations defined in (4.8) using X, its local time L, and so forth. Set $\Delta(x) = \ell(x) + m(x) + Q(x, D)$. We need the following fact.

(4.27) Lemma. *The sets which are null for the operators K_α^λ and for $K_\alpha^{*\lambda}$ are the same. We have for $g \in \mathcal{B}_+(V)$*

$$K_\alpha(g) = K_\alpha^*(g/\Delta)$$

and also

$$\ell^*\Delta = \ell, \, m^*\Delta = m, \, \Delta Q^* := Q$$

except on K null sets.

Remark. Note that if $\Delta \equiv 1$ then (4.27) implies that $K_\alpha = K_\alpha^*$ and that $\{K_\alpha^\lambda\}$ and $\{K_\alpha^{*\lambda}\}$ are (ℓ, m, Q) systems. Since such systems are determined by the operators for one pair (α, λ) it follows that $K^{*\lambda} = K^\lambda = \widetilde{U}^\lambda$ and so $\widetilde{X}$ and X^* are equal in law. Thus in this case the boundary system for the constructed process equals the system we were given to start with, and the task of constructing and identifying an extension is complete (but under the unsatisfactory condition (a6)).

Proof. By construction the potential operator U^α for X is given by

$$U^\alpha f = V^\alpha f + P_V^\alpha K_\alpha(\ell + (m + Q)\hat{H}^\alpha)f$$

and by (3.17)

$$U^\alpha f = V^\alpha f + P_V^\alpha K_\alpha^*(\ell^* + (m^* + Q^*)\hat{H}^\alpha)f.$$

Suppose f vanishes except on V and we evaluate $U^\alpha f(x)$ at a point x in V. For such an f we have $\hat{H}^\alpha f(y) = 0$ for $y \in D$ and so $Q\hat{H}^\alpha f = Q^*\hat{H}^\alpha f \equiv 0$.

We have $m\hat{H}^\alpha I_V = 0$ a.e. K_α by hypothesis and $m^*\hat{H}^\alpha I_V = 0$ a.e. K_α^* from boundary theory so for such an f, $K_\alpha(\ell f) = K_\alpha^*(\ell^* f)$, or, since $\ell \geq p$,

$$(4.28) \qquad K_\alpha(f) = K_\alpha^*(\frac{\ell^*}{\ell}f) \qquad f \in \mathcal{B}_+(V).$$

Apply this with $f = (\ell + (m + Q)\hat{H}^\alpha)g$ for $g \in \mathcal{E}_+$ and we have

$$(4.29) \qquad \begin{aligned} K_\alpha^*(\frac{\ell^*}{\ell}\{\ell + (m + Q)\hat{H}^\alpha\}g) &= K_\alpha(\ell + (m + Q)\hat{H}^\alpha)g \\ &= K_\alpha^*(\ell^* + (m^* + Q^*)\hat{H}^\alpha)g \end{aligned}$$

by the equality of the two expressions for U^α given above. Note at this point that if $N(K)$ and $N(K^*)$ denote respectively the null spaces for the families $\{K_\alpha^\lambda\}$ and $\{K_\alpha^{*\lambda}\}$ then (4.28) implies that $N(K^*) \subset N(K)$. Suppose that in (4.29) g vanishes on V. Then $K_\alpha^*(\ell^* g) = 0$ and we obtain

$$(4.30) \qquad K_\alpha^*(\frac{\ell^*}{\ell}(m + Q)\hat{H}^\alpha)g = K_\alpha^*(m^* + Q^*)\hat{H}^\alpha g.$$

Now suppose instead that g vanishes off V. Then by the remarks preceding (4.29) both sides in (4.30) equal zero. Thus (4.30) holds for any g in $\mathcal{E}_+$. In (4.30) we may now replace $\hat{H}^\alpha g$ with any function h in $\mathcal{E}_+$ since K_α^* is a kernel and by (h3) $\{\hat{H}^\alpha g | g \in \mathcal{C}(E)\}$ determines the action of a positive bounded kernel. Writing this out we have

$$(4.31) \qquad K_\alpha^*(\frac{\ell^*}{\ell}(m + Q)h) = K_\alpha^*(m^* + Q^*)h \quad h \in \mathcal{E}_+.$$

Suppose h vanishes off V. Then $Qh = Q^*h = 0$ so $K_\alpha^*(\frac{\ell^*}{\ell}mh) = K_\alpha^*(m^*h)$ and since this holds for every such h we have

$$\ell^* m = m^* \ell \qquad \text{a.e. } (K^*).$$

The same argument with h vanishing on V establishes

$$\ell^* Q = \ell Q^* \qquad \text{a.e. } (K^*).$$

Adding these with $\ell^* \ell$ we obtain

$$\ell^*(\ell + m + QI_D) = \ell.$$

From this and (4.29) the assertions of (4.27) follow immediately.

Step 3. The general case ($\ell \geq 0$).

Assume that we are given a system $(\widetilde{X}, \ell, m, Q)$ satisfying (a1) through (a5). Let $\widetilde{U}^\alpha$ denote the α potential operator for $\widetilde{X}$. Make the construction we have just described using the system $(\widetilde{X}, 1+\ell, m, Q)$, for which (a6) is indeed satisfied. We will let J_α^λ stand for the right side of (4.24) based on $(\widetilde{X}, 1+\ell, m, Q)$, so that $J^\lambda = \widetilde{U}^\lambda$, and let $\mathcal{U}_\alpha$ denote the operator

$$\mathcal{U}_\alpha = \alpha(1 + \ell + (m+Q)\hat{H}^\alpha)P_V.$$

Let Y denote the extension of the minimal process which we constructed in Steps 1 and 2, with potentials $_Y U^\alpha$, expectations $\mathcal{E}^x$, boundary system $(\overline{Y}, \overline{\ell}, \overline{m}, \overline{Q})$ local time L^Y and operators

$$\overline{J}_\alpha^\lambda f(x) = \mathcal{E}^x \int_0^\infty e^{-\alpha t} e^{-\lambda L_t^Y} f(Y_t) dL_t^Y.$$

Of course

$$_Y U^\alpha = V^\alpha + P_V^\alpha J_\alpha (1 + \ell + (m+Q)\hat{H}^\alpha).$$

Then $1 + \ell + m + QI_D = 2$ so according to (4.27)

$$\overline{\ell} = \frac{1+\ell}{2}, \quad \overline{m} = \frac{m}{2}, \quad \overline{Q} = \frac{Q}{2}$$

and

$$J_\alpha = (1/2)\overline{J}_\alpha.$$

Note that if C_α^Y and U_α^Y denote the operators C_α and U_α but defined using the boundary system of Y then

$$C_\alpha^Y = C_\alpha/2 \qquad\qquad U_\alpha^Y = \mathcal{U}_\alpha/2.$$

Define a continuous additive functional Φ of Y by $\Phi = L^Y/2$, and define for $f \in \mathcal{B}_+(V)$

$$J_\alpha'^\lambda f(x) = \mathcal{E}^x \int_0^\infty e^{-\alpha t - \lambda \Phi_t} f(Y_t) d\Phi_t.$$

Clearly we have $2J_\alpha'^\lambda = \overline{J}_\alpha^{\lambda/2}$, and using the fact from boundary theory that $\{\overline{J}_\alpha^\lambda\}$ is an $(\overline{\ell}, \overline{m}, \overline{Q})$ system it follows immediately that $\{J_\alpha'^\lambda\}$ is a

$(1 + \ell, m, Q)$ system as is $\{J_\alpha^\lambda\}$. Since such systems are determined by the action of one member and since

$$J'_\alpha(f) = \overline{J}_\alpha(f/2) = J_\alpha(f)$$

we have $J'^\lambda_\alpha = J^\lambda_\alpha$ for all λ and α, and so

$$J'^\lambda = J^\lambda = \widetilde{U}^\lambda$$

($\widetilde{U}^\lambda$ is the λ potential for $\widetilde{X}$.) Making a change of variables in the formula defining J'^λ we have for $x \in V$

$$\widetilde{U}^\lambda f(x) = \mathcal{E}^x \int_0^\infty e^{-\lambda L_t^Y/2} f(Y_t) dL_t^Y/2$$
$$= \mathcal{E}^x \int_0^\infty e^{-\lambda s} f(\overline{Y}_{2s}) ds,$$

and we conclude that $\{\overline{Y}_{2s})\}$ and the original process $\{\widetilde{X}_s\}$ are equal in law. This motivates us to try constructing the desired extension X with boundary system $(\widetilde{X}, \ell, m, Q)$ by making a random time change in the process Y. We will show that this can be done.

Step 4. A time change.

Define an additive functional Ψ of Y by

$$\Psi_t = \int_0^t \ell(Y_s) d\Phi_s + \int_0^t I_D(Y_s) ds.$$

Clearly Ψ is continuous. We will show that Ψ is strictly increasing; this is where the condition (a3) enters the argument. Let τ denote the right continuous inverse of L^Y,

$$\tau(t) = \inf\{s | L_s^Y > t\},$$

let $J = \{x | \ell(x) + m(x) > 0\}$ and let u be the function

$$u = \{1 - I_J\}^{-1} + Q(1/V^1 1).$$

First we will argue that

$$(4.32) \qquad \mathcal{E}^x \left(\int_0^t u(Y_s) d\Phi_s = \infty, t > 0 \right) = 1 \quad x \in V.$$

Since almost surely $\mathcal{E}^x L^Y$ increases immediately if x is in V we may consider instead

$$\mathcal{E}^x\left(\int_0^{\tau(t)} u(Y_s)d\Phi_s = \infty, t > 0\right)$$

$$= \mathcal{E}^x\left(\frac{1}{2}\int_0^{\tau(t)} u(Y_s)dL_s^Y = \infty, t > 0\right)$$

$$= \mathcal{E}^x\left(\int_0^{2t} u(\overline{Y}_{2r})dr = \infty, t > 0\right).$$

Since $\overline{Y}_{2s} \approx \tilde{X}_s$, by hypothesis (a3) this last probability is one, as asserted in (4.32). Next we assert that

$$(4.33) \qquad \mathcal{E}^x \int_0^{\sigma_D} e^{-t}Q(f/V^1 1)(Y_t)d\Phi_t$$

$$= \mathcal{E}^x(e^{-\sigma_D}f(Y_{\sigma_D}); Y_{\sigma_D} \in D), \quad x \in V.$$

Indeed $Qd\Phi = \overline{Q}dL^Y$ and $\overline{Q}(f/V^1 1)$ is equal to $\hat{P}^x(f(X_0); X_0 \in D)$ for the process Y; so that the left side of (4.33) is ($\hat{P}$ is the excursion kernel for Y)

$$\mathcal{E}^x \int_0^{\sigma_D} e^{-t}\hat{P}^{Y_t}(f(X_0); X_0 \in D)dL_t^Y = \mathcal{E}^x \sum_{s \in G, s \leq \sigma_D} e^{-s}f(Y_s)I_{Y_s \in D}.$$

This last sum consists of the single term $e^{-\sigma_D}fI_D(Y_{\sigma_D})$. Also we assert that

$$(4.34) \qquad \mathcal{E}^x \int_0^{\sigma_D} e^{-t}m(Y_t)d\Phi_t = 0, \quad x \in V.$$

Indeed using $md\Phi = \overline{m}dL^Y$, $\hat{H}^1 1 = 1$ and $m\hat{H}^1 1(x) = \hat{P}^x(1-e^{-\sigma}; X_0 \in V)$ for the process Y we see that the left side of (4.34) is

$$\mathcal{E}^x \sum_{s \in G, s \leq \sigma_D} e^{-s}(1 - e^{-\sigma})I_V(Y_0) \circ \theta_{\sigma_D} = \mathcal{E}^x(e^{-\sigma_D}I_{Y_{\sigma_D} \in V}\mathcal{E}^{Y_{\sigma_D}}(1 - e^{-\sigma}))$$

which is zero. Finally then we assert that if $\rho = \inf\{t|\Psi_t > 0\}$ then

$$(4.35) \qquad \mathcal{E}^y(\rho = 0) = 1 \quad y \in E.$$

Indeed $\rho \leq \sigma_D$ because of the term $\int_0^t I_D(Y_s)ds$ in the definition of Ψ. In particular (4.35) is obvious for $y \in D$. Now $\Psi = \ell\Phi$ in the time interval $[0, \sigma_D]$ and so $\int_0^\rho \ell d\Phi = \Psi(\rho) = 0$. By (4.34) $m\Phi$ is zero in the interval $[0, \rho]$ and so

$$\int_0^t (1 - I_J)^{-1}(Y_s)d\Phi_s = \Phi_t < \infty$$

for $t \leq \rho$. Finally by (4.33)

$$\mathcal{E}^x \int_0^{\sigma_D} e^{-t} Q(1/V^1 1)(Y_t) d\Phi_t \leq 1$$

and so combining the last two displays

$$\mathcal{E}^x \left(\int_0^t u(Y_s) d\Phi_s < \infty, t < \rho \right) = 1.$$

According to (4.32) this implies $\rho = 0$ almost surely relative to the process Y. From (4.35) the assertion at the beginning of the paragraph that almost surely $\mathcal{E}^x$ for all x Ψ is strictly increasing is obvious.

Let μ denote the inverse of Ψ:

$$\mu_t = \inf\{s | \Psi_s > t\}.$$

Then μ_t is strictly increasing and also is continuous in t. Let X denote the time-changed process

$$X_t = Y_{\mu_t}.$$

The process X inherits the major properties of Y, notably the path properties, the strong Markov property and (because μ_t is continuous) quasi-left continuity. Because Ψ_t increases like t when the path of Y is in D it is clear that X, like Y, is an extension of the minimal process. Let (X^*, ℓ^*, m^*, Q^*) be the boundary system of X. We now will show that it is equivalent to $(\tilde{X}, \ell, m, Q)$.

Step 5. The boundary system.

If $_X U^\alpha$ denotes the α potential operator for X then

$$_X U^\alpha f(x) = \mathcal{E}^x \int_0^\infty e^{-\alpha \Psi(t)} f(Y_t) d\Psi_t.$$

Take $x \in V$. We know from boundary theory applied to Y that

$$\mathcal{E}^x \int_0^\infty e^{-\alpha t} f I_D(Y_t) dt$$
$$= \mathcal{E}^x \int_0^\infty e^{-\alpha t} (\overline{m} + \overline{Q}) \hat{H}^\alpha f(Y_t) dL_t^Y,$$

and since $2\overline{m} = m$, $2\overline{Q} = Q$ and $2\Phi = L^Y$ the right side can be written

$$\mathcal{E}^x \int_0^\infty e^{-\alpha t}(m + Q)\hat{H}^\alpha f(Y_t)d\Phi_t.$$

Furthermore in these integrals the exponentials $e^{-\alpha t}$ may be replaced with $e^{-\alpha \Psi_t}$ as the reader should be able to argue based on the fact that over an excursion interval for Y the change in Ψ is entirely in the linear part. Using the fact that $\Psi = \ell\Phi + I_D t$ we can write

$$\mathcal{E}^x \int_0^\infty e^{-\alpha \Psi_t} f(Y_t)d\Psi_t = \mathcal{E}^x \int_0^\infty e^{-\alpha \Psi_t} f\ell(Y_t)d\Phi_t + \mathcal{E}^x \int_0^\infty e^{-\alpha \Psi_t} f I_D(Y_t)dt$$

and putting this together with what we derived just before we have for $x \in V$

$$_X U^\alpha f(x) = \mathcal{E}^x \int_0^\infty e^{-\alpha \Psi_t}(\ell + (m + Q)\hat{H}^\alpha)f(Y_t)d\Phi_t.$$

Then using the fact that X is known to be an extension of the minimal process we have on all of E

$$_X U^\alpha f = V^\alpha f + P_V^\alpha L_\alpha(\ell + (m + Q)\hat{H}^\alpha)f$$

where $\{L_\alpha^\lambda\}$ are defined by

$$L_\alpha^\lambda f(x) = \mathcal{E}^x \int_0^\infty e^{-\alpha \Psi_t - \lambda \Phi_t} f(Y_t)d\Phi_t.$$

Now the family $\{L_\alpha^\lambda\}$ is an (ℓ, m, Q) system. This is proved by an additive functionals argument similar to the ones we have presented earlier using our formula for $_X U^\alpha$ which converts integrals $d\Psi$ into integrals $d\Phi$. (Notice that $L^\lambda = J'^\lambda = \tilde{U}^\lambda$. Since $\tilde{U}^\lambda$ maps $C(V)$ into itself so does every operator L_α^λ and hence, using (a 5), so does $_X U^\alpha$. The only consequence we need from this is that $_X U^\alpha$ transforms $\mathcal{E}$ into $\mathcal{E}$, and hence X is a standard process, but we have used (a 5) for stronger purposes earlier).

Let L^X denote local time on V for the process X and set

$$L_\alpha^{*\lambda}(x) = \mathcal{E}^x \int_0^\infty e^{-\lambda L_t^X - \alpha t} f(X_t)dL_t^X, \quad x \in V.$$

By boundary theory applied to X we know that

$$_X U^\alpha f = V^\alpha f + P_V^\alpha L_\alpha^*(\ell^* + (m^* + Q^*)\hat{H}^\alpha)f$$

and combining this with our previous expression for $_XU^\alpha f$ and the fact that $P_V(x,\cdot) = \epsilon_x$ for $x \in V$ we have

$$L_\alpha(\ell + (m + Q)\hat{H}^\alpha)f = L_\alpha^*(\ell^* + (m^* + Q^*)\hat{H}^\alpha)f.$$

Take $f = \alpha = 1$. Then $\hat{H}^\alpha f = 1$ and since $\ell + m + Q1 = \ell^* + m^* + Q^*1 = 1$ we conclude that

$$(4.36) \qquad \mathcal{E}^x \int_0^\infty e^{-t}dL_t^X = \mathcal{E}^x \int_0^\infty e^{-\Psi_t}d\Phi_t$$

for all x in V and hence for all x from the growth properties of L^X and Φ. Making a change of variables on the right of (4.36) we have

$$\mathcal{E}^x \int_0^\infty e^{-t}dL_t^X = \mathcal{E}^x \int_0^\infty e^{-t}d\Phi_{\mu_t}.$$

The uniqueness theorem for continuous additive functionals tempts us to conclude that $L^X = \Phi \circ \mu$ and hence for any $f \in \mathcal{B}_+(V)$

$$
\begin{aligned}
(4.37) \qquad L_1^* f(x) &= \mathcal{E}^x \int_0^\infty e^{-t}f(X_t)dL_t^X \\
&= \mathcal{E}^x \int_0^\infty e^{-t}f(X_t)d\Phi_{\mu_t}) \\
&= \mathcal{E}^x \int_0^\infty e^{-\Psi_t}f(Y_t)d\Phi_t = L_1 f(x).
\end{aligned}
$$

and hence $L_1^* = L_1$. This is flawed because $\Phi \circ \mu$ is known only to be additive relative to $\{Y_{\mu_t}\}$. But the conclusion of (4.37) does indeed follow from (4.36) by a relatively simple approximation argument which we leave to the reader. Thus we have

$$L_1(\ell + (m + Q)\hat{H}^\alpha) = L_1(\ell^* + (m^* + Q^*)\hat{H}^\alpha).$$

According to the uniqueness statement just prior to Step 1 we conclude that

$$(4.38) \qquad \ell = \ell^*, m = m^*, Q = Q^*$$

except for sets which are null relative to L_1 and L_1^*. Finally we conclude from (4.38) that $\{L_\alpha^{*\lambda}\}$, which from boundary theory is an (ℓ^*, m^*, Q^*) system, is in fact an (ℓ, m, Q) system. Since $\{L_\alpha^\lambda\}$ is an $(\ell, m, Q)\}$ system and $L_1 = L_1^*$ we conclude that $L^\lambda = L^{*\lambda}$. Since $L^\lambda = \tilde{U}^\lambda$ and $L^{*\lambda}$ is the λ-potential operator for the process on the boundary, X^*, we have shown that X^* and $\tilde{X}$ are equal in law. This then completes the proof of Motoo's construction theorem.

Bibliography

[B, 1]   R. M. Blumenthal, "On construction of Markov processes," *Z. Wahrscheinlichkeitstheorie* **63**, (1983), 433–444.

[BG, 1]   R. M. Blumenthal and R. K. Getoor, "Markov Processes and Potential Theory," Academic Press, New York 1968.

[BJ, 1]   A. Beneveniste et J. Jacod, "Systèmes de Lévy des processus de Markov," *Invent. Math.* **21**, (1973), 183–198.

[Bo, 1]   E. S. Boylan, "Local times for a class of Markov processes," *Illinois J. Math.* **8** (1964), 19–39.

[Bu, 1]   K. Burdzy, "Multidimensional Brownian excursions and potential theory," *Pitman Research Notes in Mathematics* **164**, Longman-Wiley, New York 1987.

[DM, 1]   C. Dellacherie et P. A. Meyer, "Probabilités et Potential," Hermann, Paris 1976.

[E, 1]   K. Bruce Erickson, "Continuous extensions of skew product diffusions," *Prob. Theory Related Fields* **85** (1990), 73–89.

[F, 1]   W. Feller, "Introduction to Probability Theory and its Applications, vol. II," Wiley, New York 1966.

[I, 1]   K. Itô, "Poisson point processes attached to Markov processes," Proc. 6th Berkeley Symp. Math. Stat. Prob. vol. 3, 225–240, U. of California Press 1971.

[I, 2]   —————————, "Stochastic Processes," Lecture Note Series No. 16, Math. Inst. Aarhus Univ. 1969.

[IM, 1]   K. Itô and H. P. McKean, Jr., "Brownian motions on a half line," *Illinois J. Math.* **7** (1963), 181–231.

[IM, 2]   —————————, "Diffusion processes and their sample paths," Springer, Berlin 1965.

[JY, 1]   T. Jeulin et M. Yor, "Autour d'un théorème de Ray," Astérique 52–53, (Soc. Math. France 1978) 145–158.

[Le, 1]   J. F. LeGall, "Brownian excursions, trees and measure-valued branching processes," *Ann. of Prob.* (to appear **19**) (1991).

[Ma, 1]   B. Maisonneuve, "Exit systems," *Ann. of Prob.* **3** (1975), 395–411.

[Ma, 2]   —————————, "Temps local et dénombrements d'excursions," *Z. Wahrscheinlichkeitstheorie* **52** (1980), 109–113.

[M, 1]   P. A. Meyer, "Renaissance, recollements, mélanges, ralentissement de processus de Markov," *Ann. Inst. Fourier* **25** (1975), 465–497.

[M, 2]   ——————, "Processus de Poisson ponctuels, d'apres K. Itô," *Sem. de Prob. V, Lecture Notes in Mathematics* **191**, 177–190, Springer, Berlin 1971.

[Mo, 1] M. Motoo, "Applications of additive functionals to the boundary problem of Markov processes," Proc. 5th Berkeley Symp. Math. Stat. Prob. Vol. 2 part 2, 75–110, U. of California Press 1965.

[Ra, 1] D. B. Ray, "Sojourn times of diffusion processes," *Illinois J. Math.* **7** (1963), 615–630.

[Ro, 1] L. C. G. Rogers, "Williams' characterization of the Brownian excursion law: proof and applications," *Sem. de Prob. XV, Lecture Notes in Mathematics* **850**, 227–250, Springer, Berlin 1981.

[Ro, 2] ——————, "Itô excursion theory via resolvents," *Z. Wahrscheinlichkeitstheorie* **63** (1983), 237–255.

[S, 1]   T. S. Salisbury, "On the Itô excursion process," *Prob. Theory Related Fields* **73** (1986), 319–350.

[S, 2]   ——————, "Construction of right processes from excursions," *Prob. Theory Related Fields* **73** (1986), 351–367.

[T, 1]   H. F. Trotter, "A property of Brownian motion paths," *Illinois J. Math.* **2** (1958), 425–433.

[VW, 1] S. R. S. Varadhan and R. J. Williams, "Brownian motion in a wedge with oblique reflection," *Comm. Pure and Appl. Math.* **38** (1985), 405–433.

[V, 1]   A. D. Ventcel', "On lateral conditions for multidimensional diffusion processes," *Teor. Veroyatnostei* **4** (1959), 208–211.

[W, 1]   D. Williams, "Diffusions, Markov processes, and martingales Vol. 1," Wiley, Chichester 1979.

[W, 2]   ——————, "Path decomposition and continuity of local time for one-dimensional diffusions, 1," *Proc. London Math. Soc.* (3) **28** (1974), 738–768.

[W, 3]   ——————, "Markov properties of Brownian local time," *Bull. Amer. Math. Soc.* **75** (1969), 1035–1036.

Notation Index

Probability and Its Applications

Editors

Professor Thomas M. Liggett
Department of Mathematics
University of California
Los Angeles, CA 90024-15555

Professor Charles Newman
Courant Institute of
Mathematical Sciences
New York University
New York, NY 10012

Professor Loren Pitt
Department of Mathematics
University of Virginia
Charlottesville, VA 22903-3199

Probability and Its Applications includes all aspects of probability theory and stochastic processes, as well as their connections with and applications to other areas such as mathematical statistics and statistical physics. The series will publish research-level monographs and advanced graduate textbooks in all of these areas. It acts as a companion series to Progress in Probability, a context for conference proceedings, seminars, and workshops.

We encourage preparation of manuscripts in some form of TeX for delivery in camera-ready copy, which leads to rapid publication, or in electronic form for interfacing with laser printers or typesetters.

Proposals should be sent directly to the editors, or to: Birkhäuser Boston, 675 Massachusetts Avenue, Cambridge, MA 02139.

Series Titles

K. L. Chung/ R. J. Williams. *Introduction to Stochastic Integration,* 2nd Edition
R. K. Getoor. *Excessive Measures*
R. Carmona/ J. Lacroix. *Spectral Theory of Random Schrödinger Operators*
G. F. Lawler. *Intersections of Random Walks*
R. M. Blumenthal. *Excursions of Markov Processes*